RANDOM SUMMATION

SUMMATION

Limit Theorems and Applications

RANDOM SUMMATION

Limit Theorems and Applications

Boris V. GNEDENKO
Moscow State University
Faculty of Mechanics and Mathematics
Moscow, Russia

Victor Yu. KOROLEV
Moscow State University
Faculty of Computational Mathematics
and Cybernetics
Moscow, Russia

CRC Press
Taylor & Francis Group
Boca Raton London New York

CRC Press is an imprint of the
Taylor & Francis Group, an **informa** business

CRC Press
Taylor & Francis Group
6000 Broken Sound Parkway NW, Suite 300
Boca Raton, FL 33487-2742

First issued in paperback 2019

No claim to original U.S. Government works

ISBN-13: 978-0-367-44862-2 (pbk)
ISBN-13: 978-0-8493-2875-6 (hbk)

Visit the Taylor & Francis Web site at
http://www.taylorandfrancis.com

and the CRC Press Web site at
http://www.crcpress.com

Library of Congress Card Number 96-33765

Library of Congress Cataloging-in-Publication Data

Gnedenko, Boris Vladimirovich, 1912–
        Random summation : limit theorems and applications / Boris
V. Gnedenko, Victor Yu. Korolev.
            p. cm.
        Includes bibliographical references and index.
        ISBN 0-8493-2875-6 (alk. paper)
        1. Stochastic processes. 2. Summability theory. 3. Limit.
    theorems (Probability theory)   I. Korolev, Victor Yu. II. Title.
    QA274.G64 1996
    519.2—dc20

                                                                96-33765
                                                                CIP

Preface

A remarkable feature of the development of probability theory during the last fifty years is a considerable extension of regions covered by its results. In particular, the classical limit theorems for sums of independent random variables have been generalized in many directions. One direction is determined by the situation where the number of summands is random itself. The necessity to consider these constructions arises from numerous applied problems where they serve as mathematical models. Many various results have already been accumulated and now we can say that an entire theory of random summation has taken shape. Some fragments of this theory have already been included in textbooks in probability theory, see, e.g., (Borovkov, 1986), (Gnedenko, 1988), (Chow and Teicher, 1988). There are special monographs on random summation (Gut, 1988), (Kruglov and Korolev, 1990).

However, an explosive development of the theory of random summation in the last five to seven years displayed new deep fundamental results which describe the transformation of limit regularities for sums of independent random variables when the number of summands is replaced by a random variable. Also, new wide and important applied fields were brought into the sphere of effective applicability of the theory of random summation. These circumstances caused a need for a new book which should combine a strict exposition of the foundations of the theory of random summation with a description of widespread possibilities of its application to solving important practical problems. This was the main aim we had in mind when we were writing this book.

From the formal point of view this monograph can be regarded as a counterpart to the books (Gnedenko and Kolmogorov, 1954) and (Kruglov and Korolev, 1990), which each of us wrote in co-operation with his teacher. But the ideological motivation of writing this book is quite different and explains its distinctions from the above mentioned books as well as from (Gut, 1988). We intended not only to bring the reader up to date on recent results and foundations of the theory describing the influence of the randomness of the number of summands on the limit behavior of sums of random variables, but also to demonstrate the potentialities of probability theory in solving important applied problems, especially those which at first sight seem to have no mathematically strict solutions. The examples of these problems are prediction of software reliability from the results of debugging and determination of the regularities of stock price processes. Some engineers and programmers

i

reject the very expediency of a probabilistic approach to the first of these two problems while the second one is often compared to the search for a "philosopher's stone" in the Middle Ages.

We also wanted to show how practice interacts with theory and how new mathematical formulations of problems appear and develop. These interesting topics are usually not dealt with in mathematical monographs at all, and excellent results often attack a beginner without any attempts to explain their sources or to demonstrate a long work on formulation of more and more complicated problems being put forward by mathematicians. Of course, we describe our own experience gained when we solved some problems considered here. We will be glad if our experience turns out to be useful to the readers of this book.

In our opinion, the exposition of the asymptotic theory of random summation is particularly appealing for these goals, because the appearance and development of this theory was in many respects determined by practical problems. Chapters 1, 2 and 5 are entirely devoted to applied problems. In Chapters 2 and 5 we consider two problems from reliability theory, namely, doubling with repair and software reliability prediction. The first problem raised many formulations of problems of the theory of random summation and an appropriate solution of the second one has been gained mainly due to the results of this theory.

A word about the contents of the book:

Chapter 1 contains various examples from biology, engineering and technology, physics, communication, storage control, insurance, economics and other fields. These examples deal with phenomena whose mathematical models are sums of a random number of independent random variables (random sums). Along with examples from applied fields, there are also several examples from some sections of mathematics itself such as combinatorics and theory of branching processes which intensively exploit random summation.

Chapter 2 is entirely devoted to the problem of determining reliability characteristics of a doubled technical system in which a device that failed is repaired. This problem is used as an illustration of how some formulations of mathematical problems related to random summation appeared.

Chapter 3 contains a description of the asymptotic behavior of growing random sums. Some transfer theorems are presented on both asymptotic rapprochement of the distributions of random sums with approximating laws and their convergence. Some latest results are given that present general necessary and sufficient conditions for the weak convergence of random sums. The classes of limit laws are described. It is shown how these results being applied to some practical problems make it possible to construct basic mathematical models of real processes and phenomena. Main attention is paid to three practical problems. The first one is related to the asymptotic behavior of risk processes corresponding to the surplus of an insurance company. Answers are given to the questions what limit distributions are possible in a general situation and when normal approximation is correct. These answers have an "if and only if" form. The second problem is that of determining the distribution of increments of a process of stock prices. A sort of a base of a bridge is being built between mathematical models of this process at the micro level and those at the macro level. It is shown how limit theorems of the theory of random summation explain the observed leptokurtosity of this distribution and quite naturally lead to general models involving Wiener processes with stochastic diffusion. The third problem deals with rarefied renewal processes which play an important role in reliability theory.

In Chapter 4 nonrandomly centered random sums in the double array limit scheme are considered. This is the first time these objects are described in monographic literature. Transfer theorems are proved and the class of limit distributions is described. Partial converses of the transfer theorems as well as necessary and sufficient conditions for the convergence of random sums are given. These asymptotic results are applied to the analysis of a process of stock prices mentioned above and to the improvement of the results on the asymptotic behavior of supercritical Galton-Watson processes which model many real processes in nuclear physics, biology and epidemiology. Here we also discuss analogs of infinitely divisible and stable distributions with respect to random summation.

It is shown in Chapter 5 that limit theorems of the theory of random summation give the base for the construction of a unified mathematical theory of reliability growth of modified systems including software and hardware and the outline of this theory itself is presented. Although hundreds of reliability growth models are known which are used for prediction of reliability of software or hardware in the process of design, debugging or adjustment, in canonical monographs, textbooks or handbooks on mathematical reliability theory these problems are bypassed, possibly, due to the absence of a unified mathematical viewpoint. An attempt is made here to fill this gap. The presented theory embraces practically all existing reliability growth models and gives natural rules of their classification as well as of construction of new models. Some new classes of reliability growth models are described in detail which lead to nice mathematical constructions and unusual formulations of asymptotic statistical problems.

Appendix 1 contains some results on extremal entropy properties of probability distributions. An attempt is made to look at limit theorems of probability theory as at a manifestation of the universal principle of the non-decrease of uncertainty in closed systems and to formulate some principles of construction of probability models for real phenomena.

Appendix 2 deals with the properties of Poisson processes intensively used throughout the book. Along with well-known facts, new limit theorems are presented here for doubly stochastic Poisson processes (Cox processes) used as models in Chapters 3 and 4. These results are obtained as a consequence of one general theorem on necessary and sufficient conditions for the convergence of superpositions of independent random processes. Asymptotic results concerning generalized Cox processes given here illustrate how some theorems proved in Chapters 3 and 4 work.

In the Bibliographical Commentary we render homage to those whose works may be regarded as milestones on the way of the development of the theory of random summation and indicate the sources of the results which constitute this monograph.

When we selected the theoretical material for the book, we were guided by the desire to make the book understandable for an audience as wide as possible. Therefore we restricted ourselves to the results that would be clear to anyone who is familiar with probability theory within the limits of, say, (Feller, 1968, 1971). To provide a rational balance between a final character of the presented results and the clarity of exposition, we focused main attention on the situation in which the number of summands in a sum is assumed independent of summands. For the same reason in the double array scheme we considered only random sums of identically distributed summands. Those readers who will intend to learn more on general limit theorems for random sequences with random indices after they read this book are referred to the forthcoming monograph (Korolev and Kruglov, 1997).

The history of the work on this book is as follows. The idea to write a book on random summation came to B. V. Gnedenko more than twenty years ago. At the end of the sixties and the beginning of the seventies a special seminar was held at the Department of Probability Theory of the Mechanico-Mathematical Faculty of Moscow State University; one of the main topics was limit theorems with random indices. The research of the participants of this seminar made an essential contribution to the theory of random summation stimulating an idea of writing a special monograph. However, many principal problems remained unsolved and therefore at that time a more or less complete monograph did not appear. At the end of the eighties at the Department of Mathematical Statistics of the Faculty of Computational Mathematics and Cybernetics of Moscow State University, V. M. Kruglov and V. Yu. Korolev succeeded in solving many problems which had yet remained unsolved. In particular, necessary and sufficient conditions were found for the convergence of random sums of centered summands in the double array scheme. So, forestalling the idea of B. V. Gnedenko, the book (Kruglov and Korolev, 1990) appeared. However, beyond the limits of that book remained problems related to random sums under nonrandom centering (which are of special interest for approximations) and applications of the theory of random summation. A rather complete description of the limit behavior of nonrandomly centered random sums was obtained only recently. B. V. Gnedenko who wrote the foreword to (Kruglov and Korolev, 1990) suggested that we combine our efforts and write a new book. Our close co-operation started in the beginning of 1994. We had intensive discussions for several months to determine the ideological trend of the book and its contents. After the work started, there appeared a necessity to obtain some new results specially designated for this book, and these results, mainly related to risk processes, have been obtained.

Some fragments of the book became parts of courses read by B. V. Gnedenko at Sheffield, Rome, Warsaw and Copenhagen universities and by V. Yu. Korolev at the Faculty of Computational Mathematics and Cybernetics of Moscow State University. We think that this book may serve as a textbook for the courses in probability models, random summation and software reliability at the beginning graduate level.

In the book, usual notation is used:

$\Rightarrow$ denotes weak convergence,

$\overset{d}{=}$ denotes coincidence of distributions,

$\overset{P}{\rightarrow}$ denotes convergence in probability,

$\mathbb{R}$ is the set of real numbers,

$\mathbb{R}_+$ is the set of nonnegative numbers,

$\mathbb{R}^n$ is the set of n-dimensional vectors,

$\mathbb{R}^n_+$ is the set of n-dimensional vectors with nonnegative coordinates.

We use triple numeration for theorems, lemmas etc. For example, a reference to Theorem 3.2.1 means that to the first theorem of the second section of the third chapter. Since we managed to do almost without interchapter references to formulas, within each chapter relations are supplied with double numeration. In those very few interchapter references to formulas the third index appears. So, for example, formula (3.1) is the first formula of the third section in a given chapter and the reference to relation (1.3.2) is that to the second relation of the third section of the first chapter.

We acknowledge our friends and colleagues V. V. Kalashnikov, V. M. Kruglov and V. M. Zolotarev for fruitful and stimulating discussions. We are grateful to V. E. Bening who contributed to the book nice results concerning asymptotic expansions and concentration functions of generalized Cox processes presented in Sections A2.6 and A2.7. We also wish to thank E. V. Morozov who took the responsibility to organize the preparation of the camera-ready manuscript; S. G. Sigovtsev, B. M. Shirokov, and A. V. Kolchin who did their best to translate, typeset and edit the book. We warmly thank our guardian angel Svetlana Landau, the CRC Press acquiring editor, for her kind and careful attention to our book.

In its final stages, the work was supported in part by the Russian Foundation for Fundamental Research, project 93-011-1446 and by International Science Foundation, projects NFW000 and NFW300.

B. V. Gnedenko and V. Yu. Korolev
Moscow, September, 1995.

Contents

1 Examples   **1**
  1.1   Examples related to generalized Poisson laws   1
  1.2   A remarkable formula of queueing theory   8
  1.3   Other examples   12

2 Doubling with Repair   **17**
  2.1   Mathematical model   17
  2.2   A limit theorem for the trouble-free performance duration . . .   21
  2.3   The class of limit laws   25
  2.4   Some properties of limit distributions   30
  2.5   Domains of geometric attraction of the laws from class $\mathcal{K}$ . . .   34

3 Limit Theorems for "Growing" Random Sums   **41**
  3.1   A transfer theorem. Limit laws   41
  3.2   Necessary and sufficient conditions for convergence   49
  3.3   Convergence to distributions from identifiable families   59
  3.4   Limit theorems for risk processes   67
  3.5   Some models of financial mathematics   77
  3.6   Rarefied renewal processes   84

4 Limit Theorems for Random Sums
in the Double Array Scheme   **93**
  4.1   Transfer theorems. Limit laws   93
  4.2   Converses of the transfer theorems 102
  4.3   Necessary and sufficient conditions for the convergence of
        random sums of independent identically distributed
        random variables 113
  4.4   More on some models of financial mathematics 123
  4.5   Limit theorems for supercritical Galton–Watson processes . . . 130
  4.6   Randomly infinitely divisible distributions 137

5 Mathematical Theory of Reliability Growth.
A Bayesian Approach   **153**
  5.1   Bayesian reliability growth models 153
  5.2   Conditionally geometric models 158

5.3	Conditionally exponential models	164
5.4	Renewing models	169
5.5	Models with independent decrements of volumes of defective sets	172
5.6	Order-statistics-type (mosaic) reliability growth models	175
5.7	Generalized conditionally exponential models	180
5.8	Statistical prediction of reliability by renewing models	186
5.9	Statistical prediction of reliability by order-statistics-type models	191

Appendix 1. Information Properties of Probability Distributions   **207**

A1.1 Mathematical models of information and uncertainty 207

A1.2 Limit theorems of probability theory and the universal principle of non-decrease of uncertainty 210

Appendix 2. Asymptotic Behavior of Generalized Doubly Stochastic Poisson Processes   **217**

A2.1 General information on doubly stochastic Poisson processes . . 217

A2.2 A general limit theorem for superpositions of random processes 221

A2.3 Limit theorems for Cox processes 222

A2.4 Limit theorems for generalized Cox processes 227

A2.5 Convergence rate estimates in limit theorems for generalized Cox processes 231

A2.6 Asymptotic expansions for generalized Cox processes 239

A2.7 Estimates for the concentration functions of generalized Cox processes 249

Bibliographical Commentary   **253**

References   **255**

Index   **265**

Chapter 1

Examples

1.1 Examples related to generalized Poisson laws

EXAMPLE 1.1.1. Consider the energy received by some region of the surface of Earth from cosmic particles during time T. Let E_k be the energy of the k-th particle and N be the total number of particles arrived during time T. It is obvious that the energy received by the region is equal to

$$E = E_1 + \ldots + E_N \tag{1.1}$$

We will consider E_k, $k \geq 1$, here, as independent random variables with the same distribution function $F(x)$, and N as a random variable which is independent of the variables $\{E_k\}_{k \geq 1}$, and takes nonnegative integer values with the corresponding probabilities

$$p_n = P(N = n), \quad n = 0, 1, 2, \ldots$$

When studying these physical phenomena it is usually assumed that the random variable N has the Poisson distribution with some parameter λ, i.e.,

$$p_n = \frac{\lambda^n}{n!} e^{-\lambda}, \qquad n = 0, 1, 2, \ldots, \lambda > 0 \tag{1.2}$$

Experimental data are in good accordance with this assumption. Some more details of the mathematical premises of (1.2) are presented in the Appendix, where the properties of the Poisson flows of events are described.

Let us find the distribution function $G(x)$ of the random variable E. Since the sum (1.1) has no summands with probability p_0, then we can conclude that the function $G(x)$ has a jump at zero, which is equal to p_0. The sum (1.1) will have n summands with probability p_n and the distribution function of the variable E will be $F^{*n}(x)$, where

$$F^{*(n+1)}(x) = \int_0^x F^{*n}(x - z) \, dF(z)$$

1

and $F^{*1}(x) = F(x)$. If we denote by $F^{*(0)}(x)$ the degenerate distribution function which has the only unit jump at zero, then according to the law of total probability we can write the following relation

$$G(x) = \mathsf{P}(E < x) = \sum_{n=0}^{\infty} p_n F^{*n}(x) \tag{1.3}$$

In terms of the characteristic functions

$$g(t) = \int_0^\infty e^{itx}\, dG(x) \qquad f(t) = \int_0^\infty e^{itx}\, dF(x)$$

relation (1.3) can be written as

$$g(t) = \sum_{k=0}^{\infty} p_k f^k(t). \tag{1.4}$$

If p_k are given by (1.2), then (1.4) takes a simple form

$$g(t) = \exp\{\lambda[f(t) - 1]\}. \tag{1.5}$$

Formula (1.5) implies, in particular, that the distribution function $G(x)$ is always infinitely divisible (including the case when the distribution function $F(x)$ is not).

It is well known that under the conditions of our example the expectation and variance of the random variable E are determined as

$$\mathsf{E}E = \mathsf{E}N \cdot \mathsf{E}E_1,$$
$$\mathsf{D}E = \mathsf{E}N \cdot \mathsf{D}E_1 + \mathsf{D}N(\mathsf{E}E_1)^2.$$

Under assumption (1.2) these formulas take a very simple form

$$\mathsf{E}E = \lambda \cdot \mathsf{E}E_1, \qquad \mathsf{D}E = \lambda \cdot \mathsf{E}E_1^2.$$

This example seemed to be an interesting and useful object of the mathematical investigation. Due to these reasons the formula for the expectation of a random sum (when the number of summands is independent of the summands) was included into the textbook (Gnedenko, 1949). We should note that other authors paid their attention to this problem at the same time: (Wald, 1944), (Robbins, 1948), (Kolmogorov and Prokhorov, 1949), (Feller, 1950). The period of the accumulation of separate facts and particular results has begun.

EXAMPLE 1.1.2. In 1939 some problems of the theory of summation of independent random variables were solved providing a noticeable progress. The introduction of accompanying infinitely divisible distributions became the central idea of the method proposed by B. V. Gnedenko. New capabilities of the method of accompanying infinitely divisible distributions were demonstrated by V. M. Kruglov in (Kruglov, 1975).

The essence of this method is as follows. To every summand X_{nk} with characteristic function $f_{nk}(t)$ a random variable Y_{nk} with characteristic function $g_{nk}(t) = \exp(f_{nk}(t) - 1)$ is set in correspondence. New variables are infinitely divisible.

Consider the sum of independent random variables

$$S_n = X_{n1} + \ldots + X_{nk_n} \tag{1.6}$$

and construct a new sum of random variables with accompanying distributions

$$\bar{S}_n = Y_{n1} + \ldots + Y_{nk_n}. \tag{1.7}$$

The distribution functions of sums (1.6) converge to the limit one as $k_n \to \infty$ $(n \to \infty)$ if and only if the limit law for the distributions of sums (1.7) exists. Both limit laws coincide, if they exist. We will present an interesting proof of this fact in Chapter 4 (which differs from the one given initially).

This proof is based on the fact that

$$\exp(f_{nk}(t) - 1) = e^{-1} \sum_{j=0}^{\infty} \frac{f_{nk}^j(t)}{j!}$$

and we can see that the characteristic function of the accompanying distribution is equal to the characteristic function of the random sum of independent summands, the latter distributed according to the original law and the number of summands having the Poisson distribution with parameter 1. In particular, if for each n the summands X_{nk}, $k \geq 1$ are identically distributed, then

$$\bar{S}_n \overset{d}{=} \sum_{i=1}^{k_n} \sum_{j=1}^{N_i} Y_{nj}^{(i)} \tag{1.8}$$

where N_i are independent random variables with the same Poisson distribution with parameter 1, independent of the random variables $Y_{nj}^{(i)}$, $Y_{nj}^{(i)} \overset{d}{=} Y_{n1}$. But according to the form of characteristic function of sum (1.8) we can conclude that

$$\bar{S}_n \overset{d}{=} \sum_{j=1}^{N^{(n)}} Y_{nj}$$

where $N^{(n)}$ has the Poisson distribution with parameter k_n and is independent of Y_{nj}, $j \geq 1$.

EXAMPLE 1.1.3. When a characteristic function has the form (1.5) with $f(t)$ also being a characteristic function, the corresponding distributions are called generalized Poisson. Representatives of the set of integer generalized Poisson laws are often encountered in statistical modeling and data processing in physical, biological and some other experiments.

The range of phenomena described by these distributions is rather wide. For example, one of the first models of this type was the model of an insurance company (Lundberg, 1909), (Seal, 1969). We will not stop on this model here, because we intend to deal with it in detail later. With the help of the integer generalized Poisson distribution the numbers of fatal outcomes from scarlet fever and smallpox in Switzerland and from steam boilers explosions were described in (Eggenberger, 1924). A typical model (Pollaszek and Geiringer, 1928) supposes that breakdowns take place independently of each other and the number of the breakdowns that caused j deaths (the random variable Z_j), has the Poisson distribution with some parameter λ_j, $j = 1, 2, \ldots$ A similar model was applied to the description of the number of victims in transport accidents (Kupper, 1965). In such a model the total number of fatal outcomes resulting from all breakdowns of a year has the form

$$Z = Z_1 + 2Z_2 + 3Z_3 + \ldots \tag{1.9}$$

These models are used in epidemiology, entomology and bacteriology (Greenwood and Yule, 1920), in studying the spatial arrangement of individuals in biology and ecology (Skellam, 1952). Models (1.9) are very important in nuclear physics, in particular, in experiments aimed at the estimation of partial cross-sections of nuclear reactions. We describe these models and the problems appearing in processing of the results of such experiments in more detail using the corresponding physical terminology.

Consider a bundle of primary particles aimed cyclically on the target consisting of the investigated material placed in an accelerator. During the interaction between the particles and the nuclei reactions can take place with the generation of one, two, $\ldots k$ secondary particles. In other words we will say that a particle caused a reaction of j-th type if j secondary particles were generated as a result of this reaction ($j = 1, 2, \ldots$). All these secondary particles, the products of the reaction of any of k types, are indistinguishable. If the intensity of the primary bundle does not change from cycle to cycle, then the numbers Z_j of the reactions of j-th type are assumed to be realizations of Poisson random variables with some parameters λ_j. (This is quite usual for nuclear processes, stipulated by a small probability of an interaction and a vast number of nuclei in the target. This fact gives an opportunity to apply the Poisson theorem on rare events. This assumption was multiplicitly confirmed by experiments (Belov, Galkin and Ufimtsev, 1985)).

The parameter λ_j has the sense of the mean number of reactions of the j-th type per cycle or per time unit. The random variables Z_j are considered independent, because actually the target does not change during the irradiation. Therefore the total number of secondary particles that appeared per cycle has the form (1.9).

The indistinguishability of secondary particles that were generated by reactions of different types, what is typical for the multiple photonuclear reaction (photoneutron or photoproton), and their joint presence in the experimental output make the problem of separation of partial outputs and, hence, partial cross-sections, almost unsolvable. Under the conditions of protoneutron reactions (γ, n) and $(\gamma, 2n)$, when the reactions with the generation of one or two neutrons occur as a result of the exposition the target to the bundle of harsh γ-quanta, we can reconstruct only the "weighted" cross-sections $\sigma(\gamma, n) + 2Z(\gamma, 2n)$, not the partial ones $\sigma(\gamma, n)$ and $\sigma(\gamma, 2n)$ and the total section $\sigma(\gamma, n) + Z(\gamma, 2n)$. To find the characteristics mentioned above we should apply probability methods.

It is required to find the distribution of the number of registered particles. Under the assumption of 100-percent registration of the secondary particles the solution is as follows. As far as the characteristic function of the variable jZ_j is equal to

$$\mathsf{E}e^{it_j Z_j} = \exp\{\lambda_j(e^{it_j} - 1)\}, \qquad t \in \mathbb{R}^1,$$

and the characteristic function of the sum of independent summands equals to the product of characteristic functions of the summands, we have the following representation for the characteristic function of sum (1.9), denoting $\lambda = \sum_j \lambda_j$,

$$\mathsf{E}e^{itZ} = \prod_j \exp\{\lambda_j(e^{it_j} - 1)\} =$$

$$\exp\left\{\sum_j \lambda_j(e^{it_j} - 1)\right\} = \exp\{\lambda(f(t) - 1)\}, \tag{1.10}$$

where

$$f(t) = \sum_j \frac{\lambda_j}{\lambda} e^{it_j}$$

is the characteristic function of a random variable taking the value j with probability λ_j/λ. Thus

$$Z \overset{d}{=} X_1 + \ldots + X_N, \tag{1.11}$$

where X_j, $j \geq 1$, are independent random variables with common characteristic function $f(t)$, and N is the random variable with the Poisson distribution with parameter λ independent of $\{X_j\}$, $j \geq 1$.

If we suppose that every secondary particle is registered with some probability ε, less than 1, then the form of the function $\mathsf{E}e^{itZ}$ in (1.10) remains valid with $f(t)$ slightly changed, (Belov, Galkin and Ufimtsev, 1985), (Korolev, 1980).

Within different particular values of the maximum multiplicity k and the parameters λ_j and ε, the distributions with characteristic functions (1.10) transform into such well-known distributions as Poisson-binomial, Hermite, Neyman, Stirling-Hermite and others, to whose properties a considerable number of publications are devoted (for example, (Johnson and Kotz, 1969)).

Though representation (1.11) does not follow evidently from the physical statement of the problem, but emerges as a result of a kind of an analytical trick, we think that it may be helpful in the investigation of the structure of phenomena that advanced it.

EXAMPLE 1.1.4. Consider the classical risk process $Z(t)$ from the mathematical theory of insurance (actuarial mathematics, risk theory)

$$Z(t) = ct - \sum_{k=1}^{N(t)} X_k, \tag{1.12}$$

where $Z(0) = 0$ and $\sum_{k=1}^{0} = 0$ by definition and the process $N(t)$ and the random variables $\{X_k\}_{k\geq 1}$ are assumed independent. Here $Z(t)$ is the surplus of an insurance company during time $(0, t]$, $c > 0$ is the constant expressing the intensity of insurance premiums, $N(t)$ is the number of insurance payments during time $(0, t]$ and $\{X_k\}$ are claims. It is usual to suppose that $N(t)$ is a homogeneous Poisson process with some constant intensity α, and $\{X_k\}_{k\geq 1}$ are independent identically distributed random variables, and this makes the distribution of the random variable $Z(t)$ generalized Poisson. The risk process is explicitly studied in (Grandell, 1990). We will return repeatedly later to this example and describe some properties of the process (1.12) as well as of some of its generalizations.

EXAMPLE 1.1.5. During the last three decades the probability methods were occupying a more important place in the investigations of combinatorial nature. If we set a uniform distribution upon the ensemble of studied combinatorial objects, then we can consider the numerical characteristics of these objects as random variables and apply well-developed analytical methods of probability theory to the investigation of their properties.

Consider one of such problems, which is of some interest from the point of view of construction and analysis of computational algorithms and cryptography.

A mapping of each element a of some set of n elements into some element $\sigma(a)$ of the same set is called a transposition. Within this mapping we get all the elements of the initial set and each element only once. Therefore the notion of transposition coincides with the notion of one-to-one mapping of the set onto itself. For the pair of transpositions a multiplication operation is defined as a consecutive applying of these transpositions. Generally speaking, the multiplication of transpositions is not commutative. All the transpositions of n elements make up a group S_n, which is called a symmetrical n-degree group. Note that a transposition from S_n can be represented as an oriented graph with n vertices possessing the property that only one arc comes from each vertex and only one arc comes into each vertex. A transposition that cyclically transposes some ensemble of elements and leaves other elements on their places is called a cycle. The number of transposed elements is called a length of a cycle (in the graph presentation a cycle is presented as a contour, a length of a cycle is a length of this contour).

Consider a set $\mathbf{S}_{n,R}$ of all the n-degree transpositions, with lengths of cycles lying in some set R of natural numbers. An interest for studying such sets can be explained by their connection with simplest equations containing unknown transpositions. For example, the ensemble of all the solutions of the equation

$$\mathbf{X}^p = e$$

in the symmetrical group S_n, where e is an identical transposition and p is a prime number, coincides with the ensemble $\mathbf{S}_{n,R}$, where $R = \{1, p\}$.

If d is a composite number and $1 = d_0 < d_1 < \ldots < d_r = d$ are all the different divisors of a number d then transposition X is the solution of the equation

$$\mathbf{X}^d = e$$

if and only if the lengths of cycles of X belong to the set $R = \{d_0, d_1, \ldots, d_r\}$.

Denote the number of $\mathbf{S}_{n,R}$ elements as $a_{n,R}$. Asymptotics of the numbers $a_{n,R}$ were investigated by many authors. Quite a full bibliography is given in (A. V. Kolchin, 1993).

To the investigation of the transpositions from the set $\mathbf{S}_{n,R}$ in the work (V. F. Kolchin, 1989) the approach used earlier to the investigation of different characteristics of the random transpositions from the whole set $\mathbf{S}_{n,R}$ in (V. F. Kolchin, 1986) was applied.

Let the uniform distribution be given on $\mathbf{S}_{n,R}$ and $\nu_{n,R}$ be a number of cycles in random transposition from this set. Let us say that $b_{n,R} = (n-1)!$ if $n \in R$ and $b_{n,R} = 0$ in other cases. It is easy to see that

$$\mathsf{P}(\nu_{n,R} = j) = \frac{n!}{j! a_{n,R}} \sum_{n-1+\ldots+n_j=n} \frac{b_{n_1,R} \cdot \ldots \cdot b_{n_j,R}}{n_1! \cdot \ldots \cdot n_j!}. \tag{1.14}$$

Introduce independent identically distributed random variables $X_1, \ldots, X_j$ with

$$\mathsf{P}(X_1 = k) = \frac{b_{k,R} x^k}{k! B(x)} = \frac{x^k}{k B(x)}, \quad k \in R,$$

where

$$B(x) = \sum_{k=1}^{\infty} \frac{b_{k,R} x^k}{k!} = \sum_{k \in R} \frac{x^k}{k}, \quad 0 < x < 1.$$

Set $X_0 = 0, S_j = X_0 + X_1 + \ldots + X_j$. With the help of these values relation (1.14) can be written as

$$\mathsf{P}(\nu_{n,k} = j) = \frac{n!(B(x))^j}{j! x^n a_{n,R}} \mathsf{P}(S_j = n). \tag{1.15}$$

By summing up equalities (1.15) over j we find that

$$a_{n,R} = \frac{n! e^{B(x)}}{x^n} \mathsf{P}(S_N = n),$$

where N is a random variable with the Poisson distribution with parameter $B(x)$ and independent of $X_0, X_1, \ldots$. Thus to investigate the asymptotic behavior of the numbers $a_{n,R}$ it is sufficient to choose a suitable value of the parameter x and to get a local limit theorem for the random sum S_N which has the generalized Poisson distribution. Some results were obtained by this approach in (V. F. Kolchin, 1989) and (A. V. Kolchin, 1993).

1.2   A remarkable formula of queueing theory

In Example 1.1.3 we obtained a representation in the form of a random sum of random variables (independent and identically distributed) for the variable that seemed to be unrepresentable in such a form, if looked at from the viewpoint of the physical formulation of the problem only. We will give some more similar examples in this section. All these examples are connected with one and the same formula which is remarkable because it was rediscovered several times in connection with different applied problems.

EXAMPLE 1.2.1. The original version of this formula was proved in the works by F. Pollaczek (Pollaczek, 1930) and A. Ya. Khinchin (Khinchin, 1932). We will follow the lines of the reasoning used by Khinchin because of its extraordinary visuality. Further we will use the terminology of the original article taken from the telephone practice. Khinchin was solving the problems of the telephone service optimization in Moscow in 1930-s. Here is the problem solved by Khinchin.

A telephone operator that serves a big group of customers gets a certain average number of calls per hour. Let the probability to get precisely k calls during the time interval depend only on a length of the interval and be independent of both time of its beginning and the number of the customers waiting to talk. If a call finds the telephone operator busy, then the customer has to wait until she serves all the customers that called earlier. The duration of the operator's talk with a customer is supposed to be a random variable with density $f(t)$. Therefore the mean duration of the talk equals to

$$\mu_1 = \int_0^\infty t f(t)\, dt.$$

If n calls occur in a time unit (usually an hour) then $n\mu_1 = \alpha$ is the mean busy period of the operator. Assume that $\alpha < 1$, since otherwise the queue is growing unlimitedly.

It is required to find the distribution of the waiting time, i.e., the probability that the customer calling the operator at an arbitrary time would have to wait less than time t before he deals with the operator, $t > 0$.

The solution of this problem appears to be especially easy if we express it in terms of characteristic functions. Denote the characteristic function corresponding to the density f as $\psi(s)$,

$$\psi(s) = \int_0^\infty e^{ist} f(t)\, dt, \qquad s \in \mathbb{R}.$$

Denote the characteristic function of the desired distribution as $\phi(s)$.

Using simple but rather cumbersome calculations A. Ya. Khinchin came to the representation

$$\phi(s) = \frac{1 - \alpha}{1 - \alpha \dfrac{\psi(s) - 1}{i\mu_1 s}}, \tag{2.1}$$

and hence he had solved the problem, since (2.1) expresses $\phi(s)$ through α and $\psi(s)$. Now relation (2.1) is called Pollaczek-Khinchin formula.

Many investigators have noticed the fact that rather an unexpected result can be obtained with its help. This result is as follows. Let F be the distribution function of the duration of operator's talk with a customer.

Integrating by parts and taking account of nonnegativeness of the duration of a talk, it is easy to get the relation

$$\mu_1 = \int_0^\infty [1 - F(x)]\, dx,$$

whence it follows that the function

$$h(x) = \frac{1}{\mu_1}[1 - F(x)] \qquad (2.2)$$

is the density of some distribution. Find the characteristic function $\chi(s)$ corresponding to this density. By integrating by parts we get

$$\chi(s) = \int_0^\infty e^{isx} h(x)\, dx = \frac{1}{\mu_1}\int_0^\infty e^{isx}[1 - F(x)]\, dx =$$

$$= \frac{1}{\mu_1}\left\{ \frac{e^{isx}}{is}[1 - F(x)]\Big|_0^\infty + \frac{1}{is}\int_0^\infty e^{isx} f(x)\, dx \right\} =$$

$$= \frac{\psi(s) - 1}{i\mu_1 s}.$$

Now consider the sum $S = Y_1 + \ldots + Y_N$, where random variables $N, Y_1, Y_2, \ldots$ are independent; moreover the variables $Y_1, Y_2, \ldots$ are identically distributed and have one and the same density (2.2), and the random variable N has the geometric distribution with parameter α:

$$P(N = n) = (1 - \alpha)\alpha^n, \qquad n = 0, 1, 2, \ldots$$

Find the characteristic function of the random variable S. According to the law of total probability we have

$$Ee^{isS} = \sum_{n=0}^\infty (1 - \alpha)\alpha^n \chi^n(s) = \frac{1 - \alpha}{1 - \alpha\chi(s)},$$

that exactly coincides with the right-hand side of (2.1)! Thus we have the random variable Z, determined earlier as the waiting time for the conversation with the operator, that admits the representation

$$Z \stackrel{d}{=} Y_1 + \ldots + Y_N \qquad (2.3)$$

as the random sum of the independent random variables.

L. Takács (Takács, 1965) introduced into consideration a random process connected with the queue considered above. He called it virtual waiting time. This process equals at every time t to a length of time interval which must pass from time t to the time when the operator gets completely free from serving the calls that were received before time t. Denote this process as $Z(t)$. It is easy to see that if $t_1, t_2, \ldots$ are times of the calls occurrences, then for $t_n < t < t_{n+1}$ the process $Z(t)$ is determined as

$$Z(t) = \begin{cases} 0, & \text{if } Z(t_n) \leq t - t_n, \\ Z(t_n) - (t - t_n), & \text{if } Z(t_n) \geq t - t_n. \end{cases}$$

With the help of the methods of the theory of random processes it can be shown that characteristic function (2.1) is the characteristic function of the limit distribution for $Z(t)$ with $t \to \infty$, see, for example, (Gnedenko and Kovalenko, 1989, Sect. 4.2).

Note one interesting analogy when the same distribution serves as an adequate description of an absolutely different physical process. In times $t_1, t_2, \ldots$ water portions of volumes $Z_1, Z_2, \ldots$ correspondingly come to a reservoir as a lock gate opens. The water flow is uniform: unit volume of water flows in time unit. Then $Z(t)$ will be equal to water volume in the reservoir in time t. In particular, such problems were investigated by Moran (Moran, 1956). It is interesting that representation (2.2) was obtained by V. Beneš (Beneš, 1957) in connection with the research of a simplest queue and by D. Kendall (Kendall, 1957) in connection with problems of the storage control theory and the theory of dams. Moreover both of the authors noticed that the phenomenological interpretation of the summands Y_j and the variable N is absent in the context of the solved problems. Futhermore, Prabhu (Prabhu, 1980, p. 38) noticed that the appearance of this formula had "caused a mild surprise at the time". This formula has attracted the assiduous attention of the specialists in risk theory (actuarial mathematics) and finally the reasonable interpretation of the variables involved in representation (2.2) was found.

EXAMPLE 1.2.2. Earlier in Example 1.1.4 we introduced the classical risk process describing the surplus of an insurance company,

$$Z(t) = ct - \sum_{k=1}^{N(t)} X_k,$$

where $\{X_i\}_{i \geq 1}$ are independent nonnegative random variables with the same distribution function $F(x)$ and $\mathsf{E}X_i = \mu_1 > \infty$, and $N(t)$ is the Poisson process independent of the sequence $\{X_i\}$ with

$$\mathsf{P}(N(t) = n) = e^{-\lambda t} \frac{(\lambda t)^n}{n!}, \qquad \lambda > 0 \qquad n = 0, 1, 2, \ldots$$

This process describes the capital of the company at time t, and coincides with the process $Z(t)$ introduced in the previous example, up to the sign and the multiplier c at t. Let u be the initial capital of the company. Denote the probability of the company's bankruptcy (ruin probability) within dependence on u as $\psi(u)$,

$$\psi(u) = \mathsf{P}(u + Z(t) < 0 \quad \text{for some} \quad t > 0).$$

Introduce a new variable for the investigation of this probability. For some $\theta > 0$ set $c = (1 + \theta)\mu_1 \lambda$ in the definition of $Z(t)$. Let us call the variable

$$L = \sup_{t \geq 0} \left\{ \sum_{i=1}^{N(t)} X_i - (1 + \theta)\mu_1 \lambda t \right\} = \sup_{t \geq 0} \{-Z(t)\}$$

the maximum aggregate loss. Since $Z(0) = 0$, it is clear that $L \geq 0$. To express the ruin probability in terms of the random variable L note that

$$1 - \psi(u) = \mathsf{P}(u + Z(t) \geq 0 \quad \text{for all} \quad t > 0) =$$

$$= \mathsf{P}\left(u + (1 + \theta)\mu_1 \lambda t - \sum_{i=1}^{N(t)} X_i \geq 0 \quad \text{for all} \quad t > 0\right) = \qquad (2.4)$$

$$\mathsf{P}\left(\sum_{i=1}^{N(t)} X_i - (1 + \theta)\mu_1 \lambda t \leq u \quad \text{for all} \quad t\right) = \mathsf{P}(L \leq u).$$

Under the influence of the results of Khinchin (Khinchin, 1932) and Takács (Takács, 1965), Beekman (Beekman, 1968) got the following result. Under the above-mentioned assumptions the ruin probability which on account of (2.4) is equal to $\mathsf{P}(L > u)$, can be represented in the form

$$\mathsf{P}(L > u) = 1 - \frac{\theta}{1+\theta} \sum_{n=0}^{\infty} \frac{H^{*n}(u)}{(1+\theta)^n}, \qquad (2.5)$$

where $H^{*0}(u)$ is the distribution function degenerate at zero and for $n = 1, 2, \ldots$

$$H^{*n} = \frac{1}{\mu_1} \int_0^u H^{*(n-1)}(u - x)[1 - F(x)]\, dx, \qquad u \geq 0.$$

In this connection (2.5) implies that

$$L \overset{d}{=} L_1 + \ldots + L_N, \qquad (2.6)$$

where the random variables $N, L_1, L_2, \ldots$ are independent, N has the geometric distribution

$$\mathsf{P}(N = j) = \frac{\theta}{(1+\theta)^{j+1}}, \qquad j = 0, 1, \ldots, \qquad (2.7)$$

and variables $L_1, L_2, \ldots$ are identically distributed with density

$$h(x) = \frac{1}{\mu_1}[1 - F(x)].$$

Comparison of this result with Example 1.2.1 shows their tight connection. This circumstance even prompted Asmussen (Asmussen, 1987, p. 281) to call (2.5) Pollaczek-Khinchin formula though in actuarial mathematics it is customary to call (2.5) Beekman's convolution formula.

An elegant outline of main steps of the proof of this formula is presented in (Bowers et al., 1986, Ch. 12). This deduction is based on the consideration of moments when the risk process reaches new record values (ladder epochs). There is the probability $1 - \psi(0)$ after each record that this record will not be beaten and correspondingly there is the probability $\psi(0)$ that this record will be beaten. It follows from the fact that the generalized Poisson process $\sum_{i=1}^{N(t)} X_i$ is a homogeneous process with independent increments. In this case according to the aftereffect absence property of a Poisson process (lack of memory) it follows that the variables L_i of the previous record exceedings during i-th exceeding (ladder variables) are independent and identically distributed so that (2.6) takes place. Therefore the number of records before reaching the absolute record has the geometric distribution (2.7). For finding out the distribution of the variables L_i it is sufficient to consider L_1.

All those who want to get acquainted with the proof of formula (2.5) and its interpretation in terms of record values more explicitly are recommended to see the chapters of the book (Feller, 1971) dealing with ladder epochs.

In conclusion of this section we note that the sums of the form (2.3) or (2.6) where the number of the independent identically distributed summands has the geometric distribution and is independent of the summands, are usually called geometric random sums. We will repeatedly meet these objects in this book. In particular, they will be the main objects of the investigation in Chapters 2 and 5.

1.3   Other examples

EXAMPLE 1.3.1. Geiger-Müller counters that are used in nuclear physics and in cosmic radiation research (see Examples 1.1.1 and 1.1.3) work according to the following principle. A particle going into the counter causes a discharge that lasts for some time τ. Any particle going into the counter during the discharge gets lost and is not registered by the counter. By many researchers, τ is assumed constant. However, it is more natural to consider τ as a random variable. Assume that the lengths of time intervals $X_1, X_2, \ldots$ between the particle arrivals are independent identically distributed random variables. Let $\tau_1, \tau_2, \ldots$ be independent identically distributed random variables with the same distribution as the random variable τ and also let $\{X_j\}_{j \geq 1}$ and $\{\tau_j\}_{j \geq 1}$ be independent. It is obvious that the counter loses no particles until the discharge duration is not greater than the length of time interval between the arrivals of the successive particles. The time Z free from losses can be represented in the form

$$Z = X_1 + \ldots + X_N$$

where $N = \min\{k : X_k < \tau_k\}$.

Denote

$$\alpha = P(X_1 > \tau_1).$$

It is easy to notice that the random variable N has the geometric distribution

$$P(N = k) = \alpha^{k-1}(1 - \alpha), \qquad k = 1, 2, \ldots$$

Since N depends on the values taken by the variables $\{X_j\}$ and $\{\tau_j\}$, we should use the Wald identity to calculate the mathematical expectation of Z. According to this identity, if the event $\{N = k\}$ is independent of the events determined by the variables X_n with numbers $n > k$, then

$$EZ = EX_1 \cdot EN.$$

In this case it is obvious that $EN = \frac{1}{1-\alpha}$ so that the mean duration of the time interval Z until the first loss of a particle is equal to

$$EZ = \frac{EX_1}{1 - \alpha}.$$

Whence it follows that the more intensive the particles flow is, the less EX_1 is, and therefore the less the duration of the loss-free work is. The greater the value of α is, the greater EZ is.

EXAMPLE 1.3.2. Consider some service device and let demands for service form a queue. The demands arrive one by one. It is reasonable to consider the time between the arrivals of k-th demand and $k+1$-th demand a random variable. Consider the distribution of a busy period, i.e. the time interval between the arrival of the first demand and the moment of the queue's release from the demands. Assume that the duration Y_k of the service of the k-th demand is a random variable, $k \geq 1$.

It is obvious that the length of a busy period equals to

$$Z = Y_1 + \ldots + Y_N,$$

where

$$N = \min\{k : X_1 + \ldots + X_k > Y_1 + \ldots + Y_k\}.$$

And we again come to the necessity of considering the sum of a random number of independent identically distributed random variables.

EXAMPLE 1.3.3. Some manufacturers produce n products every day. Each product can be defective with probability p. With every defective product the manufacturers sustain losses. The value of loss depends on the type of a defect so that we should consider the value of the loss X_k with the defective product with number k as a random variable with some distribution function $F(x)$. If N is the number of the defective products produced during one day, then the total loss per day is equal to

$$L = X_1 + \ldots + X_N.$$

Under the assumption that the products become or do not become defective independently of each other, the random variable N has the binomial distribution

$$P(N = k) = C_n^k p^k (1 - p)^{n-k}, \qquad k = 0, 1, \ldots, n.$$

According to the law of total probability, under the assumption that the defects are independent, it is easy to calculate that the distribution function of the variable L is equal to

$$G(x) = P(L < x) = \sum_{k=0}^{n} C_n^k p^k (1 - p)^{n-k} F^{*n}(x),$$

If we denote the characteristic function corresponding to the distribution function $F(x)$ as $f(x)$, then the characteristic function of the random variable L becomes equal to

$$g(t) = \sum_{k=0}^{n} C_n^k p^k (1 - p)^{n-k} (f(t))^k = [1 + p(f(t) - 1)]^n.$$

Here

$$\mathsf{E}L = np\mathsf{E}X_1,$$

$$\mathsf{D}L = np[\mathsf{D}X_1 + (1 - p)(\mathsf{E}X_1)^2] = np[\mathsf{E}X_1^2 - p(\mathsf{E}X_1)^2].$$

EXAMPLE 1.3.4. Bypassing such an important and productive branch of probability theory which effectively exploits random sums of random variables as sequential statistical analysis, we will mention one more section of probability theory directly connected with random sums, namely, the theory of branching processes. Later in Chapter 3 we will see how random sums appear in the investigation of branching processes in a very unexpected way. Now we will only formulate some main definitions (moreover, branching processes are random sums by definition) and properties.

Let $\{Z_{n,j}\}_{n \geq 1, j \geq 1}$ be a double array of independent identically distributed random variables taking integer nonnegative values. From this double array construct a sequence of random variables $\{Z_n\}_{n \geq 0}$ by the following rule:

$$Z_0 = 1, \qquad Z_{n+1} = Z_{n,1} + \ldots + Z_{n,Z_n}, \qquad n \geq 1. \qquad (3.1)$$

The sequence of random variables (3.1) is called a branching process with discrete time or a Galton–Watson process. Such sequences describe many phenomena in nuclear physics, bacteriology, epidemiology and other sciences. The properties of Galton–Watson processes are described in full detail in the monographs (Sevastyanov, 1971), (Harris, 1963). We will mention only those properties which will be used further. Denote the generating function of the random variable Z_1 as $f(\cdot), f(s) = p_0 + p_1 s + p_2 s^2 + \ldots$, where $p_i = P(Z_1 = i), i \geq 0$, so that $f(s) = \mathsf{E}s^{Z_1}$. It follows from (3.1) that all random variables $Z_{n,j}$ have $f(s)$ as their generating function.

Let $m = EZ_1$. If $m \leq 1$, then (Sevastyanov, 1971, p. 19) there with probability one will exist such n_0 that $Z_n = 0$ for all $n \geq n_0$. If $m > 1$ then such probability is strictly less than one and is equal to the least root of the equation

$$f(s) = s,$$

belonging to the segment $[0, 1]$. Galton–Watson processes with $m > 1$ which are called supercritical will be the main objects of our attention in conclusive sections of Chapters 3 and 4.

Denote $f_n(s) = Es^{Z_n}$. Then $f_n(s) = {}^n f(s)$, where ${}^n f(s)$ is a n-th functional iteration of f. In particular, this implies that $EZ_n = m^n$.

Mention some properties of supercritical Galton–Watson processes that are of most interest for us. Denote $X_n = Z_n / m^n$.

THEOREM 1.3.1. *Let $\{Z_n\}_{n \geq 1}$ be a supercritical Galton–Watson process and $EZ_1 \log Z_1 < \infty$. Then with probability one $X_n \to X$ with $n \to \infty$, where X is a nondegenerate random variable taking nonnegative values.*

For the PROOF see references in (Heyde, 1970).

Denote the Laplace–Stieltjes transform of the random variable X as ϕ, $\phi(s) = E \exp\{-sX\}$, $\mathrm{Re}\, s \geq 0$.

LEMMA 1.3.1. *Let $\{Z_n\}_{n \geq 1}$ be a supercritical Galton–Watson process and $EZ_1 \log Z_1 < \infty$. Then the functions ϕ and f are connected by the equation*

$$\phi(ms) = f(\phi(s)), \qquad \mathrm{Re}\, s \geq 0, \tag{3.2}$$

$\phi'(0) = -1$. *There exists a unique characteristic function $\phi(-it), t \in \mathbb{R}$, that satisfies equation (3.2) and corresponds to a distribution with first moment equal to one.*

PROOF see in (Harris, 1963, Chapt. 1, Sect. 8.2) and references in (Heyde, 1970).

Chapter 2

Doubling with repair

2.1 Mathematical model

In this chapter we will give an example of an important practical problem that is successfully solved with the help of the apparatus of random summation. This example is of great interest since it has generated a whole sequence of asymptotic problems of the theory of random summation that were completely solved just recently.

To increase the reliability of technical systems additional (reserve) elements are used, that accept loading in the case of the main element failure. Depending on the condition of the reserve element, three types of reservation are considered: loaded (hot), unloaded (cold) and light-weight (warm). Under hot reservation the element stays in the same condition as the main one. Back axis wheels of a truck (two wheels on each side carrying the same loading) are the typical example of hot reservation, as well as a substitute wheel in the luggage compartment of a car is the example of cold reservation. Light-weight reservation supposes the reserve element to be under incomplete loading. Of course the example of the substitute wheel only approximately characterizes the situation of cold reserve, because the wheel is changing even without working: rubber is getting old, metal parts can get rusty. So in reality any cold reserve appears to be a light-weight one.

The reliability of technical systems can also be increased by repair of troubled elements. In this connection one question appears: how much does the reliability of a system increase if there are reserve elements (doubling) staying in light-weight of reservation and a troubled element is directed to the repair?

We still cannot proceed to a mathematical solution of the problem because we do not have a precise formulation yet. For this purpose, we should first construct the mathematical model of the practical problem. A mathematical model is constructed on the basis of observations of the object and determines the mathematical apparatus that should be used.

The mathematical model of the formulated engineering problem is based on the following assumptions.

17

ASSUMPTION 2.1.1. *Trouble-free performance duration of every element is a random variable.*

ASSUMPTION 2.1.2. *Instead of the troubled element a reserve element is set immediately (if there is any).*

ASSUMPTION 2.1.3. *The troubled element is immediately directed to the repair.*

ASSUMPTION 2.1.4. *Renewal duration is a random variable independent of the trouble-free performance duration.*

ASSUMPTION 2.1.5. *A repaired element immediately becomes a reserve one.*

ASSUMPTION 2.1.6. *The repair completely reconstructs all performance properties of the element (including the duration distributions of trouble-free performance and repair).*

ASSUMPTION 2.1.7. *An element does not change its properties staying in reserve.*

We shall call the moment when both elements of our system become incapacitated the moment of system trouble. Our task is to find the distribution of the trouble-free performance duration of a doubled system.

Denote the sought distribution function as $H(x)$.

Introduce the following notations:

X_{2k+1} – trouble-free performance duration of the main element after k renewals,

X_{2k+2} – trouble-free performance duration of the reserve element after k renewals,

Y_{2k+1} – repair duration of the main element after k renewals,

Y_{2k+2} – repair duration of the reserve element after k renewals,

Z – trouble-free performance duration of a doubled system.

Note that according to Assumptions 2.1.1, 2.1.4, 2.1.6 and 2.1.7 random variables $X_1, X_2, \ldots, Y_1, Y_2, \ldots$ are independent; moreover $X_1, X_2, \ldots$ have one and the same distribution function, say $F(x)$, and $Y_1, Y_2, \ldots$ also have one and the same distribution function, say $G(x)$.

Note that Z equals at least to $X_1 + X_2$ and precisely equals to that sum if the renewal of the main element continues longer than the trouble-free performance of the reserve element, that is if $X_2 < Y_1$. If $X_2 \geq Y_1$, but $Y_2 > X_3$, then

$$Z = X_1 + X_2 + X_3.$$

Continuing our reasoning further in the same way we come to the representation

$$Z = X_1 + X_2 + \cdots + X_N,$$

where the random variable N is defined as

$$N = \min\{k \geq 2 : Y_k > Y_{k+1}\}.$$

If we set

$$\alpha = \mathsf{P}(X_{k+1} \geq Y_k) = \int_0^\infty G(x)\,dx,$$

then the distribution of N takes the form

$$P(N = k) = \alpha^{k-2}(1 - \alpha), \qquad k = 2, 3, \ldots$$

Let a be the mean trouble-free performance duration of an element. Then by introducing one element into an unloaded reserve without further renewal one can only double the mean trouble-free performance duration of the system. For calculating the mean performance duration of the doubled system with repair we should use the Wald identity, according to which

$$\mathsf{E}Z = a\,\mathsf{E}N.$$

It is easy to calculate that

$$\mathsf{E}N = 1 + \frac{1}{1 - \alpha},$$

and therefore

$$\mathsf{E}Z = a\left(1 + \frac{1}{1 - \alpha}\right).$$

If we set $\alpha = 0$, then we get the system without restoration, for which $\mathsf{E}Z = 2a$ — and we already knew it. The effectiveness of reservation and repair can be defined by the formula

$$\frac{\mathsf{E}Z}{a} = 1 + \frac{1}{1 - \alpha}.$$

This formula shows that there are two ways of increasing the mean trouble-free performance duration of the system: a) to increase trouble-free performance duration of the element, that is the value of a, or b) to accelerate the repair, that is to bring α to 1. Anyone who had some practice of increasing reliability of technical objects knows how hard every percent of the increase in the mean trouble-free performance duration is. At the same time even a small increase of the parameter α can cause an essential increase of the mathematical expectation of Z. For example, by increasing a by 10% we are increasing $\mathsf{E}Z$ by 10% too. By increasing α by 10% — from 0.8 to 0.9 — we increase $\mathsf{E}Z$ from $6a$ to $11a$, that is almost twice.

Now we will dwell on the problem of finding the function $H(x)$. It would be more suitable for us to deal not with the distribution functions, but with their complements to one, that is with the functions

$$\overline{H}(x) = 1 - H(x) = P(Z \geq x), \qquad \overline{F}(x) = 1 - F(x).$$

Trouble-free performance duration of the duplicated system can be greater than x in two mutually exclusive cases:

1) The main element will perform for time greater than x;

2) The main element will fail at some moment z $(z < x)$, but the whole system will perform for time greater than x — the system consisting of the reserve element and the element failed at moment z will perform for time greater than $x - z$.

Denote the probability that the system consisting of the reserve element and the element troubled at moment 0 will perform trouble-free for time u as $P(u)$. Then by the summation theorem the following relation takes place

$$\overline{H}(x) = \overline{F}(x) + \int_0^x \overline{P}(x - z)\, dF(z). \tag{1.1}$$

Now find the function $P(u)$. Note that the system consisting of the reserve element and the element troubled at the moment $u = 0$ can perform for a time greater than u only in two mutually exclusive cases:

1) The reserve element will perform for time greater than u;

2) The reserve element will fail at a moment v ($v < u$), but the main element will be renewed by that time, and the system consisting of the repaired main element and the element failed at the moment v will perform for time greater than $u - v$.

It is easy to show that hence follows the equation

$$\overline{P}(u) = \overline{F}(u) + \int_0^u \overline{P}(u - v)G(v)\, dF(v). \tag{1.2}$$

For the unknown function $H(x)$ we have simple integral equations. Such equations are usually solved in terms of Laplace-Stieltjes transforms. Set

$$\chi(s) = \int_0^\infty e^{-sx}\, dH(x), \quad \phi(s) = \int_0^\infty e^{-sx}\, dF(x),$$

$$\gamma(s) = \int_0^\infty e^{-sx}G(x)\, dF(x), \quad \pi(s) = \int_0^\infty e^{-sx}\, dP(x).$$

In terms of the introduced Laplace-Stieltjes transforms equations (1.1) and (1.2) obtain the form

$$\chi(s) = \phi(s)\pi(s),$$
$$\pi(s) = \phi(s) - \gamma(s) + \pi(s)\gamma(s),$$

whence we have

$$\chi(s) = \phi(s)\frac{\phi(s) - \gamma(s)}{1 - \gamma(s)}.$$

The problem is solved completely. In fact, a Laplace-Stieltjes transform with the help of the inversion formula permits us to find the distribution function. Besides, the formulas

$$m_k = \int_0^\infty x^k\, dH(x) = (-1)^k \frac{d^k\chi(s)}{ds^k}\bigg|_{s=0}$$

make it possible to calculate the moments of the distribution $H(x)$ for any k. In particular, with $k = 1$ and $k = 2$ we get (using the logarithmic derivative):

$$\frac{\chi'(s)}{\chi(s)} = \frac{\phi'(s)}{\phi(s)} + \frac{\phi'(s) - \gamma'(s)}{\phi(s) - \gamma(s)} + \frac{\gamma'(s)}{1 - \gamma(s)},$$

$$\frac{\chi''(s)\chi(s) - (\chi'(s))^2}{(\chi(s))^2} = \frac{\phi''(s)\phi(s) - (\phi'(s))^2}{(\phi(s))^2} +$$

$$\frac{(\phi''(s) - \gamma''(s))(\phi(s) - \gamma(s)) + (\phi'(s) - \gamma'(s))^2}{(\phi(s) - \gamma(s))^2} +$$

$$\frac{\gamma''(s)(1 - \gamma(s)) + (\gamma'(s))^2}{(1 - \gamma(s))^2}.$$

The first of these relations gives us the formula for EZ we obtained earlier. The second one allows us to obtain additional facts.

2.2   A limit theorem for the trouble-free performance duration

The result we obtained in the previous section gives the complete solution of the problem. With its help we can find many characteristics required for theory and practice. However, this solution is not completely satisfactory since we cannot directly observe those implications that, for example, follow from the modification of the function $G(x)$, which has the meaning of the repair rate. In this section we will prove a theorem that allows us to make this question clear and besides gives a simpler expression for the solution of the original problem. In formulation of the problem we will rely on common sense and accumulated practical experience.

In most of real situations the trouble-free performance duration of the element exceeds its renewal duration. For example, a car wheel run exceeds several times the time necessary for the renewal of the inner tube. In just the same way the trouble-free performance duration of the complicated systems – computers or TV sets – as a rule exceeds many times the duration of their repair. For modern technical goods this state of things must be taken for granted.

What conclusion can be made from the fact that the renewal duration is usually less than the trouble-free performance duration of the element? We saw above that

$$\alpha = P(X_{k+1} \geq Y_k) = \int\limits_0^\infty G(x)\,dF(x).$$

Our assumption that the trouble-free performance duration is almost always greater than the repair duration can be formalized with the help of the relation

$$\alpha \to 1 \qquad \text{(but is not equal to 1)}$$

or which is the same,

$$\beta = 1 - \alpha \to 0 \quad \text{(but is not equal to 0)}.$$

Imagine that we successively improve the renewal system. Denote the distribution function of the repair duration at the n-th improval stage as $G_n(x)$, and supply with the index n all variables and functions depending on it. Thus, in particular, assume

$$\gamma_n(s) = \int\limits_0^\infty e^{-sx} G_n(x)\, dF(x), \quad \alpha_n = \int\limits_0^\infty G(x)\, dF(x).$$

Let Z_n be the trouble-free performance duration of the doubled system and $\chi_n(s)$ be its Laplace-Stieltjes transform. With these notations we have

$$\chi_n(s) = \phi(s)\frac{\phi(s) - \gamma_n(s)}{1 - \gamma_n(s)}, \quad A_n = \mathsf{E}Z_n = a\left(1 + \frac{1}{\beta_n}\right),$$

where

$$\beta_n = 1 - \alpha_n = \int\limits_0^\infty (1 - G_n(x))\, dF(x).$$

THEOREM 2.2.1. *Let in addition to the preceding assumptions the random variables X_k have finite mathematical expectation $a > 0$ and $\beta_n \to 0$ as $n \to \infty$. Then*

$$\mathsf{P}(Z_n < xA_n) \Rightarrow 1 - e^{-x} \quad (n \to \infty).$$

PROOF. According to the theorem of continuity of the correspondence between distribution functions of nonnegative random variables and their Laplace-Stieltjes transforms, a sequence of distribution functions weakly converges to its limit if and only if corresponding Laplace-Stieltjes transforms converge to the Laplace-Stieltjes transform of the limit distribution in every point. Thus to prove the theorem it suffices to show that

$$\chi_n\left(\frac{s}{A_n}\right) \to \frac{1}{1 + s}$$

as $n \to \infty$ for every $s \geq 0$ (it is easy to see that $\frac{1}{1+s}$ is the Laplace-Stieltjes transform of the exponential distribution). Later we will need an estimate of the difference

$$\beta_n - \beta_n(s),$$

where

$$\beta_n(s) = \phi\left(\frac{s}{A_n}\right) - \gamma\left(\frac{s}{A_n}\right) = \int\limits_0^\infty \exp\left\{-\frac{sx}{A_n}\right\}[1 - G_n(x)]\, dF(x).$$

It is obvious that

$$0 \le \beta - \beta_n(s) = \int\limits_0^\infty \left(1 - \exp\left\{-\frac{sx}{A_n}\right\}\right)[1 - G_n(x)]\,dF(x).$$

It is known that with $z \ge 0$,

$$1 - e^{-z} \le z,$$

so that

$$0 \le \beta_n - \beta_n(s) =$$

$$\left(\int\limits_0^{\sqrt{A_n}} + \int\limits_{\sqrt{A_n}}^\infty\right)\left(1 - \exp\left\{-\frac{sx}{A_N}\right\}\right)[1 - G_n(x)]\,dF(x) \le$$

$$\frac{s}{A_n}\int\limits_0^{\sqrt{A_n}} x[1 - G_n(x)]\,dF(x) + 2\int\limits_{\sqrt{A_n}}^\infty [1 - G(x)]\,dF(x) \le$$

$$\frac{s}{\sqrt{A_n}}\beta_n + o(\beta_n),$$

or

$$0 \le 1 - \frac{\beta_n(s)}{\beta_n} = o(1)$$

uniformly in every finite segment s.

It is easy to verify that

$$\chi_n\left(\frac{s}{A_n}\right) = \phi\left(\frac{s}{A_n}\right)\frac{1}{1 - \dfrac{\phi\left(\frac{s}{A_n}\right) - 1}{\beta_n(s)}} = \phi\left(\frac{s}{A_n}\right)\frac{1}{1 - \dfrac{\phi\left(\frac{s}{A_n}\right) - 1}{\beta_n}\dfrac{\beta_n}{\beta_n(s)}},$$

and therefore according to the preceding estimate,

$$\chi_n\left(\frac{s}{A_n}\right) = \phi\left(\frac{s}{A_n}\right)\frac{1}{1 - (1 + o(1))\dfrac{\phi\left(\frac{s}{A_n}\right) - 1}{\beta_n}}.$$

Further under conditions of the theorem on every finite segment

$$\phi\left(\frac{s}{A_n}\right) = 1 + o(1),$$

so that

$$\chi_n\left(\frac{s}{A_n}\right) = \frac{1 + o(1)}{1 - \dfrac{\phi\left(\frac{s}{A_n}\right) - 1}{\beta_n}}. \qquad (2.1)$$

Now consider the relation

$$\frac{\phi\left(\frac{s}{A_n}\right)-1}{\beta_n(s)} = \frac{\phi\left(\frac{s}{A_n}\right)-1}{s\beta_n(s)}\frac{\beta_n s}{\beta_n(s)}.$$

By virtue of the definition of a derivative,

$$\frac{\phi\left(\frac{s}{A_n}\right)-1}{\frac{s}{A_n}} \longrightarrow \phi'(0) \quad (n \to \infty).$$

But we know that $\phi'(0) = -a$ and $\beta_n = [1 + o(1)]/(A_n a)$. Therefore having gathered all the estimates together we conclude that uniformly in every finite segment of the domain of s,

$$\chi_n\left(\frac{s}{a_n}\right) \to \frac{1}{1+s} \qquad (n \to \infty),$$

Q.E.D.

We can give another formulation of the above theorem on the basis of the representation of Z_n as a random sum of identically distributed independent random variables. In this case the number of summands N_n depends both on values taken by individual summands and on the other sequence of independent random variables $\{Y_j^{(n)}\}_{j\geq 1}$; each of those has the distribution function $G_n(x)$, but the summands in the sum and the random variables $\{Y_j^{(n)}\}_{j\geq 1}$ are independent for every n. Remember that in our case the variables N_n are defined as

$$N_n = \max\{k \geq 2 : X_k < Y_{k-1}^{(n)}\}. \tag{2.2}$$

The new formulation of Theorem 2.2.1 is as follows.

THEOREM 2.2.1′. *In the above assumptions on the sequences* $\{X_j\}$ *and* $\{Y_j^{(n)}\}$, *if the variables* $X_1, X_2, \ldots$ *have a finite mathematical expectation* $a > 0$ *each and* $\beta_n \to 0$ *as* $n \to \infty$, *then*

$$P\left(\frac{1}{A_n}(X_1 + X_2 + \ldots + X_{N_n}) < x\right) \Rightarrow 1 - e^{-x} \quad (n \to \infty) \tag{2.3}$$

where the random variable N_n *is defined by relation (2.2).*

We have proved the limit theorem whose statement is similar to the classical ones. However, the randomness of the number of the summands introduces new in principle specific regularities. In fact, if, for example, in (2.3) the number of summands was nonrandom then we had to deal with the classical law of large numbers, according to which the degenerate law should be in the right-hand side of (2.3). In Chapters 3 and 4 we will see how and why the change of limit laws does happen as compared to the classical theorems.

2.3 The class of limit laws

The theorem proved in the previous section inevitably gives rise to a number of questions. And the first one is - if we stay under the conditions of identity of distributions, independence and positiveness of the summands, but relax the assumption of existence of finite mathematical expectations, then what limit distributions can appear for the sums

$$S_n = \frac{1}{B_n}(X_1 + \ldots + X_n) \tag{3.1}$$

where $\{X_j\}_{j\geq 1}$ are independent identically distributed random variables, the variables N_n are defined by relation (2.2), in which $\{Y_j^{(n)}\}_{j\geq 1}$ are independent identically distributed random variables that are independent of the sequence $\{X_j\}_{j\geq 1}$ for every n, and B_n are appropriately chosen positive functions of n? However, our reasoning will be also true under the assumption that the indices N_n are independent of $\{X_j\}_{j\geq 1}$ for every n (see (Gnedenko and Freyer, 1969)).

In this section we assume that the expectations EN_n do not increase too fast with the increase of n, namely, assume that as $n \to \infty$

$$\frac{EN_{n+1}}{EN_n} \to 1,$$

or in the notation of the previous paragraph,

$$\frac{1-\alpha_{n+1}}{1-\alpha_n} = \frac{\beta_{n+1}}{\beta_n} \to 1 \quad (n \to \infty).$$

The main aim of this paragraph is the proof of the following statement.

THEOREM 2.3.1. *Under the above conditions, a distribution function $\Psi(x)$ can be the limit one for the distribution functions of sums (3.1) within an appropriate choice of constants B_n if an only if its Laplace-Stieltjes transform has the form*

$$\psi(s) = \frac{1}{1+cs^\delta}, \tag{3.2}$$

where constants c and δ are nonnegative and $0 < \delta \leq 1$.

Preface the proof of this theorem with some auxiliary statements.

LEMMA 2.3.1. *If a distribution function $F(x)$ is such that all the integrals mentioned further exist, and functions $\phi_1(x)$ and $\phi_2(x)$ are non-decreasing and non-increasing, respectively, then the following inequality takes place*

$$\int_0^\infty \phi_1(x)\phi_2(x)\, dF(x) \leq \int_0^\infty \phi_1(x)\, dF(x) \cdot \int_0^\infty \phi_2(x)\, dF(x).$$

PROOF. Let

$$a = \int_0^\infty \phi_2(x) \, dF(x)$$

and

$$z = \sup\{x : \phi_2(x) \geq a\}.$$

Then for $x > z$ the inequality $\phi_2(x) < a$ holds. Now it is evident that

$$\int_0^\infty \phi_1(x)\phi_2(x) \, dF(x) - \int_0^\infty \phi_1(x) \, dF(x) \cdot \int_0^\infty \phi_2(x) \, dF(x) =$$

$$= \int_0^\infty \phi_1(x)[\phi_2(x) - a] \, dF(x) =$$

$$= \int_0^z \phi_1(x)[\phi_2(x) - a] \, dF(x) + \int_z^\infty \phi_1(x)[\phi_2(x) - a] \, dF(x) \leq$$

$$\leq \phi_1(z) \int_0^z [\phi_2(z) - a] \, dF(x) + \phi_1(z) \int [\phi_2(x) - a] \, dF(x) =$$

$$= \phi_1(z) \int_0^\infty [\phi_2(x) = a] \, dF(x) = 0.$$

This chain of inequalities proves our lemma.

This lemma was proved by A. D. Solov'ev in 1965. Later he noticed that it was known as far ago as to P. L. Chebyshev. The statement of the lemma becomes clear from the fact that the covariance between $\phi_1(X)$ and $\phi_2(X)$ is negative, where X is a random variable with finite values of $E\phi_1(X), E\phi_2(X)$ and $E\phi_1(X)\phi_2(X)$, which is actually elucidated by the reduced chain of inequalities.

LEMMA 2.3.2. *If the distribution functions of sums (3.1) weakly converge to a proper distribution function under the above assumptions, then*

$$B_n \to \infty \qquad (n \to \infty).$$

LEMMA 2.3.3. *If the distribution functions of sums (3.1) weakly converge to a proper distribution function under the above assumptions concerning β_n, then*

$$\frac{B_{n+1}}{B_n} \to 1 \qquad (n \to \infty).$$

LEMMA 2.3.4. *If the distribution functions of sums (3.1) weakly converge to a proper distribution function and the sequence C_n is such that*

$$\frac{B_n}{C_n} \to 1 \qquad (n \to \infty),$$

then the sequence of sums $B_n S_n / C_n$ has the same distribution as the sequence S_n.

The proofs of the last three lemmas are not given here, because they precisely repeat the proofs of the statements of Sections 10 and 29 of the monograph (Gnedenko and Kolmogorov, 1954) with the account of the representation (2.1), in which A_n are replaced by B_n.

LEMMA 2.3.5. *If $B_n \to \infty$ as $n \to \infty$, then*

$$\frac{\beta_n(s)}{\beta_n} \to 1 \qquad (n \to \infty)$$

uniformly on every finite segment $s, 0 \le s \le 1$.

PROOF. It is clear that

$$0 \le 1 - \frac{\beta_n(s)}{\beta_n} = \frac{\beta_n - \beta_n(s)}{\beta_n} =$$

$$\frac{1}{\beta_n} \left\{ \int_0^\infty [1 - G_n(x)] \, dF(x) - \right.$$

$$\left. \int_0^\infty \exp\left\{ -\frac{sx}{B_n} \right\} [1 - G_n(x)] \, dF(x) \right\} =$$

$$\frac{1}{\beta_n} \int_0^\infty \left(1 - \exp\left\{ -\frac{sx}{B_n} \right\} \right) (1 - G_n(x)) \, dF(x).$$

But according to Lemma 2.3.1 the last expression does not exceed

$$\frac{1}{\beta_n} \int_0^\infty \left(1 - \exp\left\{ -\frac{sx}{B_n} \right\} \right) dF(x) \times$$

$$\frac{1}{\beta_n} \int_0^\infty [1 - G_n(x)] \, dF(x) = 1 - \phi\left(\frac{s}{B_n} \right).$$

But as far as $B_n \to \infty$ as $n \to \infty$, the lemma is proved.

PROOF OF THEOREM 2.3.1. According to the results of Sect. 2.2., we should elucidate the form of the limit Laplace-Stieltjes transforms for

$$\chi_n\left(\frac{s}{B_n} \right) = \phi\left(\frac{s}{B_n} \right) \frac{\phi\left(\dfrac{s}{B_n} \right) - \gamma_n\left(\dfrac{s}{B_n} \right)}{1 - \gamma_n\left(\dfrac{s}{B_n} \right)} =$$

$$\phi\left(\frac{s}{B_n}\right)\frac{B_n}{\beta_n(s)+\left[1-\phi\left(\frac{s}{B_n}\right)\right]}.$$

Since by virtue of Lemma 2.3.2 $B_n \to \infty$, we have $\phi\left(\dfrac{s}{B_n}\right) \to 1$ with $n \to \infty$ for every s. Therefore if

$$\chi_n\left(\frac{s}{B_n}\right) \to \psi(s) \quad \text{with} \quad n \to \infty,$$

then necessarily

$$\frac{1-\phi\left(\dfrac{s}{B_n}\right)}{\beta_n} \to \alpha(s) \tag{3.3}$$

and

$$\psi(s) = \frac{1}{1+\alpha(s)}.$$

Thus our main task is to determine the form of the function $\alpha(s)$.

According to the preceding lemmas, for any $s > 0$ and $s^* > 0$ one can find integer numbers $m = m_n(s^*/s)$, such that with $n \to \infty$,

$$\frac{B_n \dfrac{s}{s'}}{B_m} \to 1.$$

It is evident that

$$\alpha(s') = \lim_{n\to\infty} \frac{1-f\left(\dfrac{s^*}{B_n}\right)}{\beta_n} = \lim_{n\to\infty}\left[\frac{1-f\left(\dfrac{s}{B_n\frac{s}{s^*}}\right)}{\beta_m}\cdot\frac{\beta_m}{\beta_n}\right].$$

Since the limit in the left-hand side of this equality exists and the limit of the first multiplier in the right-hand side exists also, the limit of the second multiplier, i.e. β_m/β_n, should exist. Denote it as $k(s^*/s)$. As a result we obtain

$$\alpha(s') = \alpha(s)k\left(\frac{s^*}{s}\right). \tag{3.4}$$

Since $\alpha(s)$ is an analytical function for $Res > 0$, we can use the following expansion by the Taylor formula

$$\alpha(s^*) = \alpha(s) + \alpha'(s)(s^* - s) + o(s^* - s).$$

Substituting this expression in (3.4) yields

$$\alpha'(s)(s^a st - s) = \alpha(s)\left[k\left(\frac{s^*}{s}\right) - 1\right] + o(s^a st - s). \tag{3.5}$$

Note that $k(1) = 1$. Divide (3.5) by $\alpha(s)(s^* - s)$. Get

$$\frac{\alpha'(s)}{\alpha(s)} = \frac{k\left(\dfrac{s^*}{s}\right) - k(1)}{\dfrac{s^*}{s} - 1} \cdot \frac{1}{s} + o(1).$$

Now let $s^* \to 1$. The last equality transforms into the equation

$$\frac{\alpha'(s)}{\alpha(s)} = \frac{\delta}{s},$$

where $\delta = k'(s)|_{s=1}$. The solution of this equation has the form

$$\alpha(s) = cs^\delta,$$

where c is an arbitrary constant. Thus the Laplace-Stieltjes transform of the limit distribution should have the form (3.2). And we are to clarify for which c and δ this function $\psi(s)$ will really be a Laplace-Stieltjes transform.

First of all notice that the Laplace-Stieltjes transform of a distribution function is a nonnegative function varying between 0 and 1. Hence it follows that c should be nonnegative. Really, if c is negative, then the function $\psi(s)$ has a pole at $s^\delta = -\frac{1}{c}$.

Further if $\psi(s)$ is the Laplace-Stieltjes transform of a nonnegative random variable, then its second derivative has to be nonnegative for all $s > 0$. It is easy to calculate that

$$\psi''(s) = c\delta s^{\delta-2} \frac{c(\delta + 1)s^\delta - (\delta - 1)}{(1 + cs^\delta)^3}.$$

We see that when $s < s_0$, s_0 being a root of the equation

$$cs^\delta = \frac{\delta - 1}{\delta + 1},$$

$\psi''(s)$ is negative for $\delta < 1$. Hence there should be $\delta \leq 1$. But δ cannot be negative, as if $\delta < 0$, then

$$\psi'(s) = \frac{-c\delta s^{\delta-1}}{1 + cs^\delta} > 0.$$

But this is impossible for a Laplace-Stieltjes transform. Thus we have finally determined that there should be $c > 0$ and $0 < \delta \leq 1$ in (3.2).

We have to verify that every function of the form (3.2) is the Laplace-Stieltjes transform of some distribution function when $c > 0$ and $0 < \delta \leq 1$. It follows from the theory of completely monotone functions. A function $\psi(s)$ is called completely monotone if it is defined for all $0 \leq s < \infty$, has all derivatives for $s > 0$, that satisfy for every natural n the inequality

$$(-1)^n \psi^{(n)} \geq 0, \qquad s > 0.$$

According to the famous Bernstein theorem, a function $\psi(s)$, defined for $s \geq 0$, is the Laplace-Stieltjes transform of some distribution function if and only if it is completely monotone and $\psi(0) = 1$ (see, e.g., (Feller, 1971)). It is known that if $\omega(s)$ is a completely monotone function and $\rho(s)$ is a positive function with a completely monotone derivative, then the function $\omega(\rho(s))$ is completely monotone.

Set $\omega(s) = \frac{1}{1+s}$ and $\rho(s) = cs^\delta$, where $0 < \delta \leq 1$. Then the function $\omega(s)$ is the Laplace-Stieltjes transform of the standard exponential distribution, and $\rho(s)$ is a positive function with a completely monotone derivative. Therefore $\psi(s) = \omega(\rho(s))$ is a completely monotone function. Since $\psi(0) = 1$, $\psi(s)$ is the Laplace-Stieltjes transform of some distribution function. The proof is completed.

We will denote by $\mathcal{K}$ the class of limit distributions defined by (3.2) in honor of I. N. Kovalenko, who was the first to describe it in (Kovalenko, 1965).

Note that for now the distributions from class $\mathcal{K}$, for which explicit representations are known, are exhausted by two laws:

1) Exponential distribution with distribution function $\Psi(x) = 1 - e^{-x/c}$, $x \geq 0$; this corresponds to $\psi(s) = \frac{1}{1+cs}$;

2) Distribution defined by the density

$$p(x) = \frac{1}{\sqrt{\pi x}} - \frac{2e^x}{\sqrt{x}} \int\limits_{\sqrt{x}}^{\infty} e^{-z^2}\, dz,$$

this corresponds to $\psi(s) = \frac{1}{1+\sqrt{s}}$.

2.4 Some properties of limit distributions

The absence of known explicit representations for the laws from the class $\mathcal{K}$ (with the exception of the two mentioned) makes the problem of description of the properties of the laws from this class very important. In this section we shall specify some of them.

The notion of unimodal distribution functions plays a significant part in mathematical statistics from the beginning of this century. A distribution function was called unimodal if its density $f(x) = F'(x)$ had the only maximum. This notion was generalized by A. Ya. Khinchin in 1938 in the following way.

DEFINITION 2.4.1. A distribution function $F(x)$ is called unimodal if there exists at least one value $x = a$ such that for $x < a$ the function $F(x)$ is convex and for $x > a$ it is concave.

It is easy to check that normal, exponential, uniform on segment $[a, b]$, Cauchy and Laplace distributions are unimodal.

A function $\Psi(x)$ is called convex on the halfline $x < a$ if for any $x_1 < a$ and $x_2 < a$ the following inequality is fulfilled

$$2\Psi\left(\frac{x_1 + x_2}{2}\right) \leq \Psi(x_1) + \Psi(x_2).$$

If for all $x_1 < a$ and $x_2 < a$ the opposite inequality is fulfilled, then the function Ψ is called concave.

Convex and concave functions are absolutely continuous; that is, their right-side and left-side derivatives differ in no more than a countable number of points.

A. Ya. Khinchin found an elegant necessary and sufficient condition of unimodality of a distribution $F(x)$ at the point $x = 0$. It turned out that $F(x)$ is unimodal with the vertex in the point $x = 0$ if and only if its characteristic function $\phi(t)$ is representable in the form

$$\phi(t) = \frac{1}{t}\int\limits_0^t \psi(z)\,dz,$$

where $\psi(z)$ is some characteristic function.

It can be shown (almost literally repeating Khinchin's proof) that the distribution function of a nonnegative random variable is unimodal in the point $x = 0$ if and only if its Laplace-Stieltjes transform is representable in the form

$$\phi(s) = \frac{1}{s}\int\limits_0^s v(z)\,dz, \tag{4.1}$$

where $v(s)$ is the Laplace-Stieltjes transform of some nonnegative random variable.

THEOREM 2.4.1. *All distributions from the class $\mathcal{K}$ are unimodal with the vertex in the point $x = 0$.*

PROOF. Assume that the equality (4.1) holds for the functions from the class $\mathcal{K}$ and find the form of the corresponding function $v(z)$. It is evident from (4.1) that

$$v(z) = \phi(z) + z\phi'(z) = \frac{1}{1 + cz^\delta} -$$
$$-\frac{\gamma cz^\delta}{(1 + cz^\delta)^2} = \frac{1 - \delta}{1 + cz^\delta} + \frac{\delta}{(1 + cz^\delta)^2}.$$

The first summand in the right-hand side of the last equality is a completely monotone function. The second summand is also completely monotone as the power of a completely monotone function. The sum of completely monotone functions is completely monotone. As $v(0) = 1$, then by Bernstein's theorem

$v(z)$ is the Laplace-Stieltjes transform of some distribution. The theorem is proved.

THEOREM 2.4.2. *All distribution functions from the class $\mathcal{K}$ have unbounded densities for $\delta < 1$.*

PROOF. In Chapt. 13, Sect. 4 of (Feller, 1971) a statement is presented which establishes that a distribution density is bounded by a number B if and only if for every $s > 0$ its Laplace-Stieltjes $\psi(s)$ transform satisfies the inequalities

$$0 \le \frac{(-1)^n \psi^{(n)}(s)}{n!} \le \frac{B}{s}, \qquad n = 1, 2, \ldots$$

At the same time, for the laws from the class $\mathcal{K}$ with $\delta < 1$ for $n = 1$ we have

$$0 \le \frac{-s\psi'(s)}{1!} = \frac{\delta c s^\delta}{(1 + c s^\delta)^2},$$

and therefore for s large enough (for every s satisfying the condition $s^{1-\delta} > \sqrt{2}$) the following inequality holds

$$\frac{-s\psi'(s)}{1!} > \frac{B}{s}.$$

In other words, the mentioned inequalities are not fulfilled already for $n = 1$. By virtue of Theorem 2.4.1 any distribution density from the class $\mathcal{K}$ with $\delta < 1$ infinitely increases when its argument aims for zero from the right. The theorem is proved.

Remember one important definition.

DEFINITION 2.4.2. A distribution function $F(x)$ is called infinitely divisible if for any natural n there exists a distribution function F_n, whose n-fold convolution is equal to F.

This definition can be reformulated in terms of characteristic functions. Namely, $F(x)$ is an infinitely divisible distribution function if for any natural n its characteristic function is the n-th power of some characteristic function.

THEOREM 2.4.3. *All laws from the class $\mathcal{K}$ are infinitely divisible.*

PROOF. In Chapter 13 of (Feller, 1971) a criterion is given for a Laplace-Stieltjes transform to correspond to an infinitely divisible distribution function. Namely, a function $\psi(s)$ is the Laplace-Stieltjes transform of an infinitely divisible distribution function if and only if $\psi(0) = 1$ and its logarithmic derivative is completely monotone. For functions from the class $\mathcal{K}$ we have

$$-[\log \psi(s)]' = [\log(1 + c s^\delta)]' = \frac{1}{1 + c s^\delta} \cdot \frac{c\delta}{s^{1-\delta}}.$$

For $\delta = 1$ this function is obviously completely monotone. For $\delta < 1$ it is also completely monotone as the product of two completely monotone functions. The proof is completed.

Before we pass to the formulation of the next, characteristic, property of distributions of class $\mathcal{K}$, we shall introduce a notion which will be intensively used throughout the book. Consider a sequence $X_1, X_2, \ldots$ of independent nonnegative random variables with the same distribution function $F(x)$. On the real line define the points $Y_0 = 0, Y_1 = X_1, Y_2 = X_1 + X_2, Y_3 = X_1 + X_2 + X_3, \ldots$. Let p be any number from the interval $0 < p < 1$. Construct a new sequence $\{Y_j\}_{j \geq 1}$ of random variables according to the following principle. Every point Y_j with the exception of Y_0 will be included in the new sequence independently of the others with the same probability p (at the same time we can say that each Y_j is excluded with probability $1-p$). Denote the probability distribution of the remaining point with minimal abscissa by $F_p(x)$. The operation described will be called an elementary rarefaction. This operation was considered for the first time by A. Rényi (Rényi, 1956).

This operation assumes a simple interpretation. Let Y_j be times when some events occur and in accordance with what has been said above, let the flow of these events possess the property that time intervals between successive occurrences of the events are independent and identically distributed random variables. In many practical situations the original flow becomes more and more rare. That is how the things are getting on a conveyor if defective goods are rejected after every operation. The same is going on in the misprints correction by several correctors. In both these examples the operation of rarefaction is applied succesively several times. A. Rényi in the mentioned work showed that with multiple application of elementary rarefaction, the remaining points form Poisson process in the limit. We will return to rarefied flows in a more general situation in Sect. 3.6. For now we only formulate:

DEFINITION 2.4.3. A distribution function $F(x)$ is called stable with respect to elementary rarefaction if for every $p \geq 0$ there exists an $a_p > 0$ such that for all x,

$$F_p(x) = F\left(\frac{x}{a_p}\right),$$

or in terms of Laplace-Stieltjes transforms,

$$\phi_p(s) = \phi(a_p s),$$

where ϕ_p and ϕ are Laplace-Stieltjes transforms of the points with minimal abscissas in the rarefied and non-rarefied sequences.

THEOREM 2.4.4. *A distribution function $F(x)$ belongs to the class $\mathcal{K}$ if and only if it is stable with respect to elementary rarefaction.*

PROOF. It is clear that as far as the probability that the first remaining point will have the number k in the original sequence is equal to $p^{k-1}(1 - p), k = 1, 2, \ldots$, we have

$$F_p(x) = \sum_{k-1}^{\infty} p^{k-1}(1 - p) F^{*k}(x).$$

We have already met these geometric convolutions in Chapter 1 where we made sure that the above relation can be written in terms of Laplace-Stieltjes transforms as

$$\phi_p(s) = \sum_{k=1}^{\infty} p^{k-1}(1-p)\psi^k(s) = \frac{(1-p)\psi(s)}{1-p\psi(s)}. \tag{4.2}$$

The "only if" part of the theorem is proved by the direct substitution of expression (3.2) for the laws of class $\mathcal{K}$ into (4.2):

$$\phi_p(s) = \frac{\frac{1-p}{1+cs^\delta}}{1-\frac{p}{1+cs^\delta}} = \frac{1-p}{1-p+cs^\delta} = \frac{1}{1+c((1-p)^{-1/\delta}s)^\delta} = \psi((1-p)^{-1/\delta}s).$$

We see that the constants a_p are connected with the parameter of elementary rarefaction $p = 1 - q$ by a simple relation $a_p = (1-p)^{-1/\delta}$. In particular, for $p = 0$ the equality $a_0 = 1$ takes place. This equality is intuitively evident, as far as for $p = 0$ the distribution $F(x)$ does not change under the operation of elementary rarefaction.

To prove the "if" part of the theorem we will show that every distribution that does not change under elementary rarefaction belongs to the class $\mathcal{K}$. Assume that for any p $(0 \le p < 1)$ the following identity holds

$$\phi_p(s) = \frac{(1-p)\psi(s)}{1-p\psi(s)} = \psi(a_p s).$$

Hence

$$\psi(a_p s) - \psi(s) = \frac{\psi(s)(\psi(s) - 1)}{1 - p\psi(s)}$$

or

$$\frac{\psi(a_p s) - \psi(s)}{a_p s - s} = \frac{p}{a_p - 1} \cdot \frac{\psi(s)[\psi(s) - 1]}{s[1 - p\psi(s)]}.$$

Let $s > 0$. Pass to the limit as $p \to 0$. The limit of the left-hand side of the latter equality exists and equals $\psi'(s)$. The second multiplier of the right-hand side also has the limit equal to $\frac{1}{s}\psi(s)[\psi(s) - 1]$. This means that the limit of the first multiplier in the right-hand side exists. Denote it as δ. As a result we come to the equation

$$\psi'(s) = \frac{\delta}{s}\psi(s)[\psi(s) - 1].$$

It is easy to verify that the solutions of this equation have the form (3.2). But of all the possible solutions we need only those that are Laplace-Stieltjes transforms of some distribution functions. But above we have seen that this is so if and only if $c > 0$ and $0 < \delta \le 1$. The theorem is proved.

2.5   Domains of geometric attraction of the laws from class $\mathcal{K}$

In Section 2.2 we have seen that if the summands of the sum

$$Z_n = X_1 + \ldots + X_{N_n}$$

are nonnegative, have a finite expectation different from zero, and the index N_n defined by (2.2) has the geometric distribution, then equality (2.3) holds true, i.e., the distribution function of the ratio $Z_n/\mathbf{E}Z_n$ weakly converges to $1 - e^{-x}, x \geq 0$, as $n \to \infty$.

DEFINITION 2.5.1. Assume that random variables $X_1, X_2, \ldots$ are independent and have the same distribution $F(x)$. We will say that $F(x)$ belongs to the domain of geometric attraction of the distribution $\Psi(x)$ if there exists a sequence of positive constants $\{B_n\}$ such that

$$\mathbf{P}\left(\frac{1}{B_n}(X_1 + \ldots + X_{N_n}) < x\right) \Rightarrow \Psi(x) \qquad (n \to \infty).$$

Strictly speaking, we should mention that in this definition the indexes N_n are defined by relation (2.2). However, the results formulated below also remain valid if the indexes N_n are assumed independent of $\{X_j\}_{j \geq 1}$ (see (Kruglov and Korolev, 1990, Chapter 8) or Section 4.6 of this book).

THEOREM 2.5.1. *A distribution function $F(x)$ belongs to the domain of geometric attraction of the exponential distribution if and only if*

$$\lim_{x \to \infty} \frac{x[1 - F(x)]}{\displaystyle\int_0^x [1 - F(z)]\,dz} = 0. \qquad (5.1)$$

PROOF. Introduce the notation

$$U(x) = \int_0^x [1 - F(z)]\,dz.$$

It is easy to verify that the Laplace-Stieltjes transform of the function $U(x)$ is equal to

$$u(s) = \frac{1}{s}[1 - \phi(s)],$$

where $\phi(x)$ is the Laplace-Stieltjes transform of the distribution function $F(x)$. Assume that (5.1) holds. Then

$$0 \leq \lim_{x \to \infty}\left|1 - \frac{U(bx)}{U(x)}\right| =$$

$$\lim_{x \to \infty} \frac{\left|\displaystyle\int_{bx}^x [1 - F(z)]\,dz\right|}{U(x)} \leq$$

$$|1 - b| \lim_{x \to \infty} \frac{x[1 - F(kx)]}{U(kx)},$$

where $k = 1$ if $b > 1$ and $k = b$ if $b < 1$. Hence we conclude that if (5.1) takes place, then $U(x)$ is a slowly varying function. By virtue of the well-known properties of these functions (see (Feller, 1971), for example) we conclude that in the neighborhood of the point $s = 0$,

$$1 - \phi(s) = su(1/s).$$

Hence in the same way as it was done in the previous section, we make the conclusion that $F(x)$ belongs to the domain of geometric attraction of the exponential distribution.

Now let $F(x)$ belong to the domain of geometric attraction of the exponential distribution. Then by virtue of (3.3) we have

$$\lim_{n\to\infty} \frac{1 - \phi\left(\dfrac{s}{B_n}\right)}{\beta_n} = s.$$

Hence we conclude that

$$\lim_{n\to\infty} \frac{\dfrac{B_n}{s}\left[1 - \phi\left(\dfrac{s}{B_n}\right)\right]}{B_n\left[1 - \phi\left(\dfrac{1}{B_n}\right)\right]} = 1,$$

i.e., the function

$$u(s) = \frac{1}{s}[1 - \phi(s)]$$

is slowly varying at the point $s = 0$. But then according to Sect. 5, Chapt. XIII of (Feller, 1971) the function

$$U(x) = \int_0^x [1 - F(z)]\,dz,$$

for which $u(s)$ is the Laplace-Stieltjes transform, satisfies the relation

$$\lim_{x\to\infty} \frac{u\left(\frac{1}{x}\right)}{U(x)} = 1$$

Thus the function $U(x)$ is slowly varying and therefore for any $c > 0$ and $x \to \infty$

$$\frac{U(x)}{U(cx)} - 1 \to 0.$$

Let $c < 1$. Then

$$\frac{U(x)}{U(cx)} - 1 = \frac{\displaystyle\int_{cx}^x [1 - F(z)]\,dz}{U(cx)} \geq$$

$$\geq (1 - c) \cdot \frac{x[1 - F(x)]}{U(x)} \cdot \frac{U(x)}{U(cx)} \geq 0$$

Hence it is clear that (5.1) takes place. The proof is completed.

Now let us busy ourselves with the description of domains of geometric attraction of other laws from the class $\mathcal{K}$ whose Laplace-Stieltjes transforms are given by the formula

$$\psi(s) = \frac{1}{1 + cs^\delta}, \qquad 0 < \delta < 1. \tag{5.2}$$

THEOREM 2.5.2. *A distribution function $F(x)$ belongs to the domain of geometric attraction of a distribution function $\Psi(x)$ whose Laplace-Stieltjes transform is given by (5.2) if and only if for every $c > 0$*

$$\lim_{x \to \infty} \frac{1 - F(x)}{1 - F(cx)} = c^\delta \tag{5.3}$$

PROOF. Assume that $F(x)$ belongs to the domain of geometric attraction of the law (5.2). Then according to (3.3) we have

$$\lim_{n \to \infty} \frac{1 - \phi\left(\dfrac{s}{B_n}\right)}{\beta_n} = s^\delta.$$

Hence we conclude that

$$\lim_{n \to \infty} \frac{1 - \phi\left(\dfrac{s}{B_n}\right)}{1 - \phi\left(\dfrac{1}{B_n}\right)} = 1.$$

Whence it follows that in the neighborhood of the point $s = 0$ the following equality takes place

$$1 - \phi(s) = s^\delta L\left(\frac{1}{s}\right),$$

where $L(z)$ is a slowly varying function, i.e. such that for any $c > 0$ the relation

$$\lim_{n \to \infty} \frac{L(cx)}{L(x)} = 1.$$

holds. In addition, in the neighborhood of the point $x = \infty$ according to one of Tauberian theorems (see Chapt. 13 of (Feller, 1971)) the function $F(x)$ has the form

$$1 - F(x) = \frac{1}{\Gamma(1 - \delta)} \frac{L(x)}{x^\delta}.$$

Hence by simple division we find that

$$\frac{1 - F(x)}{1 - F(cx)} = c^\delta \frac{L(x)}{L(cx} \to c^\delta \quad (x \to \infty),$$

i.e., (5.3) takes place.

Assume now that condition (5.3) is fulfilled. Then for $x \to \infty$ we have

$$1 - F(x) = \frac{L(x)}{x^\delta},$$

where $L(x)$ is a slowly varying function. But hence it follows that the asymptotic equality

$$1 - \phi(s) \sim \Gamma(1 - \delta)s^\delta L\left(\frac{1}{s}\right).$$

holds in the neighborhood of the point $s = 0$. If the sequence $\{\beta_n\}$ is given, then choosing B_n from the equality

$$B_n^\delta = \frac{\Gamma(1 - \delta)}{\beta_n} L(B_n)$$

we find that

$$\frac{1 - \phi\left(\dfrac{s}{B_n}\right)}{\beta_n} \to s^\delta$$

with $n \to \infty$. Repeating the reasoning from the proof of Theorem 2.3.1, we find that $F(x)$ belongs to the domain of geometric attraction of the law (5.2). The proof is completed.

By direct calculations it is easy to verify that the distribution $\Psi(x)$ belongs to its own domain of geometric attraction.

It can be shown that if a distribution $F(x)$ belongs to the domain of geometric attraction of the law $\Psi(x)$ with $\delta < 1$, then all the moments of orders $\mu < \delta$ are finite and the moments of orders $\mu > \delta$ are infinite.

In the conclusion of this chapter we note that some of the theorems formulated and proved here strikingly resemble the results of the classical theory of summation which deals with sums of infinitely increasing nonrandom number of independent random variables. As examples, the following analogies can be mentioned.

1) Limit distributions for the sums of identically distributed independent nonnegative summands $S_n = \frac{1}{B_n}(X_1 + \ldots + X_n)$ have Laplace-Stieltjes transforms $\psi(s) = \exp\{-cs^\delta\}, c > 0, 0 < \delta \leq 1$, (see, for example, (Feller, 1971) Chapter 13). At the same time, as we have seen above, the limit laws for the sums $Z_n = \frac{1}{B_n}(X_1 + \ldots + X_n)$ with geometrically distributed number of summands N_n also have simple Laplace-Stieltjes transforms $\psi(s) = \frac{1}{1+cs^\delta}$ with the same values of the parameters c and δ.

2) Condition (5.3) for a distribution to belong to the domain of geometric attraction coincides with that for a distribution to belong to the domain of attraction in the classical sense.

3) Distributions which belong both to the domain of geometric attraction of the law with parameter δ and to the classical domain of attraction of the law with the same parameter value have moments of orders $\mu < \delta$ and have no moments of orders $\mu > \delta$ ($\delta < 1$).

4) Condition (5.1) for a distribution to belong to the domain of geometric attraction of the exponential law coincides with the condition of relative stability of the sums $S_n = \frac{1}{B_n}(X_1 + \ldots + X_n)$. The latter one was proposed in 1936 by A. Ya. Khinchin, who made a successive attempt to find a more general form of the law of large numbers for independent positive random variables.

So many coincidences cannot be due to a pure chance. We will see in Chapters 3 and 4 that there are reasons of deep inner character, which connect two groups of problems.

But we will also see that there is still no complete analogy between the notions of the theory of random summation and those of the classical theory of summation. Instead, a complete analogy is possible only concerning problems connected with the so-called randomly infinitely divisible laws, which is rather a special case. We will consider these problems in more detail in Section 4.6.

Chapter 3

Limit theorems for "growing" random sums

3.1   A transfer theorem. Limit laws

This chapter and the next one are devoted mainly to the consideration of the general questions connected with the asymptotic behavior of random sums. We consider the scheme of "growing" random sums in this chapter and the double array scheme in the next one. Comparing limit laws appearing in these schemes, an attentive reader will notice that in contrast to the classical theory of summation the class of limit distributions for "growing" random sums is not a subclass of the class of limit laws for random sums in the double array scheme. The point is that in these two chapters we consider two different settings of problems. They take account of a new – as compared to the classical theory – source of randomness, namely, random indices, in different ways.

Let $X_1, X_2, \ldots$ be independent random variables. Denote $S_k = X_1 + \ldots + X_k, k \geq 1$. Let $\{a_k\}_{k \geq 1}$ and $\{b_k\}_{k \geq 1}$ be sequences of numbers, $b_k > 0$ $(k \geq 1)$. Denote

$$Y_k = \frac{S_k - a_k}{b_k}.$$

Denote the characteristic function of the random variable Y_k as $h_k(t), t \in \mathbb{R}$. Let $\{N_k\}_{k \geq 1}$ be integer positive random variables independent of the sequence $\{X_j\}_{j \geq 1}$ for every $k \geq 1$. Our aim is to study the asymptotic behavior of the random variables

$$Z_k = \frac{S_{N_k} - c_k}{d_k},$$

where $\{c_k\}_{k \geq 1}$ and $\{d_k\}_{k \geq 1}$ are sequences of numbers, $d_k > 0$, $k \geq 1$. Denote the characteristic function of the random variable Z_k as $f_k(t)$. Generally speaking, the sequences $\{a_k\}, \{b_k\}$ and $\{c_k\}, \{d_k\}$ guaranteeing the convergence or compactness of the random variables $\{Y_k\}$ and $\{Z_k\}$ are different.

In this section we will assume that the sequences $\{a_k\}$ and $\{b_k\}$ provide the weak convergence of centered and normalized sums

$$Y_k \Rightarrow Y \qquad (k \to \infty) \tag{1.1}$$

to some random variable Y whose characteristic function will be denoted as $h(t), t \in \mathbb{R}$. Put

$$g_k(t) = \sum_{n=1}^{\infty} P(N_k = n) \exp\left\{ it\left(\frac{a_n - c_k}{d_k}\right)\right\} h\left(t\frac{b_n}{d_k}\right).$$

LEMMA 3.1.1. *Let $b_k \to \infty, d_k \to \infty \, (k \to \infty)$. Assume that (1.1) holds and the family of the random variables $\{b_{N_k}/d_k\}_{k \geq 1}$ is weakly relatively compact. Then for every $t \in \mathbb{R}$*

$$\lim_{k \to \infty} |f_k(t) - g_k(t)| = 0.$$

PROOF. It is easy to see that according to the law of total probability,

$$f_k(t) = \sum_{n=1}^{\infty} P(N_k = n) \exp\left\{ it\left(\frac{a_n - c_k}{d_k}\right)\right\} h_n\left(t\frac{b_n}{d_k}\right).$$

Let α and β be real numbers, $0 < \alpha < \beta < \infty$, which will be specified later. Introduce the following sets

$$\begin{aligned}
\mathcal{N}_1 &= \mathcal{N}_1(k, \alpha) = \{n : b_n < \alpha d_k\}, \\
\mathcal{N}_2 &= \mathcal{N}_2(k, \alpha, \beta) = \{n : \alpha d_k \leq b_n \leq \beta d_k\}, \\
\mathcal{N}_3 &= \mathcal{N}_3(k, \beta) = \{n : b_n > \beta d_k\}.
\end{aligned}$$

If $t = 0$, then the statement is trivial. Fix arbitrary $t \neq 0$. Then

$$|f_k(t) - g_k(t)| \leq$$

$$\sum_{n=1}^{\infty} P(N_k = n) \left| \exp\left\{ it\left(\frac{a_n - c_k}{d_k}\right)\right\} \left[h_n\left(t\frac{b_n}{d_k}\right) - h\left(t\frac{b_n}{d_k}\right)\right]\right| \leq$$

$$\sum_{n=1}^{\infty} P(N_k = n) \left| h_n\left(t\frac{b_n}{d_k}\right) - h\left(t\frac{b_n}{d_k}\right)\right| =$$

$$\sum_{n \in \mathcal{N}_1} P(N_k = n) \left| h_n\left(t\frac{b_n}{d_k}\right) - h\left(t\frac{b_n}{d_k}\right)\right| +$$

$$\sum_{n \in \mathcal{N}_2} P(N_k = n) \left| h_n\left(t\frac{b_n}{d_k}\right) - h\left(t\frac{b_n}{d_k}\right)\right| +$$

$$\sum_{n \in \mathcal{N}_3} P(N_k = n) \left| h_n\left(t\frac{b_n}{d_k}\right) - h\left(t\frac{b_n}{d_k}\right)\right| \equiv$$

$$I_1(\alpha, k) + I_2(\alpha, \beta, k) + I_3(\beta, k).$$

Let $\varepsilon > 0$ be an arbitrary number. Consider $I_1(\alpha, k)$. Since the function h, being a characteristic one, is continuous at zero and t is fixed, an $\alpha_1 = \alpha_1(\varepsilon, t)$ can be chosen so that

$$\sup_{k} \sup_{n \in \mathcal{N}_1(k, \alpha_1)} \left| 1 - h\left(t\frac{b_n}{d_k}\right) \right| \leq \sup_{|\tau| \leq \alpha_1 |t|} |1 - h(\tau)| < \varepsilon. \qquad (1.2)$$

As far as (1.1) holds true, the family of the random variables $\{Y_k\}$ is weakly relatively compact and therefore according to the weak relative compactness criterion (see (Loève, 1963)) the family of characteristic functions $\{h_k\}_{k \geq 1}$ is equicontinuous at zero, i.e., for an arbitrary $\varepsilon > 0$ there exists $\delta = \delta(\varepsilon) > 0$ such that

$$\sup_{k} \sup_{|\tau| \leq \delta} |1 - h_k(\tau)| < \varepsilon. \qquad (1.3)$$

For such δ choose $\alpha_2 = \alpha_2(\varepsilon, t)$ so that

$$\alpha_2 = \alpha_2(\varepsilon, t) \leq \frac{\delta}{|t|}.$$

Then it follows from (1.3) that

$$\sup_{k} \sup_{n \in \mathcal{N}_1(k, \alpha_2)} \left| 1 - h_n\left(t\frac{b_n}{d_k}\right) \right| \leq \sup_{n} \sup_{|\tau| \leq \delta} |1 - h_n(t)| < \varepsilon. \qquad (1.4)$$

Set $\alpha = \min(\alpha_1, \alpha_2)$. Then from relations (1.2) and (1.4) we obtain

$$I_1(\alpha, k) \leq \sup_{k} \sup_{n \in \mathcal{N}_1(k, \alpha)} \left| 1 - h\left(t\frac{b_n}{d_k}\right) \right| + \sup_{k} \sup_{n \in \mathcal{N}_1(k, \alpha)} \left| 1 - h_n\left(t\frac{b_n}{d_k}\right) \right| < 2\varepsilon. \qquad (1.5)$$

Consider $I_3(\beta, k)$. Since the family $\{b_{N_k}/d_k\}_{k \geq 1}$ is weakly relatively compact by the hypothesis, for every $\varepsilon > 0$, $\beta = \beta(\varepsilon)$ can be indicated which provides the validity of the inequality

$$I_3(\beta, k) \leq 2 \sum_{n \in \mathcal{N}_3(k, \beta)} P(N_k = n) = 2P\left(\frac{b_{N_k}}{d_k} > \beta\right) \leq 2\sup_{k} P\left(\frac{b_{N_k}}{d_k} > \beta\right) < \varepsilon. \qquad (1.6)$$

Finally consider $I_2(k, \alpha, \beta)$ with α and β determined above. We have

$$I_2(k, \alpha, \beta) \leq \sum_{n \in \mathcal{N}_2} P(N_k = n) \sup_{|\tau| \leq \beta |t|} |h_n(\tau) - h(\tau)| \leq$$

$$\sup_{n \in \mathcal{N}_2(k, \alpha, \beta)} \sup_{|\tau| \leq \beta |t|} |h_n(t) - h(t)|. \qquad (1.7)$$

Condition (1.1) implies uniform on every finite interval convergence of h_k to h as $k \to \infty$. As far as $b_k \to \infty$ and $a_k \to \infty$ $(k \to \infty)$ we have

$$\inf \mathcal{N}_2(k, \alpha, \beta) \to \infty \qquad (k \to \infty).$$

This means that as $k \to \infty$, the right-hand part of (1.7) tends to zero, i.e., for every ε, $k_0 = k_0(\varepsilon)$ can be indicated such that for all $k \geq k_0$,

$$I_2(k, \alpha, \beta) < \varepsilon. \tag{1.8}$$

Unifying (1.5), (1.6) and (1.8), we obtain that for $k \geq k_0$

$$|f_k(t) - g_k(t)| < 4\varepsilon.$$

The arbitrariness of ε in this inequality proves the lemma.

Lemma 3.1.1 shows the way for the search for approximations of distributions of random sums with known indexes. It describes asymptotic rapprochement of distributions of random sums with approximating laws defined by the characteristic functions $g_k(t)$.

However, the distribution corresponding to the characteristic function $g_k(t)$ has a rather complicated form and is not suitable for practical construction of approximations. At the same time, the form of approximating distributions simplifies essentially in some cases. One of these situations is connected with the asymptotically nonrandom indices, i.e., such that

$$\frac{b_{N_k}}{d_k} \Rightarrow \text{const.} \tag{1.9}$$

This situation takes place in many applied problems, for example, in those connected with generalized Poisson processes, such as classical risk process described in Section 1.1. We will consider this example in detail in Section 3.4. Now we will prove two statements implied by Lemma 3.1.1. In both cases, convergence (1.9) is assumed.

Denote

$$w_k(t) = \mathsf{E} \exp \left\{ it \left(\frac{a_{N_k} - c_k}{d_k} \right) \right\}, \quad r_k(t) = w_k(t) h \left(t \frac{b_k}{d_k} \right), \quad t \in \mathbb{R}.$$

THEOREM 3.1.1.   *Assume that* $b_k \to \infty$, $d_k \to \infty$ $(k \to \infty)$ *and (1.1) holds. Let*

$$\frac{b_{N_k} - b_k}{d_k} \Rightarrow 0 \qquad (k \to \infty). \tag{1.10}$$

I. *If*

$$\lim_{k \to \infty} \frac{b_k}{d_k} = 0, \tag{1.11}$$

then for any $t \in \mathbb{R}$

$$\lim_{k \to \infty} |f_k(t) - w_k(t)| = 0. \tag{1.12}$$

II. *If*

$$0 < c_0 \equiv \inf_k \frac{b_k}{d_k} \leq \sup_k \frac{b_k}{d_k} \equiv c_1 < \infty, \tag{1.13}$$

then for any $t \in \mathbb{R}$

$$\lim_{k \to \infty} |f_k(t) - r_k(t)| = 0. \tag{1.14}$$

PROOF. Condition (1.10) together with (1.11) or (1.13) guarantee the weak relative compactness of the family of random variables $\{b_{N_k}/d_k\}_{k\geq 1}$. Thus by virtue of Lemma 3.1.1 in the case **I** it is necessary to verify that

$$\lim_{k\to\infty} |g_k(t) - w_k(t)| = 0, \tag{1.15}$$

and we should show in the case **II** that

$$\lim_{k\to\infty} |g_k(t) - r_k(t)| = 0. \tag{1.16}$$

We begin with the proof of (1.15). In the proof of Lemma 3.1.1 we introduced the set $\mathcal{N}_1 = \mathcal{N}_1(k, \alpha) = \{n : b_n < \alpha d_k\}$, $\alpha > 0$. If $t = 0$, then (1.15) is trivial. Fix arbitrary nonzero t. Then

$$|g_k(t) - w_k(t)| =$$

$$\left| \sum_{n=1}^{\infty} P(N_k = n) \exp\left\{ it\left(\frac{a_n - c_k}{d_k}\right) \right\} \left[h\left(t\frac{b_n}{d_k}\right) - 1 \right] \right| \leq$$

$$\sum_{n=1}^{\infty} P(N_k = n) \left| h\left(t\frac{b_n}{d_k}\right) - 1 \right| \leq$$

$$\sum_{n\in\mathcal{N}_1} P(N_k = n) \left| h\left(t\frac{b_n}{d_k}\right) - 1 \right| + P(N_k \notin \mathcal{N}_1). \tag{1.17}$$

The function h, being a characteristic one, is continuous at zero. Thus for any $\varepsilon > 0$, $\alpha = \alpha(t, \varepsilon)$ can be chosen such that

$$\sup_k \sup_{n\in\mathcal{N}_1(k,\alpha)} \left| h\left(t\frac{b_n}{d_k}\right) - 1 \right| \leq \sup_{|\tau|<\alpha|t|} |h(\tau) - 1| < \varepsilon. \tag{1.18}$$

Conditions (1.10) and (1.11) imply

$$\frac{b_{N_k}}{d_k} \Rightarrow 0 \qquad (k \to \infty).$$

Thus for α and ε chosen above, a k_0 can be found such that for all $k \geq k_0$,

$$P(N_k \notin \mathcal{N}_1(\alpha, k)) = P\left(\frac{b_{N_k}}{d_k} \geq \alpha\right) < \varepsilon. \tag{1.19}$$

Substituting (1.18) and (1.19) in (1.17) we get

$$|g_k(t) - w_k(t)| < 2\varepsilon,$$

and by virtue of the arbitrariness of ε this means that (1.15) holds true, and the last fact implies (1.12) according to Lemma 3.1.1.

Turn to item **II**. Introduce the set

$$\mathcal{N} = \mathcal{N}(k, \delta) = \{n : |b_n - b_k| \leq \delta d_k\}, \quad \delta > 0.$$

If $t = 0$, then (1.16) is trivial. Let $t \neq 0$. Then

$$|g_k(t) - r_k(t)| =$$

$$\left| \sum_{n=1}^{\infty} P(N_k = n) \exp\left\{ it\left(\frac{a_n - c_k}{d_k}\right) \right\} \left[h\left(t\frac{b_n}{d_k}\right) - h\left(t\frac{b_k}{d_k}\right) \right] \right| \leq$$

$$\sum_{n=1}^{\infty} P(N_k = n) \left| h\left(t\frac{b_n}{d_k}\right) - h\left(t\frac{b_k}{d_k}\right) \right| \leq$$

$$\sum_{n \in \mathcal{N}} P(N_k = n) \left| h\left(t\frac{b_n}{d_k}\right) - h\left(t\frac{b_k}{d_k}\right) \right| + P(N_k \notin \mathcal{N}). \qquad (1.20)$$

Since the function h is uniformly continuous being a characteristic one, for any $\varepsilon > 0$, a $\delta = \delta(t, \varepsilon) \in (0, c_0)$ can be chosen such that

$$\sup_k \sup_{n \in \mathcal{N}(k,\delta)} \left| h\left(t\frac{b_n}{d_k}\right) - h\left(t\frac{b_k}{d_k}\right) \right| \leq$$

$$\sup\{|h(t_1) - h(t_2)| : |t_1 - t_2| < \delta|t|\} < \varepsilon. \qquad (1.21)$$

But it follows from (1.10) that there exist ε and δ such that for all $k \geq k_0$

$$P(N_k \notin \mathcal{N}(k,\delta)) = P\left(\left| \frac{b_{N_k} - b_k}{d_k} \right| > \delta \right) < \varepsilon. \qquad (1.22)$$

Substituting (1.21) and (1.22) in (1.20) we obtain

$$|g_k(t) - r_k(t)| < 2\varepsilon,$$

and by virtue of the arbitrariness of ε, this means that (1.16) holds and therefore according to Lemma 3.1.1, (1.14) does so. The proof is completed.

It follows from Theorem 3.1.1 that with asymptotically nonrandom indexes in the sense of (1.9) (or (1.10)) the distribution function $F_k(x) = P(Z_k < x)$ of the random sum S_{N_k} centered and normalized by constants can be approximated by the distribution function

$$R_k(x) = W_k(x) * H\left(x\frac{d_k}{b_k}\right), \qquad (1.23)$$

where $W_k(x) = P(a_{N_k} - c_k < xd_k), H(x) = P(Y < x)$. It is easy to verify that the distribution function R_k is a discrete translation mixture of the distribution function H. Indeed, according to (1.23) $R_k(x)$ is the distribution function of the sum of two independent random variables $Y b_k / d_k$ and

$V_k = (a_{N_k} - c_k)/d_k$. Then by the formula of total probability

$$R_k(x) = \mathsf{P}(V_k + Y b_k/d_k < x) =$$

$$\sum_{n=1}^{\infty} \mathsf{P}(N_k = n) \mathsf{P}\left(\frac{a_n - c_k}{d_k} + Y\frac{b_k}{d_k} < x\right) =$$

$$\sum_{n=1}^{\infty} \mathsf{P}(N_k = n) \mathsf{P}\left(Y < \frac{x d_k + c_k - a_n}{b_k}\right) =$$

$$\sum_{n=1}^{\infty} \mathsf{P}(N_k = n) H\left(\frac{x d_k + c_k - a_n}{b_k}\right).$$

These distributions appeared for the first time as approximations for distributions of centered random sums in the work of H. Robbins (Robbins, 1948), where the sums of identically distributed summands with finite variance were considered. Robbins' result was extended later for sums of non-identically distributed summands (with finite variances) by Z. Rychlik and D. Szynal (Rychlik and Szynal, 1972, 1973). Thus Theorem 3.1.1, published in (Korolev, 1989), generalizes the cited works, since it does not assume any moment restrictions.

Lemma 3.1.1 describes asymptotic rapprochement of distributions of random sums with an approximating law which also depends on the number of the sum. The limit situation is described by the following theorem. It is customary in accordance with the paper (Gnedenko and Fahim, 1969) to call such statements transfer theorems, because they contain the description of conditions providing the transfer of convergence property from the sums of non-random number of summands (1.1) to random sums. Let $H(x) = \mathsf{P}(Y < x)$ as before.

THEOREM 3.1.2. *Let sequences of numbers* $\{a_k\}, \{b_k\}, \{c_k\}$ *and* $\{d_k\}$ *be such that* $b_k \to \infty, d_k \to \infty$ $(k \to \infty)$ *and condition (1.1) takes place. Let also*

$$\left(\frac{b_{N_k}}{d_k}, \frac{a_{N_k} - c_k}{d_k}\right) \Rightarrow (U, V) \qquad (k \to \infty) \tag{1.24}$$

for some random variables U *and* V. *Then*

$$\mathsf{P}(Z_k < x) \Rightarrow \mathsf{E}H\left(\frac{x - V}{U}\right) \qquad (k \to \infty). \tag{1.25}$$

PROOF. It is easy to see that the characteristic function

$$f(t) = \mathsf{E}h(Ut)e^{itV}, \quad t \in \mathbb{R},$$

where h is the characteristic function of the random variable Y occurring in (1.1), corresponds to the distribution function standing in the right-hand side of (1.25). That is why we will make sure that (1.25) holds, if we prove that for any $t \in \mathbb{R}$

$$\lim_{k \to \infty} f_k(t) = f(t). \tag{1.26}$$

Denote $U_k = b_{N_k}/d_k, V_k = (a_{N_k} - c_k)/d_k$. Then obviously, the characteristic function f_k can be written in the form

$$f_k(t) = \mathsf{E}h_k(U_kt)e^{itV_k},$$

where as before, h_k is the characteristic function of the random variable $Y_k = (S_k - a_k)/b_k$. Since

$$|f_k(t) - f(t)| \le |f_k(t) - g_k(t)| + |g_k(t) - f(t)|,$$

taking account of Lemma 3.1.1, it suffices to show that

$$\lim_{k\to\infty} |g_k(t) - f(t)| = 0.$$

But $g_k(t) = \mathsf{E}h(U_kt)e^{itV_k}$. Note that the function $\phi_t(x,y) = h(tx)e^{ity}$ for every fixed $t \in \mathbb{R}$ is continuous and bounded in x and y. According to the definition of weak convergence, (1.24) is equivalent to

$$\mathsf{E}\phi(U_k, V_k) \to \mathsf{E}\phi(U, V) \quad (k \to \infty) \tag{1.27}$$

for any continuous and bounded function ϕ, whence it follows that (1.27) holds true for $\phi = \phi_t$. Therefore (1.24) implies (1.26) and consequently (1.25). The proof is completed.

Note that in some cases when centering and normalizing constants are organized in a special way, it is sufficient to require the convergence of only one component of the pair (U_k, V_k). See, for example, the works (Finkelstein and Tucker, 1990) and (Finkelstein, Kruglov and Tucker, 1994), where the conditions of convergence of Z_k are formulated in terms of V_k. As an example, consider the situation, where $a_k = c_k = \alpha k, b_k = d_k = \sigma k^\gamma, \alpha \in \mathbb{R}, \sigma > 0, 0 < \gamma < 1$. Then the weak convergence of random variables $V_k = \frac{\alpha}{\sigma}\left(\frac{N_k-k}{k^\gamma}\right)$ to a proper random varible V as $k \to \infty$ implies

$$\mathsf{P}\left(\left|\left(\frac{b_{N_k}}{d_k}\right)^{1/\gamma} - 1\right| > x\right) = \mathsf{P}\left(\left|\frac{N_k}{k} - 1\right| > x\right) =$$

$$\mathsf{P}\left(\left|\frac{N_k - k}{k^\gamma}\right| > k^{1-\gamma}x\right) \to 0 \quad (k \to \infty)$$

for any $x > 0$, which means that $(b_{N_k}/d_k)^{1/\gamma} \to 1$, and therefore $U_k \to 1$ as $k \to \infty$. In aggregate with (1.1), according to Theorem 3.1.2 this means that in this case the limit law for the distribution functions of random variables Z_k has the form $(H * W)(x)$, where $W(x) = \mathsf{P}(V < x)$.

It can be seen from Theorem 3.1.2 that the class of limit distributions for "growing" random sums under nonrandom centering and normalization consists of translation-scale mixtures of the limit distribution function for the sums with nonrandom number of summands. Since according to the assumption the normalizing constants b_k and d_k are positive for any $k \geq 1$, the random variable U is nonnegative. Therefore we can assume that

$$H\left(\frac{x-v}{u}\right)\Big|_{u=0} = \mathbf{I}(v < x),$$

whatever the distribution function H is, where $\mathbf{I}(A)$ denotes the indicator function of a set A.

If Z is a random variable with distribution function $\mathsf{E}H\left(\frac{x-V}{U}\right)$, then

$$Z \overset{d}{=} YU + V, \tag{1.28}$$

where the random variable Y and the pair (U, V) are independent. Therefore we come to rather a discomforting conclusion: if there are no additional assumptions concerning possible distributions of indices or concerning normalizing and centering constants, then the class of limit laws for the distributions of "growing" random sums centered by constants coincides with the set of all distribution functions; as far as in the case $\mathsf{P}(U = 0) = 1$, the distribution of random variable Z in (1.28) is completely determined by the distribution of the random variable V which in general can be quite arbitrary.

Under additional assumptions the class of limit laws narrows. Thus in the case where $\mathsf{P}(Y = a) = 1$ for some $a \in \mathbb{R}$ (for example, in the situation described by the law of large numbers), for $\mathsf{P}(V = 0) = 1$ the distribution of the limit random variable Z coincides with that of aU. The latter one is concentrated either on the semiaxis $(-\infty, 0]$ or on the semiaxis $[0, \infty)$ in accordance with the sign of the number a, but can be arbitrary among all these distributions.

Thus if we assume that (1.1) takes place, i.e., fix the random variable Y in representation (1.28), then the limit laws can be described in more detail. If in addition to this assumption $\mathsf{P}(V = \mathrm{const}) = 1$, then the class of limit laws coincides up to a nonrandom translation with the set of scale mixtures of the distribution function $H(x) = \mathsf{P}(Y < x)$. If $\mathsf{P}(U = \mathrm{const}) = 1$, then the class of limit laws coincides up to the choice of the scale parameter with the set of translation mixtures of the distribution function H, which should be expected in accordance with Theorem 3.1.1.

A more detailed description of the class of limit laws for the distributions of "growing" random sums centered by constants seems to be impossible.

We should note that in the proofs of the statements of this section we actually did not use the assumption we made in the very beginning that the random variables S_k represent cumulative sums of independent random variables. In other words, the results of this section hold for arbitrary sequences of random variables (Korolev, 1992, 1993).

3.2 Necessary and sufficient conditions for convergence

Attempts to formulate necessary and sufficient conditions of weak convergence of "growing" random sums centered and normalized by constants come across a very serious obstacle which does not allow us to prove the necessity of conditions (1.1) or (1.24) for (1.25) in the general case. Namely, in representation (1.28), the random variable Y and the pair (U, V) independent of Y are not determined uniquely by a given distribution of the random variable Z. (Since we deal with the weak convergence, here and in what follows we identify random variables and corresponding distribution functions to simplify notation and formulations. Actually we consider classes of equivalence of random variables, each of them includes all random variables with the same distribution function.)

To pass over this obstacle, for an arbitrary random variable Z introduce the set

$$\mathcal{V}(Z) = \{(Y, U, V) : Z \stackrel{d}{=} YU + V, \quad Y \text{ and } (U, V) \text{ are independent}\}.$$

Whatever random variable Z is, the set $\mathcal{V}(Z)$ is nonempty, since always $(Y, 0, Z) \in \mathcal{V}(Z)$, where Y is an arbitrary random variable independent of Z. It also follows from this example that the set $\mathcal{V}(Z)$ always contains more than one element. The set $\mathcal{V}(Z)$ can contain different triples with nonzero U. For example, let $Z \stackrel{d}{=} W - W_1$, where W and W_1 are independent random variables with identical gamma distributions with some scale parameter $\lambda > 0$ and shape parameter $\alpha > 0$. Then the set $\mathcal{V}(Z)$ will obviously contain triples $(1, W, -W_1)$ and $(W, 1, -W_1)$ together with trivial triples $(Y, 0, W - W_1)$. Moreover, in this case $\mathcal{V}(Z)$ also contains the triple $(Y, W_2^{1/2}, 0)$, where the variable Y has the standard normal distribution and is independent of the variable W_2 which has the gamma distribution with scale parameter $\frac{1}{2}\lambda^2$ and shape parameter α. To make it sure, notice that the characteristic function of the difference $W - W_1$ is equal to

$$\mathsf{E}e^{it(W-W_1)} = \frac{1}{\left(1 - \dfrac{it}{\lambda}\right)^\alpha} \cdot \frac{1}{\left(1 + \dfrac{it}{\lambda}\right)^\alpha} = \left(\frac{\lambda^2}{\lambda^2 + t^2}\right)^\alpha. \tag{2.1}$$

Let W_2 be a gamma-distributed random variable with shape parameter α and some scale parameter μ, and let Y be an independent random variable with the standard normal distribution. Then

$$\mathsf{E}e^{itYW_2^{1/2}} = \frac{\mu^\alpha}{\Gamma(\alpha)} \int\limits_0^\infty e^{-x(\frac{1}{2}t^2 + \mu)} x^{\alpha-1}\, dx =$$
$$\frac{\mu^\alpha}{\Gamma(\alpha)\left(\frac{1}{2}t^2 + \mu\right)^\alpha} \int\limits_0^\infty e^{-y} y^{\alpha-1}\, dy = \left(\frac{2\mu}{2\mu + t^2}\right)^\alpha. \tag{2.2}$$

The right-hand sides of (2.1) and (2.2) coincide, if $\mu = \frac{1}{2}\lambda^2$, Q.E.D.

Let $L_1(\cdot, \cdot)$ and $L_2(\cdot, \cdot)$ be metrics in the spaces of one-dimesional and two-dimensional distributions, respectively, which metrize weak convergence. For example, $L_1(\cdot, \cdot)$ is the Lévy metric, $L_2(\cdot, \cdot)$ is the Lévy–Prokhorov metric, see (Zolotarev, 1986). In accordance with what has been said above, if X and

Y are random variables with distribution functions G and H, respectively, we will make no distinction between $L_1(X, Y)$ and $L_1(G, H)$. Similarly, if (X_1, X_2) and (Y_1, Y_2) are two-dimensional random variables with distribution functions G and H, respectively, then we will not distinguish $L_2(G, H)$ and $L_2((X_1, X_2), (Y_1, Y_2))$.

The following theorem plays a fundamental role in this section. All other statements in this section are its corollaries.

THEOREM 3.2.1. *Assume that*

$$N_k \xrightarrow{P} \infty \qquad (k \to \infty), \tag{2.3}$$

and the sequences $\{a_k\}$ and $\{b_k\}, b_k > 0, b_k \to \infty \ (k \to \infty)$, provide weak relative compactness of the family of random variables $\{(S_k - a_k)/b_k\}_{k \geq 1}$. For some sequences of positive numbers $\{d_k\}, d_k \to \infty \ (k \to \infty)$, and real numbers $\{c_k\}$ the convergence

$$\frac{1}{d_k} \left(\sum_{j=1}^{N_k} X_j - c_k \right) \Rightarrow Z \qquad (k \to \infty) \tag{2.4}$$

takes place if and only if there exists a weakly relatively compact sequence of triples of random variables $\{(Y_k', U_k', V_k')\}_{k \geq 1}, (Y_k', U_k', V_k') \in \mathcal{V}(Z)$ with every $k \geq 1$, for which

$$L_1 \left(\frac{1}{b_k} \left(\sum_{j=1}^{k} X_j - a_k \right), Y_k' \right) \to 0 \qquad (k \to \infty), \tag{2.5}$$

$$L_2 \left(\left(\frac{b_{N_k}}{d_k}, \frac{a_{N_k} - c_k}{d_k} \right), (U_k', V_k') \right) \to 0 \qquad (k \to \infty). \tag{2.6}$$

PROOF. Necessity. Let X_j' be a random variable independent of X_j and such that $X_j' \stackrel{d}{=} X_j$. Denote $X_j^{(s)} = X_j' - X_j$ ($X_j^{(s)}$ is a symmetrized variable X_j). Let $q \in (0, 1)$. Denote the greatest lower bound of q-quantiles of the random variable N_k as $l_k(q)$. Assuming that the variables $N_k, X_1, X_2, \ldots, X_1', X_2', \ldots$ are independent with every $k \geq 1$, define

$$T_k = \frac{1}{d_k} \sum_{j=1}^{N_k} X_j^{(s)}.$$

Using the symmetrization inequality (see (Loève, 1963)),

$$P(|X_1^{(s)}| \geq x) \leq 2P \left(|X_1 - a| \geq \frac{x}{2} \right),$$

which holds for any random variable X_1 and any $a \in \mathbb{R}, x > 0$, we get

$$P(|T_k| \geq x) = \sum_{n=1}^{\infty} P(N_k = n) P\left(\left|\frac{1}{d_k} \sum_{j=1}^{n} X_j^{(s)}\right| \geq x\right) \leq$$

$$2 \sum_{n=1}^{\infty} P(N_k = n) P\left(\left|\frac{1}{d_k}\left(\sum_{j=1}^{n} X_j - c_k\right)\right| \geq \frac{x}{2}\right) =$$

$$2P\left(\left|\frac{1}{d_k}\left(\sum_{j=1}^{N_k} X_j - c_k\right)\right| \geq \frac{x}{2}\right)$$

for every $x > 0$. Whence it follows that

$$\lim_{x \to \infty} \sup_k P(|T_k| \geq x) \leq$$

$$2 \lim_{x \to \infty} \sup_k P\left(\left|\frac{1}{d_k}\left(\sum_{j=1}^{N_k} X_j - c_k\right)\right| \geq \frac{x}{2}\right) = 0$$

by virtue of condition (2.4). According to Theorem 3 in Sect. 9, Chapt. 2 of (Gnedenko and Kolmogorov, 1954) this means that the sequence $\{T_k\}_{k \geq 1}$ is weakly relatively compact. Our further aim is to prove the compactness of the sequence $\{b_{N_k}/d_k\}_{k \geq 1}$.

Let $\mathcal{K}$ be an arbitrary sequence of natural numbers. Show that a weakly convergent subsequence can be chosen from the sequence $\{b_{N_k}/d_k\}_{k \in \mathcal{K}}$. Choose a subsequence $\mathcal{K}_1 \subseteq \mathcal{K}$ in such a way that

$$Y_k = \frac{1}{b_k}\left(\sum_{j=1}^{k} X_j - a_k\right) \Rightarrow Y \qquad (k \to \infty, k \in \mathcal{K}_1),$$

where Y is some random variable. This choice is possible on account of the condition of weak relative compactness of the family of random variables $\{(S_k - a_k)/b_k\}_{k \geq 1}$. Show that with every $q \in (0, 1)$,

$$C(q) \equiv \sup_{k \in \mathcal{K}_1} \frac{b_{l_k(q)}}{d_k} < \infty. \tag{2.7}$$

Using P. Lévy's inequality

$$P\left(\max_{1 \leq r \leq n}\left|\sum_{j=1}^{r} X_j^{(s)}\right| \geq x\right) \leq 2P\left(\left|\sum_{j=1}^{n} X_j^{(s)}\right| \geq x\right),$$

which holds true for any independent random variables $X_1, \ldots, X_n$ and any

$x > 0$ (see, for example, (Loève, 1963)), we get

$$P(|T_k| \geq x) = P\left(\left|\frac{1}{d_k} \sum_{j=1}^{N_k} X_j^{(s)}\right| \geq x\right) =$$

$$\sum_{n=1}^{\infty} P(N_k = n)P\left(\left|\frac{1}{d_k} \sum_{j=1}^{n} X_j^{(s)}\right| \geq x\right) \geq$$

$$\sum_{n > l_k(q)} P(N_k = n)P\left(\left|\frac{1}{d_k} \sum_{j=1}^{n} X_j^{(s)}\right| \geq x\right) \geq$$

$$\frac{1}{2} \sum_{n > l_k(q)} P(N_k = n)P\left(\left|\frac{1}{d_k} \sum_{j=1}^{l_k(q)} X_j^{(s)}\right| \geq x\right) \geq$$

$$\frac{1}{2}(1 - q)P\left(\left|\frac{1}{d_k} \sum_{j=1}^{l_k(q)} X_j^{(s)}\right| \geq x\right).$$

Thus the weak relative compactness of the sequence $\{T_k\}_{k \geq 1}$ stated above, implies that of the sequence of distribution functions $\{F_k^{(q)}\}_{k \geq 1}$ for any $q \in (0, 1)$ according to Theorem 3 from the book (Gnedenko and Kolmogorov, 1954) mentioned above, where

$$F_k^{(q)}(x) = P\left(\left|\frac{1}{d_k} \sum_{j=1}^{l_k(q)} X_j^{(s)}\right| \leq x\right).$$

Condition (2.3) implies that $l_k(q) \to \infty$ as $k \to \infty$ for any $q \in (0, 1)$. Therefore

$$Y_{l_k(q)} = \frac{1}{b_{l_k(q)}} \left(\sum_{j=1}^{l_k(q)} X_j - a_{l_k(q)}\right) \Rightarrow Y \tag{2.8}$$

as $k \to \infty, k \in \mathcal{K}_1$, according to the choice of the sequence $\mathcal{K}_1$. Assume that (2.7) does not take place. In this case for some $q \in (0, 1)$ and some sequence $\mathcal{K}_2 \subseteq \mathcal{K}_1$ of natural numbers we should have

$$\frac{b_{l_k(q)}}{d_k} \to \infty \qquad (k \to \infty, k \in \mathcal{K}_2). \tag{2.9}$$

Let Y' be a random variable independent of Y and such that $Y' \overset{d}{=} Y$. The distribution function $F_k^{(q)}$ can be written as

$$F_k^{(q)}(x) = P\left(\frac{1}{b_{l_k(q)}} \left(\sum_{j=1}^{l_k(q)} X_j - a_{l_k(q)}\right) - \right.$$

$$\frac{1}{b_{l_k(q)}} \left(\sum_{j=1}^{l_k(q)} X'_j - a_{l_k(q)} \right) \le \frac{d_k x}{b_{l_k(q)}} \right).$$

In this case (2.8) and (2.9) imply that for any $x \in \mathbb{R}$

$$F_k^{(q)}(x) \to \mathrm{P}(Y - Y' \le 0) \ge \frac{1}{2} \qquad (k \to \infty, k \in \mathcal{K}_2),$$

which contradicts weak relative compactness of the sequence $\{F_k^{(q)}(x)\}_{k \ge 1}$. Thus we proved that (2.7) holds for any $q \in (0, 1)$.

It is not difficult to verify that the random variable N_k has the same distribution as the random variable $l_k(W)$, where W is uniformly distributed in the interval $(0, 1)$. Therefore using (2.7), for any $x \ge 0$ and $k \in \mathcal{K}_1$ we get

$$\mathrm{P}\left(\frac{b_{N_k}}{d_k} \ge x\right) = \mathrm{P}\left(\frac{b_{l_k(W)}}{d_k} \ge x\right) = \int_0^1 \mathrm{P}\left(\frac{b_{l_k(q)}}{d_k} \ge x\right) dq \le$$

$$\int_0^1 \mathrm{P}(C(q) \ge x) dq = \mathrm{P}(C(W) \ge x).$$

Hence

$$\lim_{x \to \infty} \sup_{k \in \mathcal{K}_1} \mathrm{P}\left(\frac{b_{N_k}}{d_k} \ge x\right) \le \lim_{x \to \infty} \mathrm{P}(C(W) \ge x) = 0,$$

i.e., the sequence of random variables $\{b_{N_k}/d_k\}_{k \in \mathcal{K}_1}$ is weakly relatively compact. Thus a subsequence $\mathcal{K}_3 \subseteq \mathcal{K}_1 \subseteq \mathcal{K}$ can be chosen so that the sequence $\{b_{N_k}/d_k\}_{k \in \mathcal{K}_3}$ should be weakly convergent. But by virtue of the arbitrariness of the sequence $\mathcal{K}$ this means that the family $\{b_{N_k}/d_k\}_{k \ge 1}$ is weakly relatively compact.

Now show that the sequence $\{(a_{N_k} - c_k)/d_k\}_{k \ge 1}$ is weakly relatively compact. For an arbitrary $R > 0$, according to the formula of total probability we have

$$\mathrm{P}\left(\left|\frac{a_{N_k} - c_k}{d_k}\right| > R\right) =$$

$$\sum_{n=1}^{\infty} \mathrm{P}(N_k = n)\mathrm{P}\left(\left|\frac{S_n - c_k}{d_k} - \frac{b_n}{d_k}\left(\frac{S_n - a_n}{b_n}\right)\right| > R\right) \le$$

$$\sum_{n=1}^{\infty} \mathrm{P}(N_k = n)\mathrm{P}\left(\left|\frac{S_n - c_k}{d_k}\right| > \frac{R}{2}\right) +$$

$$\sum_{n=1}^{\infty} \mathrm{P}(N_k = n)\mathrm{P}\left(\left|\frac{S_n - a_n}{b_n}\right| > \frac{Rd_k}{2b_n}\right) \equiv I_1(k, R) + I_2(k, R). \qquad (2.10)$$

At first consider $I_2(k, R)$. For $M > 0$ denote

$$\mathcal{N}_0 = \mathcal{N}_0(k, M) = \{n : b_n \le M d_k\}.$$

We have

$$I_2(k, R) = \sum_{n \in \mathcal{N}_0(k,M)} P(N_k = n) P\left(|Y_k| > \frac{Rd_k}{2b_n}\right) +$$

$$\sum_{n \notin \mathcal{N}_0(k,M)} P(N_k = n) P\left(|Y_k| > \frac{Rd_k}{2b_n}\right) \leq$$

$$\sup_k P\left(|Y_k| > \frac{R}{2M}\right) + P\left(\frac{b_{N_k}}{d_k} \geq M\right). \tag{2.11}$$

Fix arbitrary $\varepsilon > 0$. Choose $M = M(\varepsilon) > 0$ so that

$$\sup_k P\left(\frac{b_{N_k}}{d_k} \geq M(\varepsilon)\right) < \varepsilon. \tag{2.12}$$

This choice is possible on account of the weak relative compactness of the family $\{b_{N_k}/d_k\}_{k \geq 1}$ stated above. Now choose $R_2 = R_2(\varepsilon)$ so that for any $R > R_2$,

$$\sup_k P\left(|Y_k| > \frac{R}{2M(\varepsilon)}\right) < \varepsilon. \tag{2.13}$$

This is possible on account of the weak relative compactness of the family $\{Y_k\}_{k \geq 1}$, $Y_k = (S_k - a_k)/b_k$. Thus from (2.11), (2.12) and (2.13) we get

$$I_2(k, R) < 2\varepsilon \tag{2.14}$$

for $R > R_2$ and any $k \geq 1$.

Now consider $I_1(k, R)$. It follows from (2.4) that for any $\varepsilon > 0$ there exists $R_1 = R_1(\varepsilon)$ such that for $R > R_1$ and any $k \geq 1$

$$I_1(k, R) < \varepsilon. \tag{2.15}$$

It follows from (2.10), (2.14) and (2.15) that for $R > \max(R_1, R_2)$ we have

$$\sup_k P\left(\left|\frac{a_{N_k} - c_k}{d_k}\right| > R\right) < 3\varepsilon,$$

which means that the sequence of random variables $\{(a_{N_k} - c_k)/d_k\}_{k \geq 1}$ is weakly relatively compact by virtue of the arbitrariness of ε. Together with the weak relative compactness of the family $\{b_{N_k}/d_k\}_{k \geq 1}$ this means that the family of pairs $\{(b_{N_k}/d_k, (a_{N_k} - c_k)/d_k)\}_{k \geq 1}$ is weakly relatively compact.
Denote

$$U_k = \frac{b_{N_k}}{d_k}, \quad V_k = \frac{a_{N_k} - c_k}{d_k},$$

$$\gamma_k = \inf\{L_1(Y_k, Y) + L_2((U_k, V_k,), (U, V)) : (Y, U, V) \in \mathcal{V}(Z)\}.$$

Prove that $\gamma \to 0$ as $k \to \infty$. Assume the contrary. In this case for some $\delta > 0$ and all k from some sequence $\mathcal{K}_4$ of natural numbers the inequality

$\gamma_k \geq \delta$ should hold. Choose a subsequence $\mathcal{K}_5 \subseteq \mathcal{K}_4$ so that the sequences of random variables $\{Y_k\}_{k \in \mathcal{K}_5}$ and pairs $\{(U_k, V_k)\}_{k \in \mathcal{K}_5}$ weakly converge to some random variable Y and some pair (U, V), respectively, as $k \to \infty, k \in \mathcal{K}_5$. In this connection for all $k \in \mathcal{K}_5$ large enough, the inequality

$$L_1(Y_k, Y) + L_2((U_k, V_k), (U, V)) < \delta. \tag{2.16}$$

holds.   Applying Theorem 3.1.2 to the sequences $\{Y_k\}_{k \in \mathcal{K}_5}$ and $\{(U_k, V_k)\}_{k \in \mathcal{K}_5}$ we verify that $(Y, U, V) \in \mathcal{V}(Z)$, since (2.4) takes place and therefore the limit should be the same for all convergent subsequences. But then (2.16) contradicts the assumption that $\gamma \geq \delta$ for all $k \in \mathcal{K}_5$. Thus we proved that $\gamma_k \to 0$ as $k \to \infty$.

For any $k = 1, 2, \ldots$ choose a triple (Y_k', U_k', V_k') from $\mathcal{V}(Z)$ for which

$$L_1(Y_k, Y_k') + L_2((U_k, V_k), (U_k', V_k')) \leq \gamma + \frac{1}{k}.$$

The sequence of triples (Y_k', U_k', V_k') obviously satisfies conditions (2.5) and (2.6). Its weak relative compactness follows from (2.5) and (2.6) and the fact that this property is inherent in the sequences $\{Y_k\}_{k \geq 1}$ and $\{(U_k, V_k)\}_{k \geq 1}$.

Sufficiency. Assume that the sequence $\{Z_k\}, Z_k = (S_{N_k} - c_k)/d_k$, does not converge weakly to Z. This means that for some $\delta > 0$ and all k from some sequence $\mathcal{K}$ of natural numbers the inequality $L_1(Z_k, Z) \geq \delta$ holds. Weak relative compactness of the sequence $\{(Y_k', U_k', V_k')\}_{k \geq 1}$ allows us to choose a subsequence $\mathcal{K}_1 \subseteq \mathcal{K}$ so that $Y_k' \Rightarrow Y$ and $(U_k', V_k') \Rightarrow (U, V)$ as $k \to \infty, k \in \mathcal{K}_1$, where Y, U and V are some random variables. Repeating the arguments from the proof of Theorem 3.1.2, we verify that for every $t \in \mathbb{R}$

$$\mathsf{E} \exp\{itZ\} = \mathsf{E}h_k'(tU_k') \exp\{itV_k'\} \to$$
$$\mathsf{E}h(Ut) \exp\{itV\} = \mathsf{E} \exp\{it(YU + U)\}$$

as $k \to \infty, k \in \mathcal{K}_1$, where the random variables Y and (U, V) are independent, $h_k'(t) = \mathsf{E} \exp\{itY_k'\}, h(t) = \mathsf{E} \exp\{itY\}$. This means that $(Y, U, V) \in \mathcal{V}(Z)$. It follows from the inequality

$$L_1(Y_k, Y) \leq L_1(Y_k, Y_k') + L_1(Y_k', Y)$$

and condition (2.5) that $L_1(Y_k, Y) \to 0$ as $k \to \infty, k \in \mathcal{K}_1$. Similarly the inequality

$$L_2((U_k, V_k), (U, V)) \leq$$
$$L_2((U_k, V_k), (U_k', V_k')) + L_2((U_k', V_k'), (U, V))$$

and condition (2.6) imply $L_2((U_k, V_k), (U, V)) \to 0$ as $k \to \infty, k \in \mathcal{K}_1$. Apply Theorem 3.1.2 to the sequences $\{Y_k\}_{k \in \mathcal{K}_1}$ and $\{(U_k, V_k)\}_{k \in \mathcal{K}_1}$. As a result we get $L_1(Z_k, Z) \to 0$ as $k \to \infty, k \in \mathcal{K}_1$, which contradicts the assumption that $L_1(Z_k, Z) \geq \delta$ for all $k \in \mathcal{K}_1$. The proof is completed.

The formulation of Theorem 3.2.1 contains the condition of weak relative compactness of the family of random variables $\{(S_k - a_k)/b_k\}_{k \geq 1}$. In this connection we note that, first, this condition can be relaxed in the case of identically distributed summands, as we will see in the next chapter. Second, it can be reinforced by replacing it either by the requirement of the existence of moments of summands, as will be demonstrated in Theorem 3.2.2, or by the condition of weak convergence of random variables $Y_k = (S_k - a_k)/b_k$, as we will see in Theorem 3.2.4. Both these ways lead to interesting results.

Introduce the following notation: $m_j = \mathsf{E}X_j$, $A_j = m_1 + \ldots + m_j$, $\sigma_j^2 = \mathsf{D}X_j$, $B_j^2 = \sigma_1^2 + \ldots + \sigma_j^2$.

THEOREM 3.2.2. *Assume that (2.3) is fulfilled and $\sigma_j^2 < \infty, j \geq 1$, so that $B_k^2 \to \infty (k \to \infty)$. Convergence (2.4) takes place with some sequences of positive numbers $\{d_k\}_{k \geq 1}, d_k \to \infty (k \to \infty)$, and real numbers $\{c_k\}_{k \geq 1}$ if and only if there exists a weakly relatively compact sequence of triples of random variables $\{(Y_k', U_k', V_k')\}_{k \geq 1}, (Y_k', U_k', V_k') \in \mathcal{V}(Z)$ for every $k \geq 1$, which satisfies conditions (2.5) and (2.6) with $b_k = B_k$ and $a_k = A_k$.*

PROOF is reduced to Theorem 3.2.1 since the weak relative compactness of the family $\{(S_k - a_k)/b_k\}_{k \geq 1}$ follows from the Chebyshev inequality:

$$\lim_{x \to \infty} \sup_k \mathsf{P}\left(\left|\frac{S_k - A_k}{B_k}\right| \geq x\right) \leq \lim_{x \to \infty} \frac{1}{x^2} = 0.$$

The proof is completed.

We used second moments in Theorem 3.2.2 only for simplicity. It is obvious that by defining the normalizing constants in another way, we can require the existence of any moments of random variables $\{X_j\}$ of any positive order.

Theorem 3.2.1 published in (Korolev, 1993, 1994) is a complete converse of Theorem 3.1.2. However, in practice, some situations occur when either properies of summands or those of indices provide the one of (1.1) or (1.24). Now we will give two corresponding modifications of Theorem 3.2.1.

We begin with the situation when (1.24) is fulfilled. For arbitrary random variable Z and pair (U, V) with $\mathsf{P}(U \geq 0) = 1$ introduce the set

$$\mathcal{U}(Z \mid U, V) = \{Y : Z \stackrel{d}{=} YU + V, Y \text{ and } (U, V) \text{ are independent}\}.$$

The main distinction of the situation under consideration from the one described in Theorem 3.2.1 is that for some Z, U and V the set $\mathcal{U}(Z \mid U, V)$ can turn out to be empty. In fact, let, for example, $\mathsf{P}(V = 0) = 1$ and let the random variable U have the standard exponential distribution. Let Z be a positive random variable. Then condition $\mathsf{P}(Y > 0) = 1$ is necessary for the representation $Z \stackrel{d}{=} YU$ to hold. But in this case for any $x \geq 0$

$$\mathsf{P}(YU > x) = \int_0^\infty e^{-sx} dG(s), \tag{2.17}$$

where $G(s) = \mathsf{P}(Y^{-1} < s)$. But there is the Laplace-Stieltjes transform of the distribution function G in the right-hand side of (2.17). Therefore, if as Z we take a random variable such that the function $Q(s) = \mathsf{P}(Z > s)$ is not completely monotone, then representation $Z \overset{d}{=} YU$ is impossible for any random variable Y independent of U, since according to S. N. Bernstein's theorem any Laplace-Stieltjes transform is completely monotone (see (Feller, 1971), Chapt. 13, Sect. 4).

THEOREM 3.2.3. *Assume that the sequences $\{a_k\}$, $\{b_k\}$, $\{c_k\}$, $\{d_k\}$, $b_k > 0$, $b_k \to \infty$, $d_k > 0$, $d_k \to \infty$ $(k \to \infty)$ provide (1.24) and the weak relative compactness of the family of random variables $\{(S_k - a_k)/b_k\}_{k \geq 1}$. Convergence (2.4) takes place if and only if there exists a weakly relatively compact sequence of random variables $\{Y_k'\}_{k \geq 1}, Y_k' \in \mathcal{U}(Z \mid U, V)$ for every $k \geq 1$, satisfying (2.5).*

The PROOF is reduced to Theorem 3.2.1. Note that condition (2.3) is superfluous in this case since it was required in Theorem 3.2.1 only to ascertain the weak relative compactness of the sequence $\{b_{N_k}/d_k\}_{k \geq 1}$. In the present case this property is provided automatically due to condition (1.24). The proof is completed.

To formulate one more variant of a partial converse of Theorem 3.1.2, for any two random variables Z and Y introduce the set

$$W(Z|Y) = \{(U, V) : Z \overset{d}{=} YU + V, \ Y \text{ and } (U, V) \text{ are independent}\},$$

containing all pairs of random variables (U, V) independent of Y which admit the representation $Z \overset{d}{=} YU + V$. Using the same arguments that we used when we described the properties of the set $\mathcal{V}(Z)$, we can verify that first, the set $W(Z|Y)$ is always nonempty and second, for some Z and Y the set $W(Z|Y)$ can contain more than one element. The following theorem gives an opportunity to attract the notion of identifiability of families of mixtures of probability distributions to investigate the asymptotic behavior of "growing" random sums. We will see later that it allows essential simplification of formulations.

THEOREM 3.2.4. *Assume that (2.3) takes place and (1.1) holds with some sequences of positive numbers $\{b_k\}$, $b_k \to \infty$, $(k \to \infty)$, real numbers $\{a_k\}$ and random variable Y. Convergence (2.4) takes place with some sequences of positive numbers $\{d_k\}$, $d_k \to \infty$, $(k \to \infty)$, real numbers $\{c_k\}$ and some random variable Z if and only if there exists a weakly relatively compact sequence of pairs $\{(U_k', V_k')\}_{k \geq 1}, (U_k', V_k') \in W(Z|Y)$ for every $k \geq 1$, for which (2.6) is fulfilled.*

This theorem appears to be a direct consequence of Theorem 3.2.1.

In its turn, Theorem 3.2.4 allows us to get quite simple conditions for the convergence of "growing" random sums of centered summands or which is the same, of random sums centered by random variables.

Define the subset

$$S(Z|Y) = \{(U, V) \in W(Z|Y) : P(V = 0) = 1\}$$

of the set $W(Z|Y)$, which consists of those random variables U that are independent of Y and admit representation $Z \overset{d}{=} YU$. By interchanging the variables Y and U in the reasoning we used to describe the properties of the set $\mathcal{U}(Z|U,V)$ we can verify that for some Z and Y the set $\mathcal{S}(Z|Y)$ can be empty. For some Z and Y, the set $\mathcal{S}(Z|Y)$ can contain more than one element. The set $\mathcal{S}(Z|Y)$ contains no more than one element if the family of scale mixtures of the distribution $H(x) = \mathsf{P}(Y < x)$ is identifiable (more details will be given below).

THEOREM 3.2.5. *Convergence*

$$\frac{1}{d_k}\left(\sum_{j=1}^{N_k} X_j - a_{N_k}\right) \Rightarrow Z \qquad (k \to \infty), \tag{2.18}$$

takes place under conditions of Theorem 3.2.4 with some sequence of positive numbers $\{d_k\}$, $d_k \to \infty$ $(k \to \infty)$, *and random variable* Z *if and only if there exists a weakly relatively compact sequence of random variables* $\{U'_k\}_{k \geq 1}$ *such that* $U'_k \in \mathcal{S}(Z|Y)$ *for every* $k \geq 1$ *and*

$$L_1\left(\frac{b_{N_k}}{d_k}, U'_k\right) \to 0 \qquad (k \to \infty).$$

PROOF. Assume $a_0 = 0$, $\alpha_j = a_j - a_{j-1}$, $j \geq 1$. Then by putting $X_j^* = X_j - \alpha_j$ we reduce the proof to Theorem 3.2.4 with $a_k = c_k = 0$. The proof is completed.

In the next section we will consider conditions of convergence of distributions of "growing" random sums to laws from identifiable families that are also corollaries of Theorem 3.2.1.

3.3 Convergence to distributions from identifiable families

In this section we will continue the investigation of conditions of weak convergence of "growing" random sums under the assumption that the sums of a nonrandom number of random variables, being appropriately centered and normalized, weakly converge to some random variable Y with the distribution function $H(x) = \mathsf{P}(Y < x)$. In this situation, if the family of translation/scale mixtures of the distribution function H appears to be identifiable, then conditions of convergence take especially simple form.

Recall the definition of identifiable families. It was given by H. Teicher in 1961. Let a function $G(x, y)$ be measurable with respect to y for every x and let it be a distribution function as a function of x for every y. Let $\mathcal{Q}$ be some family of random variables. Denote

$$\mathcal{F} = \{F_Q(x) = \mathsf{E}G(x, Q), x \in \mathbb{R} : Q \in \mathcal{Q}\}. \tag{3.1}$$

DEFINITION 3.3.1. The family $\mathcal{F}$ defined by the kernel G and the set $\mathcal{Q}$ is called identifiable if from the equality

$$\mathsf{E}G(x, Q_1) = \mathsf{E}G(x, Q_2), \qquad x \in \mathbb{R},$$

where $Q_1 \in \mathcal{Q}, Q_2 \in \mathcal{Q}$, it follows that $Q_1 \overset{d}{=} Q_2$.

In the case under consideration, $y = (u, v)$, $G(x, y) = H((x - v)/u)$. Therefore we can use well-known results on identifiability of families of one-dimensional distributions.

To begin with, consider the situation when mixing is carried out with respect to either the translation or the scale parameter, i.e., consider one-parameter families. For these families conditions of identifiability are well known. We will recall some of them.

DEFINITION 3.3.2. A family of distribution functions $\{G(x, y) : y > 0\}$ is called additively closed if for any $y_1 > 0, y_2 > 0$

$$G(x, y_1) * G(x, y_2) \equiv G(x, y_1 + y_2). \tag{3.2}$$

Sometimes property (3.2) of families of distributions is called the reproducibility with respect to the parameter y.

The set of normal laws with zero mean is an obvious example of an additively closed family. It is easy to see that the families of Poisson distributions, binomial distributions (with a fixed probability of success in one trial), gamma distributions with fixed scale parameter are additively closed.

A distinctive feature of additively closed families is the representability of their characteristic functions in the form

$$\int\limits_{-\infty}^{\infty} e^{itx} dG(x, y) = [\phi(t)]^y, \tag{3.3}$$

where $\phi(t)$ is some (complex-valued) function which does not depend on the parameter y.

Let $\mathcal{Q} = \{Q : P(Q > 0) = 1\}$, $P_Q(x) = P(Q < x)$, $F_Q(x) = \mathbb{E}G(x, Q)$. Set

$$f_Q(t) = \int\limits_{-\infty}^{\infty} e^{itx} dF_Q(x).$$

If the family generated by the kernel $G(x, y)$ is additively closed, then taking account of (3.3) we have

$$f_Q(t) = \int\limits_{-\infty}^{\infty} e^{itx} d\,\mathbb{E}G(x, Q) = \int\limits_{-\infty}^{\infty} e^{itx} d\left(\int\limits_{0}^{\infty} G(x, y) dP_Q(y)\right) =$$

$$\int\limits_{0}^{\infty}\left(\int\limits_{-\infty}^{\infty} e^{itx} dG(x, y)\right) dP_Q(y) = \int\limits_{0}^{\infty} [\phi(t)]^y dP_Q(y).$$

Set $z = \phi(t)$. The function $\psi_Q(z) = \int_0^\infty z^y dP_Q(y)$ is analytical at least in the domain $D = \{z : 0 < |z| < 1\}$.

Assume that random variables Q_1 and Q_2 ($Q_1 \in \mathcal{Q}, Q_2 \in \mathcal{Q}$) with different distribution functions P_{Q_1} and P_{Q_2} make up the same mixture, i.e., $F_{Q_1}(x) \equiv F_{Q_2}(x)$. But then by virtue of the fact that the correspondence between distributions and their characteristic functions is one-to-one, the functions $\psi_{Q_1}(z)$ and $\psi_{Q_2}(z)$ should coincide for all z from the set $D_\phi = \{z = \phi(t) : -\infty < t < \infty\}$. Since $D_\phi \subseteq D$ and the functions $\psi_{Q_1}(\phi(t))$ and $\psi_{Q_2}(\phi(t))$ admit analytical continuation onto the whole of domain D, they must also coincide for all points $z \in D$. Hence for all $\rho \in (0,1)$, $\psi_{Q_1}(\rho e^{it}) = \psi_{Q_2}(\rho e^{it})$. Set $\rho_n = \frac{n}{n+1}$, $n = 1,2,\ldots$. Then according to the Lebesgue's dominated convergence theorem we have

$$\lim_{n\to\infty} \int_0^\infty (\rho_n e^{it})^y dP_Q(y) = \int_0^\infty e^{ity} dP_Q(y),$$

and the equality $\psi_{Q_1}(\rho e^{it}) = \psi_{Q_2}(\rho e^{it})$ appears to be correct for $\rho = 1$ as well. But $\psi_Q(e^{it})$ is nothing but the characteristic function of the random variable Q. Therefore the equality $\psi_{Q_1}(e^{it}) = \psi_{Q_2}(e^{it}), t \in \mathbb{R}$, means the coincidence of characteristic functions of the variables Q_1 and Q_2. Hence $Q_1 \overset{d}{=} Q_2$. Thus we have proved

PROPOSITION 3.3.1. *The family of mixtures (3.1) of distribution functions* $G(x, \cdot)$ *from an additively closed set is identifiable.*

Mixtures of kernels from additively closed families by no means exhaust all cases of identifiable mixtures. Consider scale mixtures of laws concentrated on positive semiaxis.

PROPOSITION 3.3.2. *Let* $G(x,y) = G(xy)$, $y \geq 0$, $G(0) = 0$. *Assume that the Fourier transform of the function* $G^*(y) = G(e^y)$, $y \geq 0$, *does not turn into zero. Then the family of mixtures*

$$\mathcal{F} = \{F_Q(x) = \mathbb{E}G(xQ), x \geq 0 : \mathrm{P}(Q > 0) = 1\}$$

is identifiable.

PROOF. Introducing variables u and v instead of x and y, setting $x = e^u, y = e^{-v}$ and denoting $F_Q^*(u) = F_Q(e^u), P_Q^*(v) = \mathrm{P}(Q \geq e^{-v})$, write the mixture $F_Q(x)$ in the form of the convolution of the functions G^* and P_Q^*:

$$F_Q(x) = F_Q(e^u) = F_Q^*(u) = \int_{-\infty}^\infty G^*(u - v) dP_Q^*(v).$$

But the Fourier-Stieltjes transform $\psi_{F_Q^*}$ of the convolution F_Q^* is equal to the product of the Fourier-Stieltjes transforms $\psi_{G^*}(t)$ and $\psi_{P_Q^*}(t)$ of the functions G^* and P_Q^*, respectively:

$$\psi_{F_Q^*}(t) = \psi_{G^*} \cdot \psi_{P_Q^*}(t).$$

Therefore if $\psi_{G^*}(t)$ is not equal to zero in any point, then the equality $G^* * P_{Q_1}^* = G^* * P_{Q_2}^*$ implies $Q_1 \overset{d}{=} Q_2$. The proof is completed.

Some families of translation mixtures possess a similar property.

PROPOSITION 3.3.3. *The family of translation mixtures*

$$\mathcal{F} = \{F_Q(x) = EG(x - Q), x \in \mathbb{R} : Q \in \mathcal{Q}\}$$

with $\mathcal{Q}$ being the set of all real random variables is identifiable if the characteristic function of the distribution $G(x)$ does not turn into zero.

PROOF is analogous to that of the previous statement.

Proposition 3.3.1 guarantees the identifiability of scale mixtures of normal laws with zero mean, as well as scale mixtures of other symmetrical stable laws.

Proposition 3.3.3 guarantees the identifiability of translation mixtures generated by any infinitely divisible law.

Conditions of identifiability of one-parameter mixtures formulated in Propositions 3.3.1 – 3.3.3 were proved by H. Teicher (Teicher, 1961). We restricted ourselves to mentioning those examples of identifiable families which will be used below to investigate the asymptotic behavior of some special random sums. Numerous examples of other identifiable families can be found in the works (Teicher, 1961), (Teicher, 1963), (Yakowitz and Spragins, 1968).

Some necessary and sufficient conditions of the identifiability of families of mixtures are given in (Tallis, 1969), see also the review (Kruglov, 1991).

Unfortunately, in the general case, examples of identifiable families of two-parameter translation-scale mixtures are unknown. We know only examples of these families when Q is a two-dimensional discrete random vector taking only a finite number of values. In this case the family of finite translation-scale mixtures of normal laws is identifiable (Teicher, 1963).

We shall demonstrate that the family of arbitrary translation-scale mixtures of normal laws is not identifiable. Let Y_1, Y_2, U_1 and U_2 be independent random variables, and let also Y_1 and Y_2 have identical standard normal distribution, $P(U_1 = 1) = 1$ and let the variable U_2 be nondegenerate, $P(U_2 > 0) = 1$. Then the distribution of the random variable $Z = Y_1 U_2 + Y_2$ can be written both as

$$P(Z < x) = \int\limits_0^\infty \int\limits_{-\infty}^\infty \Phi\left(\frac{x - v}{u}\right) dP(Y_2 < v) dP(U_2 < u), \qquad (3.4)$$

and

$$P(Z < x) = \int\limits_0^\infty \int\limits_{-\infty}^\infty \Phi\left(\frac{x - v}{u}\right) dP(Y_1 U_2 < v) dP(U_1 < u), \qquad (3.5)$$

moreover, mixing distributions in (3.4) and (3.5) are obviously different.

During the last three decades the notion of identifiability of mixtures of probability distribution has been intensively exploited in applied problems connected with separation of mixtures in classification problems, pattern recognition and distribution identification. Bibliography on this topic is rather extensive. Here we shall use identifiability for getting simpler conditions of convergence for distributions of "growing" random sums, centered and normalized by constants. First of all, consider conditions of convergence of random sums to identifiable translation mixtures. As we have seen above, these mixtures appear as limit ones, when $b_{N_k}/d_k \Rightarrow b$ for some $b > 0$ as $k \to \infty$. Without loss of generality we will assume $b = 1$.

THEOREM 3.3.1. *Assume that $b_k \to \infty, d_k \to \infty, b_{N_k}/d_k \Rightarrow 1$ and for some sequence of real numbers $\{\alpha_j\}_{j \geq 1}$ and some random variable Y the convergence*

$$\frac{1}{b_k} \sum_{j=1}^{k}(X_j - \alpha_j) \Rightarrow Y \qquad (k \to \infty), \tag{3.6}$$

takes place. Futhermore, let the summands $\{X_j\}$ be uniformly asymptotically constant: for any $\varepsilon > 0$

$$\lim_{k \to \infty} \max_{1 \leq j \leq k} P(|X_j - \alpha_j| > \varepsilon b_k) = 0. \tag{3.7}$$

Convergence to some random variable Z

$$\frac{1}{d_k} \left(\sum_{j=1}^{N_k}(X_j - \alpha_j) - c_k \right) \Rightarrow Z \qquad (k \to \infty)$$

takes place with some sequence of real numbers $\{c_k\}_{k \geq 1}$ if and only if there exists a random variable V satisfying conditions:

1) $Z \overset{d}{=} Y + V$, *where Y and V are independent;*

2) $\dfrac{1}{d_k} \left(\sum_{j=1}^{N_k} \alpha_j - c_k \right) \Rightarrow V \qquad (k \to \infty).$

PROOF. Reduce the proof to Theorem 3.2.4. Notice that condition (2.3) appearing in the formulation of Theorem 3.2.4 was used only to prove the weak relative compactness of the sequence $\{b_{N_k}/d_k\}$. Therefore it is unnecessary in the situation under consideration, since the compactness of the mentioned sequence follows from its convergence to one. Further, condition (3.7) means that the distribution of the random variable Y in (3.6) belongs to class L introduced by P. Lévy, and therefore is infinitely divisible. But characteristic functions of infinitely divisible laws do not turn into zero anywhere (see (Gnedenko and Kolmogorov, 1954), Sect. 17, Chapt. 3) and hence according to Proposition 3.3.3 the family of translation mixtures of distribution function $H(x) = P(Y < x)$ is identifiable. Therefore in the situation under consideration, the set $W_1(Z|Y) = \{V : Z \overset{d}{=} Y + V, Y$ and V are independent$\}$ consists of no more than one element. Now the desired result follows from Theorem 3.2.4, where the set $W(Z|Y)$ should be replaced by the set $W_1(Z|Y)$ in view of condition $b_{N_k}/d_k \Rightarrow 1$. The proof is completed.

Now we turn to the investigation of conditions of convergence of distributions of "growing" random sums to identifiable scale mixtures. Here we shall consider only the case where limit laws are scale mixtures of normal distributions with zero mean. We have some reasons for doing this. First of all, conditions of convergence to other identifiable scale mixtures can be obtained from the result stated below by a simple change of notations. Second, the class of scale mixtures of normal laws is rather wide and along with others, contains symmetrized gamma distributions (as we have seen in Section 3.2., see relations (2.1) and (2.2)), including Laplace distribution (double exponential), and also Cauchy and Student distributions, symmetrical stable laws (see, for example, (Zolotarev, 1983), Theorem 3.3.1). Third, these distributions have wide practical application, for example, in metrology, financial mathematics (see Sections 3.5. and 4.4. of this book).

Following the tradition, we denote the standard normal distribution function as $\Phi(x)$, $x \in \mathbb{R}$.

THEOREM 3.3.2. *Assume that (2.3) and (3.6) take place with some b_k and α_k, $b_k \to \infty$ ($k \to \infty$), and $P(Y < x) = \Phi(x)$, $x \in \mathbb{R}$. Convergence to some random variable Z*

$$\frac{1}{d_k} \sum_{j=1}^{N_k} (X_j - \alpha_j) \Rightarrow Z \qquad (k \to \infty), \qquad (3.8)$$

takes place with some sequence of positive numbers $\{d_k\}_{k \geq 1}$, $d_k \to \infty$ ($k \to \infty$), if and only if there exists a nonnegative random variable U satisfying the following conditions:

1) $P(Z < x) = \mathsf{E}\Phi\left(\frac{x}{U}\right)$, $x \in \mathbb{R}$;

2) $b_{N_k}/d_k \Rightarrow U$ ($k \to \infty$).

PROOF. By virtue of the identifiability of the family of scale mixtures of normal laws, the set $S(Z|Y)$ introduced before Theorem 3.2.5 contains no more than one element. Therefore the desired result immediately follows from Theorem 3.2.5 with $a_k = \alpha_1 + \ldots + \alpha_k$, $k \geq 1$. The proof is completed.

The importance of this theorem for applications is stipulated by the fact that Theorem 3.2.5 is actually a natural generalization of the central limit theorem for sums with a random number of summands. It can explain deviations of experimental data distributions from the normal law in many applied problems, for example, connected with measurements processing where the normality of error distributions expected by virtue of the central limit theorem is rather an exception than a rule (see, for example, (Novitskii and Zograf, 1991)).

In the case when the random variable Z in (3.8) has a normal distribution itself, the statement of Theorem 3.3.2 can be essentially reinforced.

THEOREM 3.3.3. *Assume that (2.3) takes place and (3.6) is fulfilled for some sequences of positive numbers $\{b_k\}$, $b_k \to \infty$ ($k \to \infty$), and real numbers*

$\{\alpha_j\}$. *Let the summands $\{X_j\}$ be uniformly asymptotically constant. The convergence*

$$P\left(\frac{1}{d_k}\sum_{j=1}^{N_k}(X_j - \alpha_j) < x\right) \Rightarrow \Phi(x) \qquad (k \to \infty), \qquad (3.9)$$

takes place for some sequence of positive numbers $\{d_k\}$, $d_k \to \infty$ $(k \to \infty)$, if and only if for some $b > 0$ the following conditions are fulfilled:

1) $P(Y < x) = \Phi(bx)$;

2) $b_{N_k}/d_k \Rightarrow b$ $\quad (k \to \infty)$.

PROOF. Necessity. The condition of uniform asymptotic constancy of summands (3.7) together with (3.6) imply that the distribution function of the random variable Y from (3.6) belongs to class **L**, and therefore, is infinitely divisible. Since the summands are centered, as a consequence of (3.9), any random variable $U \in S(Z|Y)$ in this situation should satisfy the relation

$$Eh(tU) = e^{-t^2/2}, \qquad t \in \mathbb{R}, \qquad (3.10)$$

where h is the infinitely divisible characteristic function of the random variable Y. According to Lemma 1.2.1 from (Kruglov and Korolev, 1990) it follows from (3.10) that there should exist functions $a(u)$ and $\sigma^2(u)$ such that

$$h(tU) = \exp\{ita(u) - \frac{1}{2}\sigma^2(u)t^2\}, \qquad t \in \mathbb{R}, \, u > 0, \qquad (3.11)$$

and moreover $Ea(U) = 0$. Specific dependence on t and u of the argument of the function in the left-hand side of (3.11) means that $a(u) = \alpha u$, where $\alpha \in \mathbb{R}$, and $\sigma(u) = \beta u$, where $\beta \geq 0$. Therefore we have $\alpha EU = 0$ which by virtue of the nonnegativeness of the random variable U is possible only if $P(U = 0) = 1$ (but then (3.10) is violated) or if $\alpha = 0$. Thus (3.10) takes the form

$$Eh(tU) = E(\exp\{-\frac{1}{2}t^2\beta^2\})^{U^2} = e^{-t^2/2}, \qquad t \in \mathbb{R}.$$

But according to Theorem 3 from (Kruglov and Titov, 1986) this is possible only if the random variable U is degenerate at a nonzero point. Therefore any random variable $U \in S(Z|Y)$ is such that for some $b > 0$ $P(U = b) = 1$, and in addition, the distribution function of the random variable Y is $\Phi(bx)$. Now we only have to refer to Theorem 3.2.5.

The sufficiency of conditions 1) and 2) follows from Theorem 3.1.2. The proof is completed.

Along with generalizations of the central limit theorem presented in Theorems 3.3.2 and 3.3.3, Theorem 3.2.5 allows also useful extensions of the weak law of large numbers to random sums.

THEOREM 3.3.4. *Assume that (2.3) takes place. Let relation*

$$\frac{1}{b_k}\sum_{j=1}^{k}(X_j - \dot{\alpha}_j) \Rightarrow 1 \qquad (k \to \infty).$$

hold with some sequences of real numbers $\{\alpha_k\}_{k\geq1}$ and positive numbers $\{b_k\}_{k\geq1}$, $b_k \to \infty$ $(k \to \infty)$. The convergence to some random variable Z

$$\frac{1}{d_k}\sum_{j=1}^{N_k}(X_j - \alpha_j) \Rightarrow Z \qquad (k \to \infty),$$

takes place with some sequence of positive numbers $\{d_k\}_{k\geq1}$, $d_k \to \infty$ $(k \to \infty)$, if and only if

$$\frac{b_{N_k}}{d_k} \Rightarrow Z \qquad (k \to \infty).$$

PROOF. In the case under consideration we have $h(t) = \mathsf{E}\exp\{itY\} = e^{it}$. Therefore

$$\mathsf{E}e^{itZ} = \mathsf{E}h(tU) = \mathsf{E}e^{itU}, \qquad t \in \mathbb{R},$$

which is possible only if $Z \stackrel{d}{=} U$, i.e., the family of scale mixtures of degenerate laws not concentrated at zero is trivially identifiable. Now the desired result follows from Theorem 3.2.5. The proof is completed.

In its turn, one improvement of the famous Rényi–Mogyoródi theorem (Rényi, 1956), (Mogyoródi, 1971) follows from Theorem 3.3.4. It is of great significance in the investigation of rarefied flows (see Sect. 3.6 of this book) and plays a determining role in reliability growth models considered in Chapter 5.

Let $X_1, X_2,\ldots$ be independent identically distributed random variables with $\mathsf{E}X_1 = m \neq 0$. For definiteness assume that $m > 0$. As before, let integer positive random variables N_k be independent of the sequence $\{X_j\}_{j\geq1}$ for every $k \geq 1$.

THEOREM 3.3.5. Let (2.3) take place and $F(x)$ be an arbitrary distribution function. Then

$$\mathsf{P}\left(\frac{1}{k}\sum_{j=1}^{N_k}X_j < mx\right) \Rightarrow F(x) \qquad (k \to \infty),$$

if and only if

$$\mathsf{P}(N_k < kx) \Rightarrow F(x) \qquad (k \to \infty).$$

The PROOF of this result is reduced to Theorem 3.3.4, since the random variables $\{X_j\}_{j\geq1}$, possessing the above-mentioned properties, obviously satisfy the law of large numbers.

Note that we will particularly consider random sums of identically distributed summands in Chapter 4. However, the techniques we use in Chapter 4 do not allow us to obtain Theorem 3.3.5 as a corollary of the main Theorem 4.2.1, since the nondegeneratedness of the limit law for sums of a nonrandom number of summands is required for the latter theorem to be correct.

In the next section dealing with limit behavior of risk processes we will present some refinements of the theorems proved here.

3.4 Limit theorems for risk processes

Consider the random process

$$S(t) = S_0 + ct - \sum_{i=1}^{N(t)} X_i, \qquad t \geq 0, \tag{4.1}$$

where $N(t)$ is an integer non negative random process with non-decreasing trajectories $(t \geq 0)$, $\{X_j\}_{j \geq 1}$ are independent random variables, and also the random variable $N(t)$ is independent of the sequence $\{X_j\}_{j \geq 1}$ for every $t \geq 0$, $c > 0$, $S_0 \geq 0$ (by definition, $\sum_{i=1}^{0} X_j = 0$).

Let us call process (4.1) a generalized risk process. If $N(t)$ is a Poisson random process and the random variables $\{X_j\}_{j \geq 1}$ are identically distributed and have finite variance, then process (4.1) is called a classical risk process. Such a process was considered in Example 1.1.4.

Random processes of the form (4.1) play an important role in actuarial mathematics (mathematical insurance theory), having the sense of the surplus of an insurance company at time t. This interpretation becomes clear if we assume that in (4.1) S_0 is an initial ("starting") capital of the insurance company, c is a coefficient characterizing the rate of a linear increase of the capital caused by clients insurance dues (premium rate), and the random variables $\{X_j\}$ are positive and characterize the payments of the company in insurance cases (claims) whose number up to time t equals to $N(t)$. Of course this model is not comprehensive and does not take into account, for example, the nonlinear increase of the capital due to possible investment to some or other projects (stocks, etc.), and also inflation and dividend payments to the holders of the company stocks or policies. Nevertheless, processes of the form (4.1) are suitable approximations to the real situation and can be considered as a base for more detailed examination of different aspects of the activity of insurance companies.

Literature on actuarial mathematics, where processes (4.1) are studied, is extensive. We will mention only canonical guidebooks (Bowers et al., 1986) and (Grandell, 1990), where further references can be found.

It is well known that classical risk process is asymptotically normal with $t \to \infty$. We will give one more proof of this fact, based on application of Theorem 3.1.2. For simplicity, without loss of generality, instead of the process $S(t)$ with continuous time consider the process with discrete time S_n, $n = 0, 1, \ldots$, assuming $S_n = S(n)$. This assumption coordinates perfectly well with practice as time is usually measured in discrete units — in days, hours, minutes and it is very difficult to imagine that the time of a payment is fixed precisely up to seconds. Similarly, $N_n = N(n)$.

So first of all assume that the random variables $\{X_j\}_{j \geq 1}$ are identically distributed with $EX_1 = a$, $DX_1 = \sigma^2 < \infty$. Let $N(t)$ be a Poisson process with intensity $\lambda > 0$.

THEOREM 3.4.1. *The classical risk process is asymptotically normal: for any $x \in \mathbb{R}$*

$$\lim_{n \to \infty} P\left(\frac{S_n - n(c - a\lambda) - S_0}{\sqrt{n\lambda(a^2 + \sigma^2)}} < x\right) = \Phi(x).$$

PROOF. Since

$$S_n - n(c - a\lambda) - S_0 = -\left(\sum_{j=1}^{N_n} X_j - na\lambda\right), \tag{4.2}$$

we can reduce the proof to Theorem 3.1.2. In our case

$$E\sum_{j=1}^{N_n} X_j = an\lambda, \quad D\sum_{j=1}^{N_n} X_j = n\lambda(a^2 + \sigma^2).$$

Set $a_n = na$, $c_n = na\lambda$, $b_n = \sigma\sqrt{n}$, $d_n = \sqrt{n\lambda(\sigma^2 + a^2)}$. Then

$$\frac{a_{N_n} - c_n}{d_n} = \frac{a(N_n - n\lambda)}{\sqrt{n\lambda(a^2 + \sigma^2)}} = \frac{a}{\sqrt{\sigma^2 + a^2}} \cdot \frac{N_n - n\lambda}{\sqrt{n\lambda}} \Rightarrow V \ (n \to \infty), \tag{4.3}$$

where $P(V < x) = \Phi\left(x\sqrt{\frac{a^2 + \sigma^2}{a^2}}\right)$, $x \in \mathbb{R}$, by virtue of the well-known property of the Poisson distribution:

$$\lim_{n \to \infty} P\left(\frac{N_n - n\lambda}{\sqrt{n\lambda}} < x\right) = \Phi(x), \quad x \in \mathbb{R}.$$

Further

$$\frac{b_{N_n}}{d_n} = \sqrt{\frac{N_n}{n\lambda} \cdot \frac{\sigma^2}{a^2 + \sigma^2}}.$$

We have for arbitrary $\varepsilon > 0$

$$P\left(\left|\frac{N_n}{n\lambda} - 1\right| > \varepsilon\right) \leq \frac{DN_n}{\varepsilon^2 n^2 \lambda^2} = \frac{1}{\varepsilon^2 n\lambda} \to 0$$

with $n \to \infty$, so that

$$\frac{b_{N_n}}{d_n} \Rightarrow \sqrt{\frac{\sigma^2}{a^2 + \sigma^2}} = U, \quad (n \to \infty). \tag{4.4}$$

Finally, by virtue of the central limit theorem

$$\frac{1}{b_n}\left(\sum_{j=1}^{n} X_j - a_n\right) = \frac{1}{\sigma\sqrt{n}}\left(\sum_{j=1}^{n} X_j - na\right) \Rightarrow Y \ (n \to \infty), \tag{4.5}$$

where $P(Y < x) = \Phi(x)$, $x \in \mathbb{R}$. Therefore applying Theorem 3.1.2 to the random variables

$$Z_n = \frac{\sum_{j=1}^{N_n} X_j - na\lambda}{\sqrt{n\lambda(a^2 + \sigma^2)}}$$

taking account of (4.3), (4.4) and (4.5), bearing in mind the form we have got
for the distributions of the limit random variables Y, U and V, we obtain

$$\lim_{n \to \infty} P\left(\frac{\sum_{j=1}^{N_n} X_i - na\lambda}{\sqrt{n\lambda(a^2 + \sigma^2)}} < x\right) = P(YU + V < x) =$$

$$\Phi\left(x\sqrt{\frac{a^2 + \sigma^2}{\sigma^2}}\right) * \Phi\left(x\sqrt{\frac{a^2 + \sigma^2}{a^2}}\right) = \Phi(x), \quad x \in \mathbb{R}.$$

Hence taking account of (4.2) we have

$$\lim_{n \to \infty} P\left(\frac{S_n - n(c - a\lambda) - S_0}{\sqrt{n\lambda(a^2 + \sigma^2)}} < x\right) =$$

$$\lim_{n \to \infty} P\left(\frac{\sum_{j=1}^{N_n} X_j - na\lambda}{\sqrt{n\lambda(a^2 + \sigma^2)}} > -x\right) =$$

$$1 - \Phi(-x) = \Phi(x), \qquad x \in \mathbb{R},$$

Q.E.D.

The analysis of real situations shows that rather severe constraints in the
definition of a classical risk process are fulfilled very seldom. Two questions
arise in this connection. First, what limit distributions can processes of the
form (4.1) have within a weakening of the conditions determining the classical
risk process? Second, in which cases can normal approximation be used for
the distribution of the process (4.1), if we relax the conditions, determining
the classical risk process? In other words, to what extent can these conditions
be weakened with the normal approximation still possible? The remaining
part of this section contains the answers to these questions.

We will consider the situation which can be regarded as a generalization
of the one where the insurance claims $\{X_j\}_{j \geq 1}$ are assumed to be identically
distributed, though we will not formally impose any restrictions, including
moment ones, upon the random variables $\{X_j\}$ besides their independence
with the exception of particularly mentioned cases. Our additional assumptions will concern centering and normalizing sequences.

We will assume that the normalizing constants have a special form, namely,
let

$$b_n = d_n = n^{1/\alpha} B(n),$$

where $1 < \alpha \leq 2$, and $B(x), x \in \mathbb{R}$ is a slowly varying function, i.e., such that
for any $p > 0$

$$\lim_{x \to \infty} \frac{B(px)}{B(x)} = 1. \tag{4.6}$$

Concerning the centering constants assume that $a_n = c_n$; moreover, let the
sequence $\{a_n\}$ increase monotonically and

$$\lim_{n \to \infty} \frac{a_n}{n} = A \tag{4.7}$$

for some $A \in (0, \infty)$. Denote $\alpha_j = a_j - a_{j-1}$, where $a_0 = 0$ for definiteness.

THEOREM 3.4.2. *Assume that $N_k \overset{P}{\to} \infty$ as $k \to \infty$, the claims $\{X_j\}_{j\geq 1}$ are uniformly asymptotically constant: for any $\varepsilon > 0$*

$$\lim_{k\to\infty} \max_{1\leq j\leq k} \mathsf{P}(|X_j - \alpha_j| > \varepsilon b_k) = 0,$$

and their normalized sums have some limit distribution:

$$\frac{1}{b_k}\left(\sum_{j=1}^k X_j - a_k\right) \Rightarrow Y \qquad (k \to \infty). \tag{4.8}$$

A generalized risk process S_n has a limit distribution as $n \to \infty$:

$$\frac{1}{b_n}(S_n - S_0 - nc + a_n) \Rightarrow -Z \qquad (n \to \infty), \tag{4.9}$$

if and only if there exists a random variable V such that

1) $Z \overset{d}{=} Y + V$, Y *and V are independent;*

2) $\dfrac{a_{N_k} - a_k}{b_k} \Rightarrow V \qquad (k \to \infty).$

PROOF. "Only if" part. Since

$$\frac{1}{b_n}(S_n - S_0 - nc + a_n) = -\frac{1}{b_n}\left(\sum_{j=1}^{N_n} X_j - a_n\right),$$

we will reduce the proof to Theorem 3.3.1. Show that the choice of the centering and the normalizing constants and the conditions of the theorem guarantee that $b_{N_k}/b_k \Rightarrow 1$ $(k \to \infty)$. In fact, according to Theorem 3.3.1, conditions (4.8) and (4.9) together with the unbounded stochastic growth of N_k imply weak relative compactness of the sequence of pairs of random variables $\{(b_{N_k}/b_k, (a_{N_k} - a_k)/b_k)\}_{k\geq 1}$. As before, denote the greatest lower bound of q-quantiles of the random variable N_k as $l_k(q)$. Since the sequence $\{a_k\}$ is monotonically increasing, the q-quantile of the random variable $(a_{N_k} - a_k)/b_k$, which we will denote as $A_k(q)$, is equal to $(a_{l_k(q)} - a_k)/b_k$. By virtue of the weak relative compactness of the sequence of random variables $\{(a_{N_k} - a_k)/b_k\}_{k\geq 1}$ the sequence $\{A_k(q)\}_{k\geq 1}$ is uniformly bounded for any $q \in (0, 1)$:

$$\sup_k |A_k(q)| \equiv M(q) < \infty.$$

It is easy to verify this by the reasoning ad absurdum. Therefore by virtue of the definition of b_n and condition (4.7), for arbitrary $q \in (0, 1)$ we have

$$\left|\frac{a_{l_k(q)}}{a_k} - 1\right| = \left|\frac{a_{l_k(q)} - a_k}{b_k}\right| \cdot \frac{b_k}{a_k} = |A_k(q)| \cdot \frac{b_k}{a_k} \leq$$

$$M(q)\frac{k^{1/\alpha}B(k)}{Ak}\cdot\frac{Ak}{a_k}\to 0 \qquad (k\to\infty) \tag{4.10}$$

since $Ak/a_k \to 1$ according to condition (4.7) and $k^{1/\alpha}B(k)/k \to 0$ by the properties of a slowly varying function. In turn from (4.10) and (4.7) we get that

$$\frac{l_k(q)}{k} = \frac{Al_k(q)}{a_{l_k(q)}}\cdot\frac{a_{l_k(q)}}{a_k}\cdot\frac{a_k}{Ak}\to 1 \qquad (k\to\infty) \tag{4.11}$$

for any $q \in (0,1)$, since the condition of unbounded stochastic growth of N_k as $k\to\infty$ means that $l_k(q)\to\infty$ $(k\to\infty)$ for any $q\in(0,1)$. It follows from (4.11) that $N_k/k \Rightarrow 1$ $(k\to\infty)$. By virtue of the inequality

$$P(|W_1 + W_2| > \delta) \le P(|W_1| > \frac{\delta}{2}) + P(|W_2| > \frac{\delta}{2}),$$

which is valid for any random variables W_1 and W_2 and for any $\delta > 0$, for arbitrary $\varepsilon > 0$ we have

$$P\left(\left|\frac{b_{N_k}}{b_k} - 1\right| > \varepsilon\right) = P\left(\left|\left(\frac{N_k}{k}\right)^{1/\alpha}\frac{B(N_k)}{B(k)} - 1\right| > \varepsilon\right) =$$

$$P\left(\left|\left(\frac{N_k}{k}\right)^{1/\alpha}\left(\frac{B(N_k)}{B(k)} - 1\right) + \left(\frac{N_k}{k}\right)^{1/\alpha} - 1\right| > \varepsilon\right) \le$$

$$P\left(\left|\left(\frac{N_k}{k}\right)^{1/\alpha}\left(\frac{B(N_k)}{B(k)} - 1\right)\right| > \frac{\varepsilon}{2}\right). \tag{4.12}$$

Consider the first summand in the right-hand side of (4.12). The following chain of relations holds:

$$P\left(\left|\left(\frac{N_k}{k}\right)^{1/\alpha}\left(\frac{B(N_k)}{B(k)} - 1\right)\right| > \frac{\varepsilon}{2}\right) =$$

$$\sum_{n=1}^{\infty}P(N_k = n)P\left(\left|\frac{B(n)}{B(k)} - 1\right|\frac{\varepsilon k^{1/\alpha}}{2n^{1/\alpha}}\right) =$$

$$\sum_{n:|\frac{n}{k}-1|\le\frac{1}{2}}P(N_k = n)P\left(\left|\frac{B(n)}{B(k)} - 1\right| > \frac{\varepsilon k^{1/\alpha}}{2n^{1/\alpha}}\right) +$$

$$\sum_{n:|\frac{n}{k}-1|>\frac{1}{2}}P(N_k = n)P\left(\left|\frac{B(n)}{B(k)} - 1\right| > \frac{\varepsilon k^{1/\alpha}}{2n^{1/\alpha}}\right) \le$$

$$\sum_{n:|\frac{n}{k}-1|\le\frac{1}{2}}P(N_k = n)P\left(\left|\frac{B(n)}{B(k)} - 1\right| > \frac{2^{1-1/\alpha}\varepsilon}{3^{1/\alpha}}\right) + P\left(\left|\frac{N_k}{k} - 1\right| > \frac{1}{2}\right) \le$$

$$\sum_{n:|\frac{n}{k}-1|\le\frac{1}{2}} P(N_k=n)P\left(\sup_{\frac{1}{2}\le p\le\frac{3}{2}}\left|\frac{B(kp)}{B(k)}-1\right|>\frac{2^{1-1/\alpha}\varepsilon}{3^{1/\alpha}}\right)+P\left(\left|\frac{N_k}{k}-1\right|\frac{1}{2}\right)\le$$

$$P\left(\sup_{\frac{1}{2}\le p\le\frac{3}{2}}\left|\frac{B(kp)}{B(k)}-1\right|>\frac{2^{1/\alpha-1}\varepsilon}{3^{1/\alpha}}\right)+P\left(\left|\frac{N_k}{k}-1\right|>\frac{1}{2}\right). \qquad (4.13)$$

According to Theorem 1.1 in (Seneta, 1976) convergence (4.6) is uniform in every closed segment of values of p. Hence a $k_0=k_0(\varepsilon)$ can be found such that for all $k\ge k_0$

$$\sup_{\frac{1}{2}\le p\le\frac{3}{2}}\left|\frac{B(kp)}{B(k)}-1\right|<\frac{2^{1/\alpha-1}\varepsilon}{3^{1/\alpha}}. \qquad (4.14)$$

Thus according to (4.13) and (4.14) for all $k\ge k_0$ we have

$$P\left(\left|\left(\frac{N_k}{k}\right)\left(\frac{B(N_k)}{B(k)}-1\right)\right|>\frac{\varepsilon}{2}\right)\le P\left(\left|\frac{N_k}{k}-1\right|>\frac{1}{2}\right),$$

and therefore according to what has already been proved,

$$\lim_{k\to\infty}P\left(\left|\left(\frac{N_k}{k}\right)^{1/\alpha}\left(\frac{B(N_k)}{B(k)}-1\right)\right|>\frac{\varepsilon}{2}\right)\le$$

$$\lim_{k\to\infty}P\left(\left|\frac{N_k}{k}-1\right|>\frac{1}{2}\right)=0. \qquad (4.15)$$

Consider the second summand in the right-hand side of (4.12). Since $N_k/k\Rightarrow 1$ $(k\to\infty)$, $N_k/k\overset{P}{\to}1$ and therefore $(N_k/k)^{1/\alpha}\overset{P}{\to}1$, i.e.,

$$\lim_{k\to\infty}P\left(\left|\left(\frac{N_k}{k}\right)^{1/\alpha}-1\right|>\frac{\varepsilon}{2}\right)=0. \qquad (4.16)$$

Now the required relation $b_{N_k}/b_k\Rightarrow 1$ $(k\to\infty)$ follows from (4.12), (4.15) and (4.16). And we only have to refer to Theorem 3.3.1.

"If" part. As before verify that condition 2) implies $b_{N_k}/b_k\Rightarrow 1$ $(k\to\infty)$, which allows reducing the proof to Theorem 3.1.2. The proof is completed.

COROLLARY 3.4.1. *Assume that $N_k\overset{P}{\to}\infty$ and the normalized sums of claims have some limit distribution, i.e., (4.8) takes place. A generalized risk process is asymptotically normal, i.e.,*

$$P\left(\frac{1}{b_k}(S_k-S_0-kc+a_k)<x\right)\Rightarrow\Phi(x) \qquad (k\to\infty), \qquad (4.17)$$

if and only if there exist numbers $\mu\in\mathbb{R}$ and $0\le\sigma^2\le 1$ such that

1) $P(Y<x)=\Phi\left(\frac{x-\mu}{\sigma}\right),\quad x\in\mathbb{R};$

2) $P\left(\dfrac{a_{N_k} - a_k}{b_k} < x\right) \Rightarrow \Phi\left(\dfrac{x + \mu}{\sqrt{1 - \sigma^2}}\right) \quad (k \to \infty)$.

PROOF. By Theorem 3.4.2, the limit standard normal random variable Z should satisfy the relation $Z \overset{d}{=} Y + V$ with Y and V independent. Then according to the Lévy–Cramér theorem on decomposability of a normal law only into normal components both Y and V should have normal distributions, which is infinitely divisible. Thus the condition of uniform asymptotic constancy of claims is unnecessary in this case. The desired result follows from Theorem 3.4.2. The corollary is proved.

We have assumed in Theorem 3.4.2 and Corollary 3.4.1 that the known properties of claims provide the convergence of the distributions of their normalized and centered sums to some law and the conditions for the convergence of distributions of a generalized risk process were formulated in terms of conditions imposed on the process $N(t)$. Now assume that the asymptotic properties of the process $N(t)$ with $t \to \infty$ are known, and formulate the conditions for convergence of distributions of a generalized risk process in terms of normalized sums of claims.

THEOREM 3.4.3. *Assume that* $N_k \overset{P}{\to} \infty$ *as* $k \to \infty$, *the convergence*

$$\frac{a_{N_k} - a_k}{b_k} \Rightarrow V \qquad (k \to \infty) \tag{4.18}$$

to some random variable V *takes place, and the sequence of random variables* $\{Y_k\}_{k \geq 1}$,

$$Y_k = \frac{1}{b_k}\left(\sum_{j=1}^{k} X_j - a_k\right), \qquad k \geq 1,$$

is weakly relatively compact. A generalized risk process S_n *satisfies relation (4.9) if and only if there exists a weakly relatively compact sequence of random variables* $\{Y_k'\}_{k \geq 1}$ *such that*

1) $Z \overset{d}{=} V + Y_k'$, *where* V *and* Y_k' *are independent;*

2) $L_1(Y_k, Y_k') \to 0 \qquad (k \to \infty)$.

PROOF. As in the proof of Theorem 3.4.2 we make sure that condition (4.18) implies $b_{N_k}/b_k \Rightarrow 1$ with $k \to \infty$, i.e., the random variables U_k' from the triples (Y_k', U_k', V_k') mentioned in Theorem 3.2.1, should be equal to one with probability one. Now the required statement follows from Theorem 3.2.1. The proof is completed.

COROLLARY 3.4.2. *Let in addition to the assumptions of Theorem 3.4.3 the family of translation mixtures of the distribution function* $P(V < x)$, $x \in \mathbb{R}$, *be identifiable. Then (4.9) takes place if and only if there exists a random variable* Y *such that*

1) $Z \overset{d}{=} Y + V$, *where* Y *and* V *are independent;*

2) $Y_k \Rightarrow Y$ $(k \to \infty)$.

PROOF. By virtue of the condition of identifiability of the family of translation mixtures of the distribution function of the random variable V, no more than one random variable Y satisfies the relation $Z \overset{d}{=} Y+V$ with independent summands in the right-hand side. Reference to Theorem 3.4.3 completes the proof.

COROLLARY 3.4.3. *Under the conditions of Theorem 3.4.3, a generalized risk process is asymptotically normal (4.17) if and only if there exist numbers $\mu \in \mathbb{R}$ and $0 \leq \sigma^2 \leq 1$ such that*

1) $P(V < x) = \Phi\left(\dfrac{x - \mu}{\sigma}\right), \quad x \in \mathbb{R};$

2) $P(Y_k < x) \Rightarrow \Phi\left(\dfrac{x + \mu}{\sqrt{1 - \sigma^2}}\right) \quad (k \to \infty).$

PROOF. By virtue of Theorem 3.4.3 the limit random variable Z should be representable in the form $Z \overset{d}{=} Y_k' + V$, where Y_k' and V are independent, $k \geq 1$. But by virtue of normality of Z, for every $k \geq 1$ according to the Lévy-Cramér theorem the representation $Z \overset{d}{=} Y_k' + V$ with independent summands in the right-hand side is possible only if both Y_k' and V are normal. Moreover, since the family of translation mixtures of the normal distribution function of the random variable V is identifiable, the normal distribution function of the random variable Y_k' does not depend on k. The corollary is proved.

Finally consider the conditions for the convergence of generalized risk processes without any assumptions about the convergence of the process $N(t)$ or of the normalized sums of claims.

THEOREM 3.4.4. *Assume that $N_k \overset{P}{\to} \infty$ with $k \to \infty$ and the sequence $\{Y_k\}_{k \geq 1}$ is weakly relatively compact. A generalized risk process S_n converges (4.9) to some random variable Z if and only if there exist weakly relatively compact sequences of random variables $\{Y_k'\}_{k \geq 1}$ and $\{V_k'\}_{k \geq 1}$ such that*

1) $Z \overset{d}{=} Y_k' + V_k'$, *where Y_k' and V_k' are independent, $k \geq 1$;*

2) $L_1(Y_k, Y_k') \to 0 \quad (k \to \infty);$

3) $L_1\left(\dfrac{a_{N_k} - a_k}{b_k}, V_k'\right) \to 0 \quad (k \to \infty).$

PROOF of this theorem is reduced to Theorem 3.2.1 by virtue of the relation $b_{N_k}/b_k \Rightarrow 1$ $(k \to \infty)$, which has been established in the proof of Theorem 3.4.2. The proof is completed.

COROLLARY 3.4.4. *Under the conditions of Theorem 3.4.4, a generalized risk process is asymptotically normal (4.17) if and only if there exist sequences of numbers $\{\beta_k\}_{k \geq 1}$ and $\{\sigma_k^2\}_{k \geq 1}$ such that $\sup_k |\beta_k| < \infty$, $0 \leq \sigma_k^2 \leq 1$, $k \geq 1$,* and

1) $L_1\left(P(Y_k < x), \Phi\left(\frac{x - \beta_k}{\sigma_k}\right)\right) \to 0 \quad (k \to \infty);$

2) $L_1\left(P\left(\frac{a_{N_k} - a_k}{b_k} < x\right), \Phi\left(\frac{x + \beta_k}{\sqrt{1 - \sigma_k^2}}\right)\right) \to 0 \quad (k \to \infty).$

PROOF. By virtue of the Lévy–Cramér theorem on decomposability of a normal law only into normal components each of the random variables Y_k' and V_k', appearing in condition 1) of Theorem 3.4.4 should have normal distribution. Furthermore, the conditions $\sup_k |\beta_k| < \infty$ and $0 \le \sigma^2 \le 1$ are necessary and sufficient for the weak relative compactness of the sequences $\{Y_k'\}$ and $\{V_k'\}$. Now the desired result follows from Theorem 3.4.4. The corollary is proved.

As an illustration consider the situation when the counting process $N(t)$, i.e., the number of the insurance claims received by time t, is a doubly stochastic Poisson process (Cox process) which is obtained from a homogeneous Poisson process with the help of a random change of time. Namely, let $N_1(t)$ be a homogeneous Poisson process with unit intensity and $\Lambda(t), t \ge 0$, be a process independent of $N_1(t)$ possessing the following properties: $\Lambda(0) = 0, P(\Lambda(t) < \infty) = 1$ for any $t > 0$, the trajectories of $\Lambda(t)$ do not decrease and are right-continuous. Then define the doubly stochastic Poisson process $N(t)$ as

$$N(t) = N_1(\Lambda(t)).$$

In this case we will say that the Cox process $N(t)$ is guided or controlled by the process $\Lambda(t)$. Some properties of doubly stochastic Poisson processes are presented in Appendix 2. These processes are quite adequate models of many real processes and are often used in actuarial mathematics. We will also meet them in Section 3.5. Denote $\Lambda_n = \Lambda(n)$ for natural n.

THEOREM 3.4.5. Assume that insurance claims $\{X_j\}_{j \ge 1}$ are independent and identically distributed with $EX_j = a \ne 0$ and $DX_j = \sigma^2 < \infty$, and $N(t)$ is a Cox process controlled by the process $\Lambda(t)$. A risk process S_n has some limit distribution:

$$\frac{1}{\sigma\sqrt{n}}(S_n - S_0 - n(c - a)) \Rightarrow -Z \quad (k \to \infty), \qquad (4.19)$$

if and only if there exists a random variable V such that

1) $Z \overset{d}{=} \sqrt{\frac{a^2 + \sigma^2}{\sigma^2}} W + \frac{a}{\sigma} V$, where W and V are independent and W has the standard normal distribution;

2) $\frac{\Lambda_n - n}{\sqrt{n}} \Rightarrow V \quad (n \to \infty).$

PROOF. Apply Theorem 3.4.2 with $a_n = na_n$, $b_n = \sigma\sqrt{n}$. According to this theorem, (4.19) takes place if and only if

$$\frac{a(N_n - n)}{\sigma\sqrt{n}} \Rightarrow V_0 \quad (n \to \infty), \qquad (4.20)$$

where V_0 is a random variable such that $Z \overset{d}{=} W_1 + V_0$, W_1 and V_0 are independent and $P(W_1 < x) = \Phi(x)$, $x \in \mathbb{R}$. But according to Theorem A2.3.2, (4.20) takes place if and only if there exists a random variable V such that $V_0 \overset{d}{=} \frac{a}{\sigma}(W_2 + V)$, where W_2 and V are independent, $P(W_2 < x) = \Phi(x)$, $x \in \mathbb{R}$, and 2) is fulfilled. Thus taking account of the normality and independence of W_1 and W_2 we come to representation 1). The proof is completed.

COROLLARY 3.4.5. *Under the conditions of Theorem 3.4.5, a risk process is asymptotically normal*

$$P\left(\frac{1}{\sigma\sqrt{n}}(S_n - S_0 - n(c - a)) < x\right) \Rightarrow \Phi\left(\frac{x - \mu}{\delta}\right) \qquad (n \to \infty)$$

with some $\mu \in \mathbb{R}$ and $\delta^2 < \infty$, if and only if $\delta^2 \geq 1$ and

$$P\left(\frac{\Lambda_n - n}{\sqrt{n}} < x\right) \Rightarrow \Phi\left(\frac{ax - \sigma\mu}{\sigma\sqrt{\delta^2 - 1}}\right) \qquad (n \to \infty).$$

This statement follows from Theorem 3.4.5 and the Lévy–Cramér theorem on decomposability of a normal random variable only into normal components.

Thus the answers to the questions posed above can be formulated in the following way.

Limit distributions for generalized risk processes under the mentioned special choice of the normalizing and centering constants have the form of convolutions of the limit law for the distribution functions of the normalized indices and of that for the normalized sums of insurance claims.

A normal approximation for the distributions of generalized risk processes is adequate if and only if both the normalized sums of insurance claims and the normalized indices are asymptotically normal.

We did not use the property of positiveness of the random variables $\{X_j\}$, which follows from their interpretation as insurance claims, in any of the statements of this section. Thus all the statements of this section are limit theorems for "growing" random sums of arbitrary independent random variables within a special choice of normalizing and centering constants.

The choice of normalizing constants as $b_n = n^{1/\alpha}B(n)$ where $B(x)$ is a slowly varying function, is typical for the case of identically distributed independent summands $\{X_j\}$ and provides the convergence of the distributions of their normalized sums to the stable law with index α (Tucker, 1968). Therefore, not imposing formally the condition of identity of distributions of the summands, we nevertheless have considered the case of "almost identically distributed" in the mentioned sense summands.

The conditions imposed upon centering and normalizing constants in this section were introduced in order that instead of convergence of pairs of randomly indexed centering and normalizing constants, the convergence of only one component of these pairs, characterizing translations, can be considered at the expense of the asymptotic degeneracy of the other component characterizing scale modification. This conditions allowed us to get stronger and simpler statements. If we consider risk processes in a general situation not imposing any conditions upon normalizing and centering constants, then the limit theorems for risk processes will coincide up to terminology with the ones presented in Sections 3.1-3.3. Unfortunately, in a general situation we cannot get as simply conclusions about the conditions of the applicability of the normal approximation as we managed to do within a special choice of constants.

We obtained the limit theorems presented in this section as implications of Theorems 3.2.1 and 3.3.1. However some particular variants of these statements were proved earlier in a direct way. In particular, the statement of Theorem 3.4.1 in the case of identically distributed insurance claims under the assumption of the finiteness of variances of the claims and the process $N(t)$ can be obtained from one result by V. M. Kruglov (Kruglov, 1988) on the convergence of moments of random sums, where the case $a_n = c_n - na$, $b_n = d_n = \sigma\sqrt{n}$ was considered. M. Finkelstein and H. Tucker (Finkelstein and Tucker, 1990) have independently obtained the same result with the same constants without assumptions about moments of the random variable N_n. Finkelstein, Kruglov and Tucker (Finkelstein, Kruglov and Tucker, 1994) generalized this result for random sums of nonidentically distributed variables with uniformly bounded variances and equal expectations. Thus, the statements of this section can be regarded as variations on the theme of the mentioned results by Kruglov, Tucker and Finkelstein.

3.5 Some models of financial mathematics

In this section we will describe an example of application of the results of asymptotic theory of random summation to the search for regularities in the process of the stock price variation which at first sight may seem absolutely chaotic. This problem attracted and continues to attract attention of many mathematicians. From all the variety of the problems appearing here we will stop only on one whose essence is to construct a model for the distributions of increments of the stock price process. We will speak about stock prices though our models are quite universal. Further we will discuss what "tuning" of a model for each particular situation means.

Let $P(t)$ be a stock price at time t. It is clear that since it is impossible to predict precisely the price behavior in the future, it is reasonable to consider $P(t)$ as a random process. The investigation of real data shows that the increments of this process $P(t + \tau_1) - P(t)$ and $P(t) - P(t - \tau_2)$ can be assumed to be uncorrelated for all t and for all τ_1, τ_2 which exceed some small positive number. It may be assumed as well that $E(P(T + \tau) - P(t)) = 0$ for all t and τ. These assumptions can be easily obtained with the help of the following practical reasoning: if it had not been so, then attentive traders having noticed deviations from noncorrelatedness and stationarity of the increments of the process $P(t)$, would surely have tried to play on it, by the same token promoting the leveling of the situation, i.e., its return to noncorrelatedness and stationarity.

One of the first models of the stock price process was suggested by L. Bachelier (Bachelier, 1900). Having made a start from the assumptions of noncorrelatedness and stationarity of increments, he had come to the same construction as the model of Brownian motion suggested later by A. Einstein (Einstein, 1905), namely, to the one-dimensional diffusion equation. As is known, this construction leads to Wiener random processes as a model for trajectories of Brownian motion of a particle and therefore for the stock prices process.

The increments of a Wiener process have normal distributions. Wiener process is a limit model and is closely connected with the central limit theorem, whose application to the analysis of the distributions of the stock prices increments is reduced to the following reasoning. Dividing the time interval $[t, t + \tau]$ into a sufficiently large number of subintervals, we notice that the increment $P(t + \tau) - P(t)$ is represented as the sum of the increments of the process $P(t)$ on subintervals. Under the assumption of independence of increments on subintervals which we have discussed above, according to the central limit theorem the distribution of the increment $P(t + \tau) - P(t)$ approaches normal with the increase of the number of subintervals, if the Lindeberg condition is fulfilled. This condition can be interpreted in the following way: no one of the summands plays a dominating role in the total increment. If the subintervals have identical length, then we can assume that the distributions of the increments on the subintervals are identical. In this case the Lindeberg condition is reduced to the requirement of finiteness of variances of the elementary increments.

However, the statistical analysis of real data shows that the increments of the stock price process on relatively short (up to 2-3 weeks) time intervals have distributions different from normal. The works in which the attention is paid to this fact began to appear as far back as in 1915. Very serious statistical analysis was performed by M. Kendall (Kendall, 1953). This phenomenon is universal: non-normality of the distributions of increments becomes apparent for practically any trade objects at any exchanges. Therefore Wiener processes appear to be rather questionable models for the stock price processes.

The mentioned non-normality of the increment distributions reveals itself in that there are essentially more very small and very large values of increments in real data than there should be under a normal distribution. In other words, the observed distribution of increments on time intervals of identical moderate length is more leptokurtic than the normal.

A serious attempt to explain the observed leptokurtosis of the considered distributions was launched by Mandelbrot (Mandelbrot, 1963a, 1963b, 1969). One of the central ideas of his reasoning is as follows. The deviation of the distributions of increments from the normal means that the central limit theorem is not applicable here. Therefore its conditions are violated. But which of them are violated? If we do not want to fall outside the frames of the general structure of the model and the assumption of identity of distributions of the summands, then the finiteness of variances appears to be the only condition which might be violated. So Mandelbrot suggested using a limit theorem about the behavior of sums of identically distributed summands with infinite variance instead of the central limit theorem. In this case stable distributions are the limit laws for the sums of identically distributed summands. Mandelbrot's investigations have caused an outburst of a purely mathematical interest to the properties of stable laws (Zolotarev, 1986). Along with numerous favorable features, stable laws possess one unpleasant property: with the exception of four cases (normal law, Cauchy and Lévy laws and the law symmetrical to Lévy distribution) explicit expressions for their densities are unknown.

Unfortunately, the mentioned deficiency (practically unessential with computer analysis) is not a single circumstance making us consider the perspectives of the use of stable laws for the description of the distributions of stock prices increments with doubt. The main premise for their application in this capacity is the assumption about the infiniteness of the variances of price increments on time intervals of arbitrarily small length. But the variances are infinite if the mentioned elementary increments can be arbitrarily large by absolute value, what appears to be impossible in practice.

Many investigators paid their attention to the fact that the intensity of stock trades is irregular and have tried to explain the leptokurtosity of the increment distributions making a start from this irregularity. Apparently, P. Clark (Clark, 1970, 1973) was the first who investigated the heterogeneity of time at stock exchanges. Instead of ordinary Wiener processes he suggested using subordinated Wiener processes as models of the stock prices processes, i.e. Wiener processes with random time of the type $W(X(t))$, where $W(t)$, $t \geq 0$ is Wiener process and $X(t)$ is a process with non-decreasing trajectories coming from zero.

According to Clark, when using the central limit theorem as a basis for the price process model construction, it is not the finiteness of the variances of elementary increments that becomes violated but the other structural assumption, namely, that of nonrandomness of the number of summands.

Based on Clark's conjecture we will show how to construct a more realistic and quite simple model of the stock price process on time intervals of a small length and then how to pass to the models of the form $W(X(t))$ on macrolevel with the help of the central limit theorem for random sums. Actually these models are suitable approximations.

Consider the stock price process on a time interval of small length. The formal expression of the stock price exists only in the deal (contract) concluded on a stock exchange. Therefore we assume that the price remains constant on time intervals between successive deals and is equal to the one stated in the last contract concluded. Then the variation of the stock price during the considered time interval is described by the marked point process $\{(T_i, P_i)\}_{i \geq 1}$, where T_i is the time of the i-th deal and P_i is the price of the i-th deal. Hence $P(t) = P_i$ for $T_i \leq t < T_{i+1}$. Let $N(t)$ be the number of deals concluded by time t. Consider $t = 0$ to be the beginning of the countdown. Then

$$P(t) - P(0) = \sum_{j=1}^{N(t)} (P_j - P_{j-1}), \quad t \geq 0, \quad P_0 = P(0). \tag{5.1}$$

Denote $X_j = P_j - P_{j-1}, j \geq 1$. Assume that $\{X_j\}_{j \geq 1}$ are independent random variables. We shall consider the behavior of the distribution of the random variable $P(t) - P(0)$ with the growth of $N(t)$. Note that we do not assume that necessarily $t \to \infty$. The time interval $[0, t]$ may remain fixed. For the convenience of the construction of approximations consider a sequence of natural-valued random variables $\{N_k\}_{k \geq 1}$ and assume that for every $k \geq 1$ the random variables N_k and $\{X_j\}_{j \geq 1}$ are independent.

From Lemma 3.1.1 we immediately obtain the following result. Let $f_k(s)$ be the characteristic function of the random variable $P(t) - P(0)$, $s \in \mathbb{R}$, with $N(t)$ replaced by N_k in (5.1).

LEMMA 3.5.1. *Let for some* $b_k > 0, b_k \to \infty \ (k \to \infty)$,

$$\mathsf{P}\left(\frac{1}{b_k}\sum_{j=1}^{k} X_j < x\right) \Rightarrow \Phi(x) \qquad (k \to \infty),$$

and for some $d_k > 0, d_k \to \infty \ (k \to \infty)$, *let the family of random variables* $\{b_{N_k}/d_k\}$ *be weakly relatively compact. Then for every* $s \in \mathbb{R}$,

$$\lim_{k \to \infty}\left|f_k\left(\frac{s}{d_k}\right) - \int_0^\infty \exp\left\{-\frac{1}{2}x^2\sigma^2\right\} dA_k(x)\right| = 0,$$

where $A_k(x) = \mathsf{P}(b_{N_k} < xd_k)$.

It is not difficult to see that if we denote $U_k = b_{N_k}/d_k$, then

$$\int_0^\infty \exp\{-\frac{1}{2}x^2\sigma^2\}dA_k(x) = \mathsf{E}\exp\{is(XU_k)\}, \qquad s \in \mathbb{R}$$

where $\mathsf{P}(X < x) = \Phi(x)$; moreover X and U_k are independent for every $k \geq 1$. Thus with the help of Lemma 3.5.1 we come to the following conclusion. If the number of deals concluded during time interval $[0, t]$ is sufficiently large and the price variations in successively concluded deals satisfy the conditions of the central limit theorem, then for some nonnegative random variable U

$$\mathsf{P}(P(t) - P(0) < x) \approx \mathsf{P}(XU < x) = \mathsf{E}\Phi\left(\frac{x}{U}\right), \qquad x \in \mathbb{R}, \qquad (5.2)$$

where $\mathsf{P}(X < x) = \Phi(x)$ and X and U are independent. In other words, the distributions approximating the distributions of stock prices increments should be searched for among scale mixtures of normal laws.

The type of the mixing distribution in the right-hand side of (5.2) is defined by the limit behavior of the random variable b_{N_k}/d_k. In this connection Theorem 3.3.2 implies that knowing the distribution of U we can also try to reconstruct the distribution of U_k, since the condition $U_k \Rightarrow U$ is not only sufficient but also necessary for the substitution of the approximate equality in (5.2) by the precise one with $N_k \xrightarrow{P} \infty$ to be possible under suitable normalization of the variable $P(t) - P(0)$. The correctness of these reasonings will be demonstrated by one important particular example.

Consider a sequence $\{N^{(k)}(t)\}_{k\geq 1}$ of doubly stochastic Poisson processes (Cox processes), i.e., $N^{(k)}(t) = N_1(\Lambda_k(t))$, where N_1 is a homogeneous Poisson process with unit intensity and Λ_k are processes independent of N_1 with non-decreasing finite and right-continuous trajectories coming from zero (see Sect. 3.4 and Appendix 2). Let the random variables $\{X_j\}_{j\geq 1}$ be independent and identically distributed with $\mathsf{E}X_j = 0, \mathsf{D}X_j = \sigma^2 < \infty$. Assume that for every $k\geq 1$ the process $N^{(k)}(t)$ is independent of the sequence $\{X_j\}_{j\geq 1}$.

THEOREM 3.5.1. *Assume that* $\Lambda_k(t) \xrightarrow{P} \infty$ *and the sequence of random variables* $\{\Lambda_k(t)/d_k^2\}_{k\geq 1}$ *is weakly relatively compact for some* $d_k > 0$. *Then*

$$\frac{1}{d_k}(P(t) - P(0)) \Rightarrow (\text{some}) \ Z \qquad (k \to \infty)$$

if and only if there exists a nonnegative random variable U *such that*

1) $\mathsf{P}(Z < x) = \mathsf{E}\Phi\left(\frac{x}{U}\right), \qquad x \in \mathbb{R};$ (5.3)

2) $\dfrac{\sigma\sqrt{\Lambda_k(t)}}{d_k} \Rightarrow U \qquad (k \to \infty).$ (5.4)

PROOF is the combination of Theorems 3.3.2 and A2.3.1.

Theorem 3.5.1 serves as an illustration of the realization of the stock exchange time heterogeneity principle suggested by Clark. In fact $\Lambda_k(t)$ has the meaning of the mean number of the deals concluded by time t. If the trajectories of $\Lambda_k(t)$ are differentiable, then $\lambda_k(t) = \Lambda_k'(t)$ determines the trade intensity at time t. In the considered model it is natural to identify the function $\lambda_k(t)$ with the stock volatility at time t.

Let us see how much the results of Lemma 3.5.1 and Theorem 3.5.1. agree with experimental data. The investigation of real data shows that the effect of non-normality of the distributions of the increments $P(t) - P(0)$ is fading away with the growth of t. With rather large t (beginning from 3-4 weeks) these distributions are practically indistinguishable from the normal one. With the help of Theorem 3.5.1 we can suggest the following explanation of the non-normality of the distributions of $P(t) - P(0)$ for moderate t and the "normalization" of these distributions with the growth of t.

On the one hand, if t is sufficiently large to consider $\Lambda_k(t)$ large, but at the same time it is sufficiently small so that the dispersion of the values of $\Lambda_k(t)/d_k^2$ on the nonintersecting intervals of length t is large, then it is natural to consider the distribution of the variable U in Theorem 3.5.1 to be different from degenerate.

On the other hand, pay attention that an auxiliary parameter k has not still been formally connected with time. We can consider integer-valued time (as in the previous section), set $k = t$ in Theorem 3.5.1 and consider one process $\Lambda(t)$ instead of the sequence $\{\Lambda_k(t)\}_{k\geq 1}$ and set $\Lambda_k = \Lambda(k)$. In this case the constants d_k obviously depend on time. In practice the fluctuations of the process $\Lambda(t)$ on the intervals of large length are small in the sense that $\Lambda_k/k \xrightarrow{P} \text{const}$ with $k \to \infty$. Considering $d_k^2 = \sigma^2 k$ we come to the conclusion that the random variable U in (5.4) turns into constant, i.e., the distribution of the normalized increment $(P(t) - P(0))/(\sigma\sqrt{t})$ turns into normal with $t \to \infty$.

As we have mentioned, the class of scale mixtures $E\Phi\left(\frac{x}{U}\right)$ of normal laws is very wide. There are very many publications devoted to the choice of the specific variables U in the analysis of specific data about stock prices. Theorem 3.5.1 can give an explanation to the fact that there is no universal random variable U describing the behavior of all the prices on all the stock exchanges. Really condition (5.4) and the statement right after Theorem 3.5.1 connect this variable with the behavior of the intensities of trade or volatilities which certainly depend on the stock and on the place and time of the trade. In other words for each particular stock its own variable U (i.e., its distribution) should be chosen.

One curious paradox is connected with mixtures $E\Phi\left(\frac{x}{U}\right)$. As we have already said the true distributions of stock price increments are always noticeably more leptokurtic than the normal law. At the same time, at first sight it is not clear whether the mixtures $E\Phi\left(\frac{x}{U}\right)$ suggested as models for the distributions of stock prices increments are more leptokurtic than normal laws. Indeed,

$$E\Phi\left(\frac{x}{U}\right) = P(XU < x), \qquad x \in \mathbb{R},$$

where X and U are independent and $P(X < x) = \Phi(x)$. We know that normal laws with large variances have more gently sloping density with a more flat vertex than normal laws with small variances. Therefore it seems that multiplying the normal variable X by a variable U whose values are concentrated around zero, we get more leptokurtic densities than normal, and multiplying X by a variable U with values essentially greater than one we can get less leptokurtic densities than normal. Fortunately, this reasoning seems to be correct only until it is verified by calculations. Consider the excess coefficient which is traditionally used as a characteristic of leptokurtosity. Recall that for a random variable Y with $EY^4 < \infty$ the excess coefficient (kurtosis) $æ(Y)$ is defined as

$$æ(Y) = E\left(\frac{Y - EY}{\sqrt{DY}}\right)^4.$$

If $P(X < x) = \Phi(x)$, then $æ(X) = 3$. Densities with sharper vertices (and respectively heavier tails) than the normal density, have $æ > 3$, and $æ < 3$ for densities with more flat vertices. The following lemma destroys the "arguments" based on the "common sense" presented above.

LEMMA 3.5.2. *Let X and U be independent random variables with finite fourth moments; moreover $EX = 0$ and $P(U \geq 0) = 1$. Then $æ(XU) \geq æ(X)$. Furthermore, $æ(XU) = æ(X)$ if and only if $P(U = \text{const}) = 1$.*

PROOF. The independence of X and U implies that

$$æ(XU) = E\left(\frac{XU - EXU}{\sqrt{DXU}}\right)^4 = \frac{E(XU - EXU)^4}{(E(XU - EXU)^2)^2} =$$

$$\frac{E(XU - EXEU)^4}{(E(XU - EXEU)^2)^2} = \frac{EX^4EU^4}{(EX^2)^2(EU^2)^2} =$$

$$\text{æ}(X) \cdot \frac{EU^4}{(EU^2)^2}. \tag{5.5}$$

But according to Jensen's inequality $EU^4 \geq (EU^2)^2$. Therefore the right-hand side of (5.5) always is not less than $\text{æ}(X)$. Furthermore, it is equal to $\text{æ}(X)$ if and only if $EU^4 = (EU^2)^2$, which is obviously possible only if $P(U = \text{const}) = 1$. The proof is completed.

Thus scale mixtures of normal laws are always more leptokurtic than normal laws themselves.

Sometimes there are reasons to assume that the summands X_j in Lemma 3.5.1 are not centered. Nevertheless Lemma 3.5.1 and Theorem 3.5.1 provide a suitable possibility for analyzing the trends of the stock prices. Trend indicators are very popular among traders, e.g., see (Elder, 1993). Let $M(t)$ be a trend (general tendency) that is followed by the stock price at time t. The price deviations from $M(t)$ in particular contracts are stipulated by the behavior of particular sides of the trade and therefore can be considered "more random" than $M(t)$. From a formal viewpoint consider $M(t)$ as a nonrandom function. Let $X_j^* = P_j - P_{j-1}$ be the price increment from the $(j-1)$-th to the j-th deals and let T_j be the time of the j-th deal. Set $\alpha_j^* = M(T_j) - M(T_{j-1})$. Then we can assume that the random variables $X_j = X_j^* - \alpha_j^*$ are centered and Lemma 3.5.1 gives the method to construct approximations for the distributions of the centered increments $Z(t) = (P(t) - M(t)) - (P(0) - M(0))$. In this connection we get a simple criterion to verify the correctness of the choice of a trend: if the observed distribution of the variable $Z(t)$ is symmetrical and leptokurtic, then the trend is constructed correctly.

The deviation of the distribution of the variable $Z(t)$ from symmetrical can be caused by a wrong choice of scale. Above we have considered the behavior of the process $P(t)$. However, as is mentioned by J. Hull (Hull, 1989), the key aspect of the stock prices is the fact that the investor is interested mainly in the stock return expressed in percent irrespective of the stock price. In this connection the process $\log(P(t)/P(0))$ should be considered instead of the process $P(t) - P(0)$. All the reasonings from this section obviously remain true for the increments of this process. Thus Lemma 3.5.1 and Theorem 3.5.1 give a method to test the correctness not only of the adjustment of the trend but of the choice of the scale by the symmetry and leptokurtosity of the increments of the considered process.

In this section we have assumed that the random variables $\{X_j\}_{j \geq 1}$ are independent. However, as it follows from Theorem A2.2.1 this condition is not essential. Our main assumptions are the independence of the indices (the number of deals) of the sequence $\{X_j\}_{j \geq 1}$ and asymptotic normality of the sums of centered and normalized variables $\{X_j\}_{j \geq 1}$.

We will return once more to the problem of the construction of the approximations for the distributions of stock price increments in Section 4.4, where we will consider this problem from somewhat different positions in the double array scheme where the approximations implicitly taking account of a trend can be proposed.

3.6   Rarefied renewal processes

In Section 2.4 we have introduced the operation of elementary rarefaction. This section is devoted to rarefaction models of a more general type and that have useful practical applications. This section is based on the results of the publications (Rényi, 1956), (Kovalenko, 1965), (Gnedenko and Freyer, 1969), (Mogyoródi, 1971).

Let $T_0 \equiv 0 \le T_1 \le T_2 \le \ldots$ be a renewal process, i.e., the random variables $X_j = T_j - T_{j-1}$ are independent and identically distributed. Interpreting $T_1, T_2, \ldots$ as moments of occurrences of some events it is customary to say that the sequence $T_1, T_2, \ldots$ determines the flow of events. Our nearest aim is to determine the general rarefaction model for the sequence $T_0, T_1, T_2 \ldots$.

Let $\{Z_{n,j}\}_{j \ge 1}, n = 1, 2, \ldots$ be a double array of row-wise independent identically distributed random variables taking integer nonnegative values. Assume that the random variables $\{T_j\}_{j \ge 0}$ and $\{Z_{n,j}\}_{j \ge 1, n \ge 1}$ are jointly independent. Set

$$N_{n,j} = Z_{n,1} + Z_{n,2} + \ldots + Z_{n,j},$$

$$T_{n,0} \equiv 0, \quad T_{n,j} = T_{N_{n,j}}, \quad j \ge 1, \quad n \ge 1.$$

With fixed n, we will call the sequence of the random variables $\{T_{n,j}\}_{j \ge 0}$ the rarefaction of the original sequence $\{T_j\}_{j \ge 0}$ with respect to the random variables $\{Z_{n,j}\}$.

It is obvious that $\{T_{n,j}\}_{j \ge 0}$ with every n is also a renewal process; in this case the distribution of the random variables $X_{n,j} = T_{n,j} - T_{n,j-1}$ is determined by the formula

$$P(X_{n,j} < x) = \sum_{k=1}^{\infty} P(Z_{n,j} = k) F^{*k}(x),$$

where $F(x) = P(X_j < x)$. Moreover if $EX_j = m < \infty$ and $EZ_{n,j} = M_n < \infty$, then as we have seen,

$$EX_{n,j} = mM_n.$$

Since $X_{n,j}$ is a random sum, the following statement can be easily deduced from Theorems 3.1.1 and 3.3.5.

THEOREM 3.6.1. *Let there exist a sequence $\{b_n\}_{n \ge 1}$ of positive numbers such that $b_n \to \infty$ with $n \to \infty$ and for some distribution function $W(x)$ the relation*

$$P(Z_{n,j} < b_n x) \Rightarrow W(x) \qquad (n \to \infty), \quad j = 1, 2, \ldots \qquad (6.1)$$

holds. Then

$$P(X_{n,j} < mb_n x) \Rightarrow W(x) \qquad (n \to \infty), \quad j = 1, 2, \ldots \qquad (6.2)$$

Furthermore, if $Z_{n,j} \to \infty$ in probability with $n \to \infty$ $(j = 1, 2, \ldots)$ then (6.1) is not only sufficient for (6.2) but is also necessary. In this connection any distribution function whose all points of growth are concentrated on the *nonnegative semiaxis, can be a limit distribution function for the random variables* $X_{n,j}/(mb_n)$ *with* $n \to \infty$.

Theorem 3.6.1 leads to the main conclusion. It turns out that the asymptotic behavior of the rarefied sequence is determined not by the distribution of the original sequence but by the distribution of random variables with respect to which the rarefaction is performed. This conclusion generalizes one result by A. Rényi who had shown that within an elementary rarefaction the normalized flow converges to the Poisson process irrespective of the distribution of the original flow.

The circumstance that any distribution function of a nonnegative random variable can be a limit one in the model of general rarefaction, makes its detailed investigation essentially difficult. Therefore further we will focus on the investigation of the so-called special rarefaction model (according to the terminology proposed by J. Mogyoródi).

Let $\{K_{n,j}\}_{j\geq 1, n\geq 1}$ be independent identically distributed random variables taking only integer positive values and independent of the original sequence $\{T_j\}_{j\geq 0}$. Assume that

$$1 \leq \mathsf{E}K_{n,j} = M < \infty.$$

We will perform the rarefaction successively and each time accompany it with the corresponding scale change. On the first step rarefy the sequence $\{T_j\}$ with respect to the random variables $\{K_{1,j}\}_{j\geq 1}$. Then the random variables $X_{1,j} = T_{1,j} - T_{1,j-1}$, $j \geq 1$, are independent and identically distributed; moreover

$$\mathsf{P}(X_{1,j} < x) = \sum_{k=1}^{\infty} \mathsf{P}(K_{1,j} = k)F^{*k}(x)$$

and $\mathsf{E}X_{1,j} = mM$. Now change the scale by the compression of the rarefied sequence. Set

$$\tilde{T}_{1,0} \equiv 0, \quad \tilde{T}_{1,j} - \tilde{T}_{1,j-1} = \frac{1}{M}(T_{1,j} - T_{1,j-1}), \quad j = 1, 2, \ldots$$

Thus we have got a new sequence $\{\tilde{T}_{1,j}\}_{j\geq 1}$, defining a new renewal process for which $\mathsf{E}\tilde{X}_{1,j} = m$, where $\tilde{X}_{1,j} = \tilde{T}_{1,j} - \tilde{T}_{1,j-1}$, $j \geq 1$. Further act in the following way: from the process $\{\tilde{T}_{k,j}\}_{j\geq 0}$ obtain its rarefaction $\{T_{k+1,j}\}_{j\geq 0}$ with respect to the random variables $\{K_{k+1,j}\}_{j\geq 1}$ and then define

$$\tilde{T}_{k+1,0} \equiv 0, \ \tilde{T}_{k+1,j} - \tilde{T}_{k+1,j-1} = \frac{1}{M}(T_{k+1,j} - T_{k+1,j-1}), \ j = 1, 2, \ldots$$

and so on. In this case every time we will have $\mathsf{E}\tilde{X}_{k,j} = m$, where $\tilde{X}_{k,j} = \tilde{T}_{k,j} - \tilde{T}_{k,j-1}$, $j = 1, 2, \ldots$; $k = 1, 2, \ldots$. The described procedure is the generalization of the elementary rarefaction described in Chapter 2, where the random variables $\{K_{n,j}\}_{j\geq 1, n\geq 1}$ have identical geometric distributions.

The rarefied process $\{\tilde{T}_{n,j}\}_{j\geq 0}$ can be described in terms of the process $\{T_j\}_{j\geq 0}$ and the variables $\{K_{n,j}\}$ in the following way. Let $Z_{1,j} = K_{1,j}$ and let random variables $Z_{n,j}$ be defined with the help of the recursive relations

$$Z_{n,1} = \sum_{j=1}^{K_{n,1}} Z_{n-1,j}, \quad Z_{n,2} = \sum_{j=K_{n,1}+1}^{K_{n,1}+K_{n,2}} Z_{n-1,j}, \ldots \quad (n \geq 2). \qquad (6.3)$$

Then setting $N_{n,j} = Z_{n,1} + Z_{n,2} + \ldots + Z_{n,j}$ and

$$T_{n,j}^* = T_{N_{n,j}}, \quad j = 1, 2, \ldots; \quad n = 1, 2, \ldots, \qquad (6.4)$$

we notice that

$$\tilde{T}_{n,0} \equiv 0, \quad \tilde{T}_{n,j} - \tilde{T}_{n,j-1} = \frac{1}{M^n}(T_{n,j}^* - T_{n,j-1}^*). \qquad (6.5)$$

It is easy to see that $T_{n,j}^* - T_{n,j-1}^*$ is a random sum of independent random variables $X_j = T_j - T_{j-1}$ where the number of the summands is equal to $Z_{n,j}$, i.e.,

$$\tilde{T}_{n,j} - \tilde{T}_{n,j-1} = \sum_{k=U_1(j,n)}^{U_2(j,n)} \frac{X_j}{M^n},$$

where

$$U_1(j,n) = \sum_{k=1}^{j-1} Z_{n,k} + 1, \quad U_2(j,n) = \sum_{k=1}^{j} Z_{n,k}.$$

Let $f(s)$ be a common generating function of the random variables $\{K_{n,j}\}$:

$$f(s) = \sum_{k=1}^{\infty} P(K_{n,j} = k)s^k, \quad |s| \leq 1, \quad j = 1, 2, \ldots; n = 1, 2, \ldots$$

The independence of the random variables $\{K_{n,j}\}_{j\geq 1, n\geq 1}$ and the definition of the random variables $Z_{n,j}$ (6.3) imply that $f_n(s) = \mathbb{E}s^{Z_{n,j}}$ is the n-th functional iteration of the function f. This fact allows us to use the apparatus of the theory of branching processes.

Under the assumption made above, $M > 1$, representation (6.5) and Theorem 1.3.1 imply the following statement.

THEOREM 3.6.2. Let $\mathbb{E}[K_{n,j} \log K_{n,j}] < \infty$, $n \geq 1$, $j \geq 1$. Then for the special rarefaction model the following relation takes place

$$\frac{1}{m}(\tilde{T}_{n,j} - \tilde{T}_{n,j-1}) \Rightarrow X \qquad (n \to \infty); \qquad (6.6)$$

furthermore, the distribution function G of the random variable X belongs to class $\mathcal{W}$ of the limit distributions for supercritical branching processes.

We have mentioned some properties of the laws from class $\mathcal{W}$ in Section 1.3. Here we will additionally mention that if $DK_{n,j} = b^2 < \infty$, then the corresponding distribution function $G \in \mathcal{W}$ is absolutely continuous everywhere except the point $x = 0$. Along with this, $EX = 1$, $DX = b^2/(M^2 - M)$. In our case $f(0) = 0$ by the definition of the random variables $K_{n,j}$ and therefore the function G is absolutely continuous everywhere.

Relation (6.6) can be considered as an analog of the law of large numbers for the rarefied sequences in the special model. Looking ahead we will give an analog of the central limit theorem. According to the method of its proof it would be reasonable to include it into the next chapter, but we will formulate it now in order not to return again to the rarefied flows. Denote $\sigma^2 = D(T_j - T_{j-1})$.

THEOREM 3.6.3. *Let* $E[K_{n,j} \log K_{n,j}] < \infty$, $n \geq 1$, $j \geq 1$; $\sigma^2 < \infty$. *Then for any* $j \geq 1$

$$\lim_{n\to\infty} P\left(\frac{M^{n/2}}{\sigma}(\tilde{T}_{n,j} - \tilde{T}_{n,j-1} - mZ_{n,j}) < x\right) =$$

$$\int_0^\infty \Phi\left(\frac{x}{\sqrt{u}}\right) dG(u), \quad x \in \mathbb{R}.$$

PROOF. It is sufficient to mention that the difference $\tilde{T}_{n,j} - \tilde{T}_{n,j-1} - mZ_{n,j}$ can be represented as a random sum of centered summands and to refer to Theorem 4.1.2.

The existence of variance of the random variable X_j in Theorem 3.6.3 was assumed in order to get the possibility of using Φ as the limit distribution function of the normalized sum of nonrandom number of random variables. This assumption is not of principle. Within its absence the class of limit laws for the normalized differences $\tilde{T}_{n,j} - \tilde{T}_{n,j-1} - mZ_{n,j}$ will consist of the distributions whose characteristic functions have the form $Eg^X(t)$, where $g(t)$ is a stable characteristic function, as it follows from the results of the next chapter.

Now consider limit laws for noncentered rarefied renewal processes when the existence of finite mathematical expectations for the random variables $X_j = T_j - T_{j-1}$ is not assumed. Let F be a distribution function and ϕ be the Laplace-Stieltjes transform of the random variable $X_j, j \geq 1$.

THEOREM 3.6.4. *Let* $E[K_{n,j} \log K_{n,j}] < \infty$. *A sequence of constants* $\{\delta_n\}_{n\geq 1}$ *providing the existence of a limit distribution for the random variables*

$$\delta_n(T_{n,j}^* - T_{n,j-1}^*) \tag{6.7}$$

as $n \to \infty$ ($j = 1, 2, \ldots$) *with* $T_{n,j}^*$ *defined by relations (6.4) and (6.5), exists if and only if for* $s > 0$

$$\lim_{n\to\infty} M^n(1 - \phi(\delta_n s)) = 0$$

or

$$\lim_{n\to\infty} M^n(1 - \phi(\delta_n s)) = cs^\alpha,$$

where $0 < \alpha \leq 1$ *and* $c > 0$.

PROOF. Consider the Laplace-Stieltjes transform $\phi_n(s)$ of random variables (6.7). Iterating equation (3.2) n times we make sure that

$$\phi_n(s) = f_n(\phi(\delta_n s)), \tag{6.8}$$

where f_n is the n-th functional iteration of f. For $0 < z \leq 1$ the function $f(z)$ has the inverse one, $u(z)$. In this connection u_n (n-th functional iteration of u) will be the inverse function for f_n. The point $z = 1$ will be the only fixed point of $u(z)$ on $(0,1]$. It can be verified that $u(z)$ satisfies all the conditions under which the limit

$$\psi(z) = \lim_{n \to \infty} M^n(u_n(z) - 1).$$

exists. The function $\psi(z)$ is the solution of the equation

$$M\psi(u(z)) = \psi(z), \tag{6.9}$$

which is known as a Poincaré equation. We will meet with these equations in the closing section of the next chapter.

The function $\psi(z)$ which is the solution of equation (6.9) is strictly increasing on $(0,1]$, twice differentiable, $\psi(1) = 0, \psi'(1) = 1$. Now from (6.8) we have

$$u_n(\phi_n(s)) = \phi(\delta_n s), \quad s > 0,$$

whence it follows that

$$M^n(u_n(\phi_n(s)) - 1) = M^n(\phi(\delta_n s) - 1). \tag{6.10}$$

According to the assumption $\phi(s)$ converges to the Laplace-Stieltjes transform of the limit distribution, which we denote as $h(s)$. Further,

$$\lim_{n \to \infty} M^n(u_n(h(s)) - 1) = \psi(h(s)).$$

Therefore by virtue of the continuity of $\psi(z)$ and $h(s)$ have

$$\lim_{n \to \infty} M^n(u_n(\phi_n(s)) - 1) = \psi(h(s)),$$

whence taking account of (6.10) it follows that

$$\lim_{n \to \infty} M^n(\phi(\delta_n s) - 1) = \psi(h(s)), \quad s > 0.$$

Consider two cases separately. At first assume that $h(s) \equiv 1, s \geq 0$. In this case $\psi(h(s)) = \psi(1) = 0$, so that

$$\lim_{n \to \infty} M^n(1 - \phi(\delta_n s)) = 0,$$

i.e., the necessity of the first condition is established. In another case $h(s_0) < 1$ for some $s_0 > 0$. Since the function $\psi(h(s))$ is nonpositive and decreases as s grows, then for $s = s_0$ we have

$$\lim_{n \to \infty} M^n(\phi(\delta_n s_0) - 1) = \psi(h(s_0)) < 0,$$

as $h(s_0) < 1$. Comparing the last two relations we notice that there exists the limit

$$\lim_{n\to\infty} \frac{1 - \phi(\delta_n s)}{1 - \phi(\delta_n s_0)} = \lim_{n\to\infty} \frac{1 - \phi(\delta_n s_0 \frac{s}{s_0})}{1 - \phi(\delta_n s_0)} = \frac{\psi(h(s))}{\psi(h(s_0))}. \tag{6.11}$$

On the other hand the function $v(s) = 1 - \phi(s)$ satisfies the condition of a lemma from (Feller, 1971), according to which

$$\lim_{n\to\infty} \frac{1 - \phi(\delta_n s)}{1 - \phi(\delta_n s_0)} = \left(\frac{s}{s_0}\right)^{\alpha}; \qquad -\infty < \alpha < \infty. \tag{6.12}$$

Relations (6.11) and (6.12) imply that

$$\psi(h(s)) = \psi(h(s_0))s^{\alpha} = c^* s^{\alpha}, \tag{6.13}$$

where $c^* = \psi(h(s_0))s_0^{\alpha} < 0$. It follows from (6.13) that

$$h(s) = \psi^{-1}(c^* s^{\alpha}). \tag{6.14}$$

Now notice that $h(s)$ being a Laplace-Stieltjes transform, is monotonically decreasing and convex with $s > 0$. Thus (6.13) implies that $\alpha \geq 0$. It follows from (6.14) that

$$h'(s) = \frac{c^* \alpha s^{\alpha-1}}{\psi'(h(s))}.$$

If $\alpha > 1$, then $h'(0) = 0$, which is impossible because in this case $h(s) \equiv 1$, for $s > 0$. This contradicts the fact that $h(s_0) < 1$. Therefore $0 \leq \alpha \leq 1$. Moreover, the case $\alpha = 0$ is also impossible since if $\alpha = 0$, then (6.13) implies that

$$h(s) = \psi^{-1}(c^*), \qquad s \geq 0.$$

But $c^* < 0$ and ψ^{-1} is strictly monotone. Hence $h(s) = \psi^{-1}(c^*) < 1$ which contradicts the fact that h is a Laplace-Stieltjes transform. Let $c = -c^*$. Then

$$\lim_{n\to\infty} M^n(1 - \phi(\delta_n s)) = cs^{\alpha}, \tag{6.15}$$

where $c > 0$ and $0 < \alpha \leq 1$. The necessity is proved.

Sufficiency. If (6.15) takes place, then with $n \to \infty$,

$$\phi(\delta_n s) = 1 - cs^{\alpha} M^{-n}(1 + o(1)).$$

Therefore for fixed $s \geq 0$ we have

$$|1 - cs^{\alpha} M^{-n}(1 + o(1)) - \exp\{-cs^{\alpha} M^{-n}(1 + o(1))\}| = o\left(\frac{1}{M^n}\right),$$

which means that

$$\phi(\delta_n s) = \exp\{-cs^{\alpha} M^{-n}(1 + o(1))\} + o\left(\frac{1}{M^n}\right).$$

Thus the Laplace-Stieltjes transform of the random variables (6.7) has the form

$$f_n(\phi(\delta_n s)) = f_n\left(\exp\left\{-cs^\alpha M^{-n}(1+o(1)) + o\left(\frac{1}{m^n}\right)\right\}\right).$$

According to the Lagrange formula,

$$|f_n(\exp\{-cs^\alpha M^{-n}(1+o(1)) + o(M^{-n})\}) - f_n(\exp\{-cs^\alpha M^{-n}\})| =$$

$$|f_n'(\theta(s))||\exp\{-cs^\alpha M^{-n}\}(\exp\{-cs^\alpha M^{-n}o(1) - 1) + o\left(\frac{1}{M^n}\right)| \le$$

$$M^n(o(1)M^{-n}) + o(M^{-n}) = o(1).$$

Thus if the limit

$$\lim_{n\to\infty} f_n(\exp\{-cs^\alpha M^{-n}\}), \tag{6.16}$$

exists, then there exists the limit of Laplace-Stieltjes transforms of the variables (6.7). To verify the existence of (6.16), consider the Laplace-Stieltjes transform of the random variable $M^{-n}Z_{n,j}$, where $Z_{n,j}$ is determined in (6.3). The value of this Laplace-Stieltjes transform at the point cs^α ($s \ge 0$) is precisely equal to

$$f_n(\exp\{-cs^\alpha M^{-n}\}).$$

According to the assumption, Theorem 1.3.1 is applicable; therefore the limit

$$\lim_{n\to\infty} f_n(\exp\{-cs^\alpha M^{-n}\}) = h(s),$$

exists, which is the Laplace-Stieltjes transform of some distribution function. If $\phi(\delta_n s) = s + o(M^{-n})$, then the reasoning remains the same. In this case $h(s) \equiv 1$. The proof is completed.

In Section 2.5 (Theorems 2.5.1 and 2.5.2) we have verified that (6.15) takes place if and only if the distribution function $F(x) = P(T_j - T_{j-1} < x)$ satisfies the condition

$$\lim_{x\to\infty} \frac{1 - F(kx)}{1 - F(x)} = \frac{1}{k^\alpha} \tag{6.17}$$

for any $k > 0$ if $0 < \alpha < 1$ and

$$\lim_{x\to\infty} \frac{x(1 - F(x))}{\int_0^x [1 - F(z)]dz} = 0, \tag{6.18}$$

if $\alpha = 1$. Taking this into account we can summarize it all. Under the conditions of Theorem 3.6.4 the Laplace-Stieltjes transform of the limit distribution function for random variables (6.7) can have either the form

$$h(s) \equiv 1, \qquad s \ge 0,$$

or the form

$$h(s) = \psi^{-1}(cs^\alpha)$$

where $c > 0, 0 < \alpha \leq 1$, and $\psi^{-1}(z)$ is the function inverse to the solution of equation (6.9). The Laplace-Stieltjes transform of the limit distribution function has the form

$$h(s) = \psi^{-1}(cs), \qquad c > 0$$

if and only if the distribution function F of the original renewal process satisfies (6.18). In this case the limit distribution function belongs to class W of the limit distributions for supercritical branching processes. The Laplace-Stieltjes transform of the limit distribution function has the form

$$h(s) = \psi^{-1}(cs^\alpha), \qquad c > 0, \qquad 0 < \alpha < 1,$$

if and only if F satisfies (6.17) for every $k > 0$. Finally, $h(s) \equiv 1$ if and only if $1 - \phi(\delta_n s) = o(m^{-n})$ as $n \to \infty$.

THEOREM 3.6.5. If $\delta_n = M^{-n}$ then the limit distribution for random variables (6.7) exists if and only if $\mathsf{E}(T_j - T_{j-1}) = m < \infty$.

PROOF. If $m < \infty$, then the existence of the limit distribution function follows from Theorem 3.6.2.

Otherwise if the random variables $M^{-n}(T^*_{n,j} - T^*_{n,j-1})$ have the limit distribution function with $n \to \infty$, then according to Theorem 3.6.4 either

$$\lim_{n\to\infty} M^n(1 - \phi(sM^{-n})) = 0,$$

or

$$\lim_{n\to\infty} M^n(1 - \phi(sM^{-n})) = cs^\alpha, \qquad c > 0, \qquad 0 < \alpha \leq 1.$$

In the first case the Laplace-Stieltjes transform of the limit distribution function has the form $h(s) \equiv 1$ and the statement is evident. The second condition is fulfilled if

$$\lim_{n\to\infty} \int_0^\infty \frac{1 - \exp\{-sxM^{-n}\}}{sM^{-n}} dF(x) = cs^\alpha.$$

According to the Fatou lemma this implies that

$$m = \int_0^\infty x dF(x) \leq \lim_{n\to\infty} \int_0^\infty \frac{1 - \exp\{-sxM^{-n}\}}{sM^{-n}} dF(x) = cs^\alpha, \qquad s > 0.$$

The proof is completed.

In the conclusion of this section consider the following problem. It is required to describe renewal processes which are invariant with respect to rarefaction with the corresponding scale change according to (6.3), (6.4) and (6.5). By invariance we mean the coincidence of the distributions of the random variables $T_j - T_{j-1}$ and $\tilde{T}_{n,j} - \tilde{T}_{n,j-1}$ for all n and j.

THEOREM 3.6.6. Let the supercritical Galton-Watson process normalized by the variables M^n and defined by the generating function $f(s) = \mathsf{E}s^{K_{1,1}}$

have a limit distribution function G as $n \to \infty$. Then the renewal process $\{T_j\}_{j \geq 0}$ is invariant with respect to rarefaction with the corresponding change of scale if and only if $F(x) \equiv P(T_j - T_{j-1} < x) = G(x/m)$, $x \in \mathbb{R}$, where m is an arbitrary finite positive number.

PROOF. Let g be the Laplace-Stieltjes transform of the distribution function G. Then the Laplace-Stieltjes transform of the random variable $\tilde{T}_{1,j} - \tilde{T}_{1,j-1}$ has the form $f(g(sM^{-1}))$. Taking account of Lemma 1.3.1 we conclude that the renewal processes with the distribution function $G(x/m)$ are invariant.

Otherwise let F be the distribution function of the invariant renewal process and ϕ be its Laplace-Stieltjes transform. The invariance of the process implies that

$$\phi(s) = f(\phi(sM^{-1})).$$

Hence it follows that

$$\phi(s) = f_n(\phi(sM^{-n})), \qquad n = 1, 2, \ldots,$$

and the limit

$$\lim_{n \to \infty} f_n(\phi(sM^{-n})) = \phi(s).$$

exists. Moreover Theorem 3.6.5 implies that $\mathsf{E}(F) = m < \infty$. Now according to Theorem 3.6.2, $F(x) = G(x/m)$. The proof is completed.

Chapter 4

Limit theorems for random sums in the double array scheme

4.1 Transfer theorems. Limit laws

In this and the next two sections we will not be concerned with applied problems but will concentrate our attention on some general problems. In this connection everywhere in this chapter we will restrict ourselves to the consideration of the case when the summands in a random sum have identical distributions. This case illustrates well regularities appearing within the transfer from sums of nonrandom numbers of summands to random sums and it does not require the attraction of a refined apparatus of the theory of random processes by which the general case can be investigated in full detail only. Those readers who would like to extend their knowledge after acquaintance with this chapter should refer to the monograph (Kruglov and Korolev, 1990) or to the paper (Korolev and Kruglov, 1993), where the detailed presentation of the theory of random summation in the double array scheme of nonidentically distributed random variables is contained.

Let $\{X_{n,j}\}_{j \geq 1}$ be a double array of row-wise independent identically distributed random variables and let $\{N_n\}_{n \geq 1}$ be a sequence of integer-valued positive random variables such that for every $n \geq 1$ the variables $\{X_{n,j}\}_{j \geq 1}$ and N_n are independent. For natural k set

$$S_{n,k} = X_{n,1} + \ldots + X_{n,k}.$$

LEMMA 4.1.1. *Let sequences of numbers $\{a_n\}_{n \geq 1}$ and $\{k_n\}_{n \geq 1}$ (a_n are real, k_n are natural) provide weak convergence of centered sums $S_{n,k_n} - a_n$ as $n \to \infty$ to some random variable Y:*

$$S_{n,k_n} - a_n \Rightarrow Y \quad (n \to \infty), \tag{1.1}$$

93

and let the family of random variables $\{N_n/k_n\}_{n\geq 1}$ be weakly relatively compact. Let $\{c_n\}_{n\geq 1}$ be a sequence of real numbers. Then for any $t \in \mathbb{R}$

$$\lim_{n\to\infty} |\mathsf{E}\exp\{it(S_{n,N_n} - c_n)\} - g_n(t)| = 0, \qquad (1.2)$$

where

$$g_n(t) = \int_0^\infty h^u(t)\exp\{it(ua_n - c_n)\}\, dA_n(u),$$

$$A_n(u) = \mathsf{P}(N_n < k_n u), \quad h(t) = \mathsf{E}\exp\{itY\}.$$

PROOF. Denote $f_n(t) = \mathsf{E}\exp\{itX_{n,1}\}$. Then

$$\Psi_n(t) \equiv \mathsf{E}\exp\{it(S_{n,k_n} - a_n)\} = f_n^{k_n}(t)\exp\{ita_n\}.$$

Condition (1.1) implies that for any $T \in (0,\infty)$

$$\lim_{n\to\infty} \sup_{|t|<T} |\Psi(t) - h_n(t)| = 0, \qquad (1.3)$$

and since exponential function is continuous, the following relation takes place for every $u \geq 0$

$$\lim_{n\to\infty} \sup_{|t|<T} |\Psi_n^u(t) - h^u(t)| = 0. \qquad (1.4)$$

Further, according to the law of total probability,

$$\mathsf{E}\exp\{it(S_{n,N_n} - c_n)\} = \sum_{k=1}^\infty \mathsf{P}(N_n = k)\mathsf{E}\exp\{it(S_{n,N_n} - c_n)\} =$$

$$\sum_{k=1}^\infty \mathsf{P}(N_n = k)f_n^k(t)\exp\{-itc_n\} =$$

$$\int_0^\infty f_n^{uk_n}(t)\exp\{-itc_n\}\, dA_n(u) =$$

$$\int_0^\infty \Psi_n^u(t)\exp\{it(ua_n - c_n)\}\, dA_n(u).$$

Therefore

$$|\mathsf{E}\exp\{it(S_{n,N_n} - c_n)\} - g_n(t)| \leq \int_0^\infty |\Psi_n^u(t) - h^u(t)|\, dA_n(u). \qquad (1.5)$$

At first assume that $\mathcal{N}$ is a sequence of natural numbers such that $A_n \Rightarrow A$ with $n \to \infty, n \in \mathcal{N}$, where A is some distribution function. Let $l_n(s)$ and $r_n(s)$ be the greatest lower and least upper bounds of s-quantiles of the distribution function A_n, respectively, $l(s)$ and $r(s)$ be the greatest lower and least upper bounds of s-quantiles of the distribution function A. In (Kruglov and Korolev, 1990) there is Theorem 1.1.1, according to which the weak convergence of A_n to A as $n \to \infty, n \in \mathcal{N}$, implies that for every $s \in (0,1)$,

$$l(s) \leq \liminf_{\substack{n \to \infty \\ n \in \mathcal{N}}} l_n(s) \leq \limsup_{\substack{n \to \infty \\ n \in \mathcal{N}}} r_n(s) \leq r(s).$$

Therefore if $l(s) = r(s)$ for some $s \in (0,1)$, then for this s the condition $A_n \Rightarrow A$ $(n \to \infty, n \in \mathcal{N})$ implies

$$l_n(s) \to l(s) \qquad (n \to \infty, n \in \mathcal{N}). \tag{1.6}$$

It is not difficult to verify that whatever the distribution function A is, the set $\{s \in (0,1) : l(s) \neq r(s)\}$ is no more than countable and therefore has zero Lebesgue measure. Rewrite the right-hand side of (1.5) as

$$\int_0^\infty |\Psi_n^u(t) - h^u(t)| \, dA_n(u) = \int_0^1 |\Psi_n^{l_n(s)}(t) - h^{l_n(s)}(t)| \, ds. \tag{1.7}$$

For every $s \in (0,1)$ satisfying (1.6), $l_n(s) = l(s) + o(1)$ with $n \to \infty, n \in \mathcal{N}$. Therefore it follows from (1.4) that

$$|\Psi_n^{l_n(s)}(t) - h^{l_n(s)}(t)| \leq$$
$$\sup_{|\tau| \leq T} |\Psi_n^{l(s)+o(1)}(\tau) - h^{l(s)+o(1)}(\tau)| \to 0$$

with $n \to \infty, n \in \mathcal{N}$. Hence according to the Lebesgue dominated convergence theorem the integral in the right-hand side of (1.7) tends to zero as $n \to \infty, n \in \mathcal{N}$, and therefore in correspondence with (1.5), for $n \in \mathcal{N}$ (1.2) takes place.

Now assume that in the general case (1.2) does not hold. In other words, assume that there exist an $\varepsilon > 0$ and a sequence $\mathcal{N}_1$ of natural numbers such that

$$|\mathrm{E} \exp \{it(S_{n,N_n} - c_n)\} - g_n(t)| > \varepsilon \tag{1.8}$$

for all $n \in \mathcal{N}_1$. But according to the condition, the family of the distribution functions $\{A_n\}_{n \geq 1}$ is weakly relatively compact. Therefore from the sequence $\mathcal{N}_1$ a subsequence $\mathcal{N}_2 \subseteq \mathcal{N}_1$ can be chosen for which $A_n \Rightarrow A^*$ with $n \to \infty, n \in \mathcal{N}_2$, where A^* is some distribution function. However, we have verified earlier that if $A_n \Rightarrow A^* (n \to \infty, n \in \mathcal{N}_2)$, then for all $n \in \mathcal{N}_2$ beginning from some, the inequality contrary to (1.8) should be fulfilled. We had come to the contradiction with the definition of the sequence $\mathcal{N}_1$ which completes the proof.

This lemma shows the way for the investigations of approximations for distributions of random sums of independent identically distributed random variables with the known prelimit distributions of random indices. The limit situation is described by the following transfer theorem.

THEOREM 4.1.1. *Let sequences $\{k_n\}_{n \geq 1}, \{a_n\}_{n \geq 1}$ and $\{c_n\}_{n \geq 1}$ (k_n are natural, a_n and c_n are real) be such that (1.1) takes place and for some pair* of random variables (U, V)

$$\left(\frac{N_n}{k_n}, \; a_n \frac{N_n}{k_n} - c_n\right) \Rightarrow (U, V) \qquad (n \to \infty). \tag{1.9}$$

Then

$$S_{n,N_n} - c_n \Rightarrow Z \qquad (n \to \infty), \tag{1.10}$$

where Z is a random variable with characteristic function

$$f(t) = \mathsf{E}[h^U(t) \exp\{itV\}]. \tag{1.11}$$

PROOF. Denote

$$U_n = \frac{N_n}{k_n}, \quad V_n = a_n \frac{N_n}{k_n} - c_n.$$

Then as is easy to see,

$$g_n(t) = \mathsf{E}[h^{U_n}(t) \exp\{itV_n\}]$$

Therefore taking account of Lemma 4.1.1 we are to verify that

$$|g_n(t) - f(t)| \to 0$$

with $n \to \infty$ for all $t \in \mathbb{R}$. Fix an arbitrary $t \in \mathbb{R}$ and consider the function $\phi_t(x, y) = h^x(t)e^{ity}$. This function is continuous and bounded in x and y. According to the definition of weak convergence, relation (1.9) means that

$$\mathsf{E}\phi(U_n, V_n) \to \mathsf{E}\phi(U, V) \qquad (n \to \infty)$$

for any continuous and bounded function ϕ including $\phi(x, y) \equiv \phi_t(x, y)$. The proof is completed.

In the considered case the structure of the class of limit laws can be described rather explicitly.

We shall say that a pair of random variables (U, V) belongs to class $\mathcal{K}_0$ if, first, $\mathsf{P}(U \geq 0) = 1$ and second, either one of the random variables U or V is degenerate or for some real numbers α and β

$$\mathsf{P}(V = \alpha U + \beta) = 1.$$

If the summands are asymptotically small, i.e., for any $\varepsilon > 0$

$$\lim_{n \to \infty} \mathsf{P}(|X_{n,1}| > \varepsilon) = 0,$$

then Khinchin's theorem about convergence of types (Khinchin, 1938, p. 87) implies that the class of the limit distributions for the centered random sums of independent identically distributed random variables consists of those distribution functions whose characteristic functions are representable in the form (1.11), where h is an infinitely divisible characteristic function and the pair of random variables (U, V) belongs to the class $\mathcal{K}_0$.

If in this connection the random variable V is degenerate, then for some $\beta \in \mathbb{R}$

$$f(t) = \exp\{it\beta\}\, \mathsf{E}h^U(t), \tag{1.12}$$

i.e., the limit characteristic function is a power mixture of infinitely divisible characteristic function to within a nonrandom translation. This situation will be considered in more detail later.

If both of the random variables U and V are nondegenerate, then for some α and β

$$f(t) = \mathsf{E}[h^U(t)\exp\{it\alpha U\}\exp\{it\beta\}] = \exp\{it\beta\}\, \mathsf{E}h_\alpha^U(t),$$

where $h_\alpha = e^{it\alpha}h(t)$ and since h_α is an infinitely divisible characteristic function, situation (1.12) takes place again.

When considering random sums centered by nonrandom variables we should mention that the situation when the random variable U is degenerate plays a special role. In this case for some $\gamma \geq 0$

$$f(t) = h^\gamma(t)\mathsf{E}\exp\{itV\}, \tag{1.13}$$

and the distribution function of the random variable Z from (1.10) takes the form

$$F(x) = \int_{-\infty}^{\infty} H_\gamma(x-z)\,dG(z) = (H_\gamma * G)(x), \tag{1.14}$$

where H_γ is the distribution function corresponding to the characteristic function h^γ and $G(z) = \mathsf{P}(V < z)$, i.e., $F(x)$ is a translation mixture of the infinitely divisible distribution function $H_\gamma(x)$.

Thus if the summands $\{X_{n,j}\}$ are asymptotically small, then the limit distributions for nonrandomly centered random sums of independent identically distributed random variables are corresponded by characteristic functions which have the form either $e^{it\beta}\mathsf{E}g^U(t)$, or $q(t)g(t)$, where $\mathsf{P}(U \geq 0) = 1$, $\beta \in \mathbb{R}$, g is infinitely divisible and q is an arbitrary characteristic function.

The special role of situation (1.13) (or (1.14)) shows itself in the following. If the random variable U is degenerate, then (1.13) implies the possibility of representation of the limit random variable Z as $Z = Y_\gamma + V$, where Y_γ is a random variable with the characteristic function h^γ independent of V. Hence if the array $\{X_{n,j}\}$ and the sequence $\{N_n\}$ are not fixed, then any distribution function can be a limit one for the distribution functions of the random variables $S_{n,N_n} - c_n$, since the degenerate random variable $Y_\gamma = 0$ is infinitely divisible and any distribution function can be represented as a weak limit of lattice distributions.

This situation illustrates the principal difference between the centering by constants of sums with random and nonrandom number of summands. In fact, in the classical summation theory within the consideration of the double array scheme the centering constants are not even written in an explicit form because it is assumed that they can simply be shared among summands. Therefore the centering by constants of sums and the centering of summands are equivalent in the classical situation. Within the consideration of random sums the centering of each summand inevitably leads to the situation where the sum itself will be centered by a random variable since the number of summands and consequently, of the constants centering each of them is random.

The above Theorem 4.1.1 was proved in (Korolev and Kruglov, 1993). It generalizes a transfer theorem for random sums of centered summands (or noncentered random sums) proved in (Gnedenko and Fahim, 1969). Of course, the centering of random sums by constants is more appropriate for solving problems connected with construction of approximating distributions. However, the investigation of noncentered random sums is important for some problems. Corresponding examples will be presented later. Therefore we will also formulate the transfer theorem for noncentered random sums.

THEOREM 4.1.2. *Let a sequence* $\{k_n\}_{n\geq1}$ *of natural numbers be such that the relations*

$$S_{n,k_n} \Rightarrow Y \qquad (n \to \infty), \qquad (1.15)$$

$$\frac{N_n}{k_n} \Rightarrow U \qquad (n \to \infty) \qquad (1.16)$$

take place. Then

$$S_{n,N_n} \Rightarrow Z \qquad (n \to \infty), \qquad (1.17)$$

where the characteristic function of the random variable Z *has the form*

$$f(t) = \mathsf{E}h^U(t) = \int_0^\infty h^u(t)\, dA(u), \qquad (1.18)$$

$h(t)$ *is the characteristic function of the random variable* Y *and* $A(u)$ *is the distribution function of the random variable* U.

PROOF appears to be a simple implication of Theorem 4.1.1.

REMARK 4.1.1. If the random variables $X_{n,k}$ are nonnegative, then it is natural to use the apparatus of Laplace-Stieltjes transforms instead of the apparatus of characteristic functions. In this case the statement of Theorem 4.1.2 remains unchanged, only relation (1.18) should be replaced by

$$\phi(s) = \mathsf{E}\chi^U(s) = \int_0^\infty \chi^z(s)\, dA(z), \qquad (1.19)$$

where ϕ and χ are the Laplace-Stieltjes transforms of the random variables Z and Y, respectively.

The class of possible limit laws for the distributions of noncentered random sums of independent identically distributed random variables is more restricted than that for centered random sums, as it can be seen from the comparison of (1.18) and (1.11). The class of laws (1.18) at the same time appears to be more extensive than the class of infinitely divisible distributions. To make it sure mention that any infinitely divisible distribution with characteristic function $h(t)$ is contained in the class of laws (1.18) with $A(u)$ having a single jump at the point $u = 1$.

Infinitely divisible laws can also appear as the limit ones in Theorem 4.1.2 in other cases. As it follows from one statement from (Feller, 1971), if the distribution function $A(u)$ in (1.18) is infinitely divisible, then the characteristic function $f(t)$ is also infinitely divisible.

The comparison of the classes of limit laws for "growing" random sums (see Section 3.1) and for random sums of identically distributed summands in the double array scheme shows that in contrast to the classical summation theory the scheme of "growing" random sums is not a particular case of the double array scheme. The point is that the presence of a new source of randomness, namely, random indices, brings new in principle elements to the settings of the appearing problems. The randomness of the indices is considered in different ways in this section and in Chapter 3. So the results of Chapter 3 and of this section are simply related to different settings of problems.

The classical summation theory and transfer theorems give us the possibility to obtain very simply sufficient conditions for the convergence of distributions of random sums of independent identically distributed random variables to the limit law for fixed sequences $\{a_n\}, \{k_n\}$ and $\{c_n\}$. We only have to replace condition (1.1) by the equivalent classical conditions of convergence to a prescribed infinitely divisible law (e.g., see (Gnedenko and Kolmogorov, 1954, Sect. 25)).

The structure of limit laws (1.11) makes it possible to simplify the conditions of Theorem 4.1.1. In order to do so, based on Khinchin's theorem on convergence of types (Khinchin, 1938), we formulate the following statement.

PROPOSITION 4.1.1. *Theorem 4.1.1 remains valid, if condition (1.9) of convergence of joint distributions of N_n/k_n and $a_n N_n/k_n - c_n$ to that of a pair (U, V) is replaced by two conditions of convergence of marginal distributions:*

$$\frac{N_n}{k_n} \Rightarrow U \qquad (n \to \infty), \tag{1.20}$$

$$a_n \frac{N_n}{k_n} - c_n \Rightarrow V \qquad (n \to \infty). \tag{1.21}$$

PROOF. Denote $\Delta_n(t) = |g_n(t) - f(t)|$, $t \in \mathbb{R}$, where g_n is defined in Lemma 4.1.1 and f is defined by (1.11). In the proof of Theorem 4.1.1 condition (1.9) was used only to prove that

$$\Delta_n(t) \to 0 \qquad (n \to \infty) \tag{1.22}$$

for each $t \in \mathbb{R}$. So we are to make sure that (1.22) remains valid with (1.9) replaced by (1.20) and (1.21).

We will separately consider three cases: (i) both U and V are nondegenerate; (ii) $\mathsf{P}(U = a) = 1$ for some $a \in \mathbb{R}$; (iii) $\mathsf{P}(V = \gamma) = 1$ for some $\gamma \in \mathbb{R}$. Recall that we denote $U_n = N_n/k_n, V_n = a_n U_n - c_n$.

(i). In this case, as we have already noted, according to Khinchin's convergence of types theorem there exist $a \in \mathbb{R}$ and $b \in \mathbb{R}$ such that $V \overset{d}{=}$

$aU + b$, $0 < |a| < \infty$, $|b| < \infty$, and (1.20) together with (1.21) imply $a_n \to a$, $-c_n \to b \, (n \to \infty)$. We have

$$\Delta_n(t) \leq \left| \int_0^\infty h^u(t)[\exp\{it(a_n u - c_n)\} - \exp\{it(au + b)\}]\, dA_n(u) \right| +$$

$$\left| \int_0^\infty h^u(t) \exp\{it(au + b)\}\, d[A_n(u) - A(u)] \right| \equiv I_1(t, n) + I_2(t, n)$$

where as above, $A_n(u) = P(U_n < u)$, $A(u) = P(U < u)$. Consider $I_1(t, n)$. For any $R > 0$ we have

$$I_1(t, n) \leq \int_0^\infty |\exp\{it(au + b)\} - \exp\{it(a_n u - c_n)\}|\, dA_n(u) \leq$$

$$\int_0^R |\exp\{it(au + b)\} - \exp\{it(a_n u - c_n)\}|\, dA_n(u) + 2P(U_n > R). \quad (1.23)$$

Let $\varepsilon > 0$ be arbitrary. Condition (1.20) implies that

$$\lim_{R \to \infty} \sup_n P(U_n > R) = 0$$

Therefore there exists an $R = R(\varepsilon)$ such that

$$2P(U_n > R(\varepsilon)) < \varepsilon \qquad (1.24)$$

for all $n \geq 1$. Conditions (1.20), (1.21) and Khinchin's convergence of types theorem imply that for each $u \in [0, R]$ we have $a_n u - c_n \to au + b \, (n \to \infty)$ and since the limit function $w(u) = au + b$ is continuous and bounded for $u \in [0, R]$, this convergence is uniform, i.e., for any $\delta' > 0$ there exists an $n_0 = n_0(\delta')$ such that for all $n \geq n_0$

$$\sup_{0 \leq u \leq R} |(a_n u - c_n) - (au + b)| < \delta'. \qquad (1.25)$$

Further, for each fixed $t \in \mathbb{R}$ the function $w_t(v) = \exp\{itv\}$ is uniformly continuous, i.e., for each $\varepsilon' > 0$ there exists a $\delta = \delta(\varepsilon') > 0$ such that

$$\sup_{v_1, v_2 : |v_1 - v_2| < \delta} |w_t(v_1) - w_t(v_2)| < \varepsilon'. \qquad (1.26)$$

Put $\varepsilon' = \varepsilon$ where ε is the same as in (1.24). Then from (1.25) and (1.26) it follows that there exists an $n_0 = n_0(\delta(\varepsilon))$ such that

$$\sup_{0 \leq u \leq R} |\exp\{it(au + b)\} - \exp\{it(a_n u - c_n)\}| < \varepsilon$$

for all $n \geq n_0(\delta(\varepsilon))$ whence in its turn it follows that

$$\int_0^{R(\varepsilon)} |\exp\{it(au+b)\} - \exp\{it(a_n u - c_n)\}|\, dA_n(u) < \varepsilon. \tag{1.27}$$

Since ε is arbitrary, from (1.23), (1.24) and (1.27) it follows that $I_1(t,n) \to 0$ as $n \to \infty$. The convergence of $I_2(t,n)$ to zero as $n \to \infty$ can be established by the reasoning based on the definition of weak convergence similar to that which was used to prove Theorem 4.1.1. So the convergence of $\Delta(t)$ to zero in case (i) is proved.

(ii). In this case we have

$$\Delta(t) = |Eh^{U_n}(t)\exp\{itV_n\} - h^a(t)E\exp\{itV\}| \leq$$
$$|Eh^{U_n}(t)\exp\{itV_n\} - h^a(t)E\exp\{itV_n\} +$$
$$|E\exp\{itV_n\} - E\exp\{itV\}| \equiv J_1(t,n) + J_2(t,n). \tag{1.28}$$

For any $\delta > 0$ we have

$$J_1(t,n) \leq \int_0^{\infty} |h^u(t) - h^a(t)|\, dA_n(u) \leq$$
$$\int_{a-\delta}^{a+\delta} |h^u(t) - h^a(t)|\, dA_n(u) + 2P(|U_n - a| > \delta).$$

Let $\varepsilon > 0$ be arbitrary. Choose a $\delta = \delta(\varepsilon)$ so that $|h^u(t) - h^a(t)| < \varepsilon$ if $|u - a| < \delta$, which is possible since the exponential function is continuous. Then

$$\int_{a-\delta}^{a+\delta} |h^u(t) - h^a(t)|\, dA_n(u) < \varepsilon$$

for all $n \geq 1$. For $\delta = \delta(\varepsilon)$ chosen in such a way, now choose an $n_1 = n_1(\delta, \varepsilon)$ so that $P(|U_n - a| > \delta) < \varepsilon$ for $n \geq n_1$, which is possible due to (1.20) and the assumption $P(U = a) = 1$. Then we have $J_1(t,n) < 3\varepsilon$ for $n \geq n_1$ which means that $J_1(t,n) \to 0$ $(n \to \infty)$ since ε is arbitrary. At the same time $J_2(t,n) \to 0$ $(n \to \infty)$ by virtue of (1.21). Therefore by (1.28), $\Delta(t) \to 0$ $(n \to \infty)$ in the case $P(U = a) = 1$

(iii). In this case we have

$$\Delta(t) = |Eh^{U_n}(t)\exp\{itV_n\} - \exp\{it\gamma\}\,Eh^{U}(t)| \leq$$

$$\left| \int_0^{\infty} h^U(t)[\exp\{it(a_n u - c_n)\} - \exp\{it\gamma\}]\, dA_n(u) \right| +$$

$$\left| \int_0^{\infty} h^U(t)\, d[A_n(u)' - A(u)] \right| \equiv Q_1(n,t) + Q_2(n,t).$$

Consider $Q_1(n,t)$. Let δ be a positive number to be specified later. Denote

$$M_n(\delta) = \{u \geq 0 : |a_n u - c_n - \gamma_n| \leq \delta\}.$$

Whatever $\delta > 0$ is, we have

$$Q_1(n,t) \leq \int\limits_0^\infty |\exp\{it(a_n u - c_n)\} - \exp\{it\gamma\}|\, dA_n(u) =$$

$$\int\limits_{M_n(\delta)} |\exp\{it(a_n u - c_n)\} - \exp\{it\gamma\}|\, dA_n(u) +$$

$$\int\limits_{\mathbb{R}_+ \setminus M_n(\delta)} |\exp\{it(a_n u - c_n)\} - \exp\{it\gamma\}|\, dA_n(u)$$

With a fixed t, the function $\exp\{itx\}$ is uniformly continuous in x. Therefore from the definition of the sets $M_n(\delta)$ it follows that for any $\varepsilon > 0$ there exists a $\delta = \delta(\varepsilon) > 0$ such that for all $n \geq 1$

$$\int\limits_{M_n(\delta(\varepsilon))} |\exp\{it(a_n u - c_n)\} - \exp\{it\gamma\}|\, dA_n(u) < \varepsilon.$$

At the same time condition (1.21) together with the assumption that $P(V = \gamma) = 1$ implies the existence of an $n_1 = n_1(\varepsilon)$ such that for all $n \geq n_1(\varepsilon)$

$$\int\limits_{\mathbb{R}_+ \setminus M_n(\delta(\varepsilon))} |\exp\{it(a_n u - c_n)\} - \exp\{it\gamma\}|\, dA_n(u) \leq$$

$$2P(U_n \in M_n(\delta(\varepsilon))) = 2P(|V_n - \gamma| > \delta(\varepsilon)) < \varepsilon,$$

whence it follows that $Q_1(n,t) \to 0$ $(n \to \infty)$ since ε is arbitrary. The convergence of $Q_2(n,t)$ to zero follows from (1.20), because with any fixed t, the function $\xi_t(u) = h^u(t)$ is continuous and bounded in u. The proposition is completely proved.

Theorem 4.1.1 plays a key role in what follows and therefore will be mentioned many times. Proposition 4.1.1 gives us the right to replace the condition of weak convergence of pairs of random variables by the weaker condition of convergence of components of these pairs in each reference to Theorem 4.1.1. Nevertheless everywhere in what follows we will consider the weak convergence (or "weak rapprochement") of pairs. We will do so to simplify notations and to preserve the unity of the formulations of the results of this and preceding chapters having in mind that actually we deal with component-wise weak convergence of (U_n, V_n).

4.2   Converses of the transfer theorems

Our nearest aim is to prove the statements that are partial converses of Theorems 4.1.1 and 4.1.2 containing necessary and sufficient conditions of weak convergence of random sums of identically distributed random variables under the assumption that either (1.1) or (1.9) takes place.

Denote $F(x) = P(Z < x)$, $H(x) = P(Y < x)$. For arbitrary distribution functions F and H introduce the set

$$\mathcal{M}_1 = \mathcal{M}_1(F \mid H) = \{(U, V) \in \mathcal{K}_0 : f(t) = E[h^U(t) \exp\{itV\}]\}$$

of those pairs (U, V) from $\mathcal{K}_0$ with which the characteristic function f corresponding to the distribution function F can be represented in the form (1.11), the characteristic function h corresponding to the distribution function H being fixed. At first note that the set $\mathcal{M}_1(F \mid H)$ is nonempty whatever the distribution functions F and H are. Indeed, the set $\mathcal{M}_1(F \mid H)$ always contains the pair $(0, V)$ where $P(V < x) = F(x)$. Second, the set $\mathcal{M}_1$ can contain more than one pair. Let, for example, $H = \Phi$ (where as before Φ is the standard normal distribution function; we will also use the notation $\Phi_{a,\sigma^2}(x) = \Phi\left(\frac{x-a}{\sigma}\right)$), and $F(x) = P(W^{(s)} < x)$, where $W^{(s)} = W - W'$ and W and W' are independent random variables with identical gamma distribution. Then along with the pair $(0, W^{(s)})$ the set $\mathcal{M}_1(F \mid \Phi)$ will contain the pair $(\sqrt{W}, 0)$ since it can be shown that

$$W^{(s)} \stackrel{d}{=} Y \cdot \sqrt{W},$$

where Y and W are independent, $P(Y < x) = \Phi(x)$ (e.g., see (Bagirov, 1988) and Sect. 3.2) and in this situation

$$E \exp\left\{itY\sqrt{W}\right\} = Eh^{\sqrt{W}}(t).$$

Let as before $L_1(\cdot, \cdot)$ and $L_2(\cdot, \cdot)$ be metrics, metrizing weak convergence in the spaces of one-dimensional and two-dimensional distributions, respectively. As before, if X and Y are random variables with distribution functions G and H respectively, then we will make no difference between the notations $L_1(X, Y)$ and $L_1(G, H)$. Similarly if (X_1, X_2) and (Y_1, Y_2) are two-dimensional random variables with distribution functions G and H, then we will not distinguish $L_2((X_1, X_2), (Y_1, Y_2))$ and $L_2(G, H)$.

THEOREM 4.2.1. *Let there exist sequences $\{k_n\}_{n\geq 1}$ and $\{a_n\}_{n\geq 1}$ (k_n are natural, a_n are real) such that (1.1) takes place; moreover let the distribution function H of the random variable Y be nondegenerate. The convergence of random sums (1.10) takes place with some sequence of real numbers $\{c_n\}_{n\geq 1}$ if and only if there exists a weakly relatively compact sequence of pairs of random variables $\{(U'_n, V'_n)\}_{n\geq 1}$, $(U'_n, V'_n) \in \mathcal{M}_1(F \mid H)$, satisfying the condition*

$$L_2\left(\left(\frac{N_n}{k_n}, a_n\frac{N_n}{k_n} - c_n\right), (U'_n, V'_n)\right) \to 0 \quad (n \to \infty). \tag{2.1}$$

PROOF. Necessity. Show that the sequence of pairs $\{(N_n/k_n, a_nN_n/k_n - c_n\}_{n\leq 1}$ is weakly relatively compact. At first consider the sequence of random variables $\{N_n/k_n\}_{N\geq 1}$. Let $\{X'_{n,j}\}_{j\geq 1}$ be independent random variables. Let each of them have the same distribution as $X_{n,1}$, and moreover let the sequences $\{X_{n,j}\}_{j\geq 1}$ and $\{X'_{n,j}\}_{j\geq 1}$ be independent. Set $X^{(s)}_{n,j} = X_{n,j} - X'_{n,j}$, $Z_{n,k} = X^{(s)}_{n,1} + \ldots + X^{(s)}_{n,k}$. According to the law of total probability we have

$$P(|Z_{n,N_n}| > x) = \sum_{k=0}^{\infty} P(N_n = k)P(|Z_{n,k}| > x). \qquad (2.2)$$

Assume that the sequence $\{N_n/k_n\}$ is not weakly relatively compact. In this case there exists an $\varepsilon_0 > 0$ such that for any integer r there is an index n_r for which

$$P(N_{n_r}/k_{n_r} \geq r) \geq \varepsilon_0.$$

It is obvious that $n_r \to \infty$ with $r \to \infty$. According to condition (1.1), (1.3) holds. Therefore for every $t \in \mathbb{R}$,

$$E\exp\{itZ_{n,k_n}\} = |f_n^{k_n}(t)e^{-ita_n}|^2 = |f_n(t)|^{2k_n} \to |h(t)|^2 \quad (n \to \infty).$$

Hence

$$|f_n(t)|^{2k_{n_r}} \to |h(t)|^2 \qquad (r \to \infty).$$

Therefore the sequence $\{|f_n(t)|^{2rk_{n_r}}\}_{r\geq 1}$ should converge to zero with $r \to \infty$ in all points t in which $|h(t)| \neq 1$. This means that any convergent subsequence of this sequence cannot converge to a continuous function. Hence the sequence $\{Z_{n_r,rk_{n_r}}\}_{r\geq 1}$ is not weakly relatively compact, i.e., there exists an $\varepsilon_1 > 0$ such that for all $x > 0$

$$\limsup_{r\to\infty} P(|Z_{n_r,rk_{n_r}}| > x) \geq \varepsilon_1. \qquad (2.3)$$

Using Lévy's inequality

$$P(|X^{(s)}_1 + \ldots + X^{(s)}_j| > x) \leq 2P(|X^{(s)}_1 + \ldots + X^{(s)}_k| > x),$$

which is correct for any symmetrical random variables $X^{(s)}_1, X^{(s)}_2, \ldots, k \geq j$ and any $x > 0$ (e.g., see (Loève, 1963)) we make sure that (2.3) implies the inequality

$$P(|Z_{n_r,k}| > x) \geq \frac{1}{2}P(|Z_{n_r,rk_{n_r}}| > x),$$

which is correct for $k \geq rk_n$ and any $x > 0$. Therefore taking account of (2.2) we have

$$\limsup_{r\to\infty} P(|Z_{n_r,N_{n_r}}| > x) \geq$$

$$\limsup_{r\to\infty} \sum_{k>rk_{n_r}} P(N_{n_r} = k)P(|A_{n_r,k}| > x) \geq$$

$$\frac{1}{2}\limsup_{r\to\infty} P(|Z_{n_r,rk_{n_r}}| > x)P\left(\frac{N_{n_r}}{k_{n_r}} \geq r\right) \geq \frac{1}{2}\varepsilon_0\varepsilon_1.$$

I.e., the sequence $\{Z_{n,N_n}\}_{n\geq 1}$ is not weakly relatively compact, which is impossible since according to the symmetrization inequality

$$P(|X - a| \geq x) \geq \frac{1}{2}P(|X^{(s)}| \geq 2x),$$

which is valid for any random variable X, any real a and any $x > 0$ ($X^{(s)}$ is a symmetrization of the random variable X, $X^{(s)} \stackrel{d}{=} X - X'$, where X and X' are independent and identically distributed), (e.g., see (Loève, 1963)) for any $x > 0$ we have

$$P(|Z_{n,N_n}| \geq x) = \sum_{k=0}^{\infty} P(N_n = k)P(|Z_{n,k}| \geq x) \leq$$

$$2\sum_{k=0}^{\infty} P(N_n = k)P\left(\left|\sum_{j=1}^{k} X_{n,j} - c_n\right| \geq \frac{x}{2}\right) =$$

$$2P(|S_{n,N_n} - c_n| \geq \frac{x}{2}), \tag{2.4}$$

and by virtue of (1.10) the right-hand side of (2.4) should tend to zero as $x \to \infty$ uniformly in n. The obtained contradiction proves the weak relative compactness of the sequence $\{N_n/k_n\}_{n \geq 1}$.

Now consider the sequence $\{a_n N_n/k_n - c_n\}_{n \geq 1}$. By virtue of what has been proved earlier from an arbitrary sequence $\mathcal{N}$ of natural numbers we can choose a subsequence $\mathcal{N}_1 \subseteq \mathcal{N}$ so that

$$\frac{N_n}{k_n} \Rightarrow U_1 \qquad (n \to \infty, n \in \mathcal{N}_1) \tag{2.5}$$

for some random variable U_1. Rewriting condition (1.1) as

$$S_{n,k_n} - a = \left(X_{n,1} - \frac{a_n}{k_n}\right) + \ldots + \left(X_{n,k_n} - \frac{a_n}{k_n}\right) \Rightarrow Y \quad (n \to \infty), \tag{2.6}$$

note that according to the transfer Theorem 4.1.2, (2.5) and (2.6) imply weak relative compactness of the sequence $\{a_n N_n/k_n - c_n\}_{n \in \mathcal{N}_1}$ since the sequence $\{S_{n,N_n} - c_n\}_{n \geq 1}$ is weakly relatively compact on account of (1.10). Therefore a subsequence $\mathcal{N}_2 \subseteq \mathcal{N}_1$ can be chosen from $\mathcal{N}_1$ so that the sequence $\{a_n N_n/k_n - c_n\}_{n \in \mathcal{N}_2}$ will be weakly convergent. Thus due to the arbitrariness of $\mathcal{N}$ the sequence $\{a_n N_n/k_n - c_n\}_{n \geq 1}$ is weakly relatively compact which means the weak relative compactness of the sequence of pairs of random variables $\{(N_n/k_n, a_n N_n/k_n - c_n)\}_{n \geq 1}$.

Denote

$$\beta_n = \inf\left\{L_2\left(\left(\frac{N_n}{k_n}, a_n \frac{N_n}{k_n} - c_n\right), (U, V)\right) : (U, V) \in \mathcal{M}_1(F \mid H)\right\}.$$

Show that $\beta_n \to 0$ with $n \to \infty$. Assume the contrary. In this case there exist a $\delta > 0$ and a sequence $\mathcal{N}_3$ of natural numbers such that $\beta_n \geq \delta$ for all $n \in \mathcal{N}_3$. Choose a subsequence $\mathcal{N}_4 \subseteq \mathcal{N}_3$ so that the pairs of random variables $(N_n/k_n, a_n N_n/k_n - c_n)$ weakly converge some pair (U, V) with $n \to \infty, n \in \mathcal{N}_4$ (this can be done by virtue of the weak relative compactness of the sequence $\{(N_n/k_n, a_n N_n/k_n - c_n)\}$ stated above). Hence and from condition (1.1) according to Theorem 4.1.1 (in which $n \in \mathcal{N}_4$) we make sure that $(U, V) \in \mathcal{M}_1(F \mid H)$. But this contradicts the assumption that $\beta_n \geq \delta$ for all $n \in \mathcal{N}_3$. Thus $\beta \to 0$ with $n \to \infty$.

For every $n = 1, 2, \ldots$ choose a pair (U'_n, V'_n) from $\mathcal{M}_1(F \mid H)$ for which

$$L_2\left(\left(\frac{N_n}{k_n}, a_n\frac{N_n}{k_n} - c_n\right), (U'_n, V'_n)\right) \leq \beta_n + \frac{1}{n}.$$

For these pairs (U'_n, V'_n) condition (2.1) is obviously fulfilled. Weak relative compactness of the sequence of the pairs $\{(U'_n, V'_n)\}_{n \geq 1}$ follows from (2.1) and from the fact that this property is inherent to the sequence $\{(N_n/k_n, a_nN_n/k_n - c_n)\}_{n \geq 1}$. Necessity is proved.

Sufficiency. Assume that the sequence of the distribution functions

$$F_n(x) = \mathsf{P}(S_{n, N_n} - c_n < x)$$

does not weakly converge to F. In this case there exist a sequence $\mathcal{N}$ of natural numbers and a $\delta > 0$ such that for $n \in \mathcal{N}$ the inequality $L_1(F_n, F) \geq \delta$ holds. Choose a subsequence $\mathcal{N}_1 \in \mathcal{N}$ so that the pairs (U'_n, V'_n) weakly converge as $n \to \infty, n \in \mathcal{N}_1$, to some pair (U, V). Just as in the proof of Theorem 4.1.1 we make sure that for any $t \in \mathbb{R}$,

$$f(t) = \mathsf{E}[h^{U'_n}(t) \exp\{itV'_n\}] \to \mathsf{E}h^U(t) \exp\{itV\} \quad (n \to \infty, n \in \mathcal{N}_1),$$

i.e., the limit pair (U, V) belongs to $\mathcal{M}_1(F \mid H)$. The inequality

$$L_2\left(\left(\frac{N_n}{k_n}, a_n\frac{N_n}{k_n} - c_n\right), (U, V)\right) \leq$$

$$L_2\left(\left(\frac{N_n}{k_n}, a_n\frac{N_n}{k_n} - c_n\right), (U'_n, V'_n)\right) + L_2((U'_n, V'_n), (U, V))$$

and condition (2.1) imply that

$$L_2\left(\left(\frac{N_n}{k_n}, a_n\frac{N_n}{k_n} - c_n\right), (U'_n, V'_n)\right) \to 0 \quad (n \to \infty, n \in \mathcal{N}_1).$$

Apply Theorem 4.1.1 to the sequences $\{S_{n, k_n} - a_n\}_{n \in \mathcal{N}_1}$ and $\{(N_n/k_n, a_nN_n/k_n - c_n\}_{n \in \mathcal{N}_1}$. As a result we obtain $L_1(F_n, F) \to 0$ with $n \to \infty$, $n \in \mathcal{N}_1$, which contradicts the assumption that $L_1(F_n, F) \geq \delta$ for all $n \in \mathcal{N}$. This contradiction completes the proof of the theorem.

The statement of Theorem 4.2.1 includes the condition of the nondegeneracy of the distribution function H. Show that this condition cannot be omitted if the sequences $\{a_n\}, \{c_n\}$ and $\{k_n\}$ are not fixed. Let the random variables $X_{n,j}$ have normal distribution with variance $\frac{1}{n}$, $\mathsf{P}(X_{n,j} < x) = \Phi(x\sqrt{n}), j \geq 1, n \geq 1$. Set $k_n = [\sqrt{n}]$, where $[a]$ means the integer part of a number a, $a_n = c_n = 0$. Let $\mathsf{P}(N_n = n) = 1 - \frac{1}{n}, \mathsf{P}(N_n = 1) = \frac{1}{n}$. Then obviously $\mathsf{P}(S_{n,k_n} < x) \Rightarrow \mathcal{E}_0(x)$, where $\mathcal{E}_0(x)$ is the distribution function degenerate at zero, but

$$\frac{N_n}{k_n} \overset{P}{\to} \infty.$$

In common words, we can say that degeneracy of the random variable Y (or the distribution function H) reduces to nothing the structural dependence of the problem on the summation of independent random variables, as can be seen from the representation of the limit law. Therefore the condition of non-degeneracy of the limit random variable in (1.1) is substantially reasonable. In this connection note also that in the scheme of "growing" random sums considered in Chapter 3, the condition of nondegeneracy of the limit random variable for the sums with nonrandom number of summands does not play such a fundamental role, since the results of Chapter 3, generally speaking, remain correct not only for sums, but for arbitrary random sequences with random indices as well, see (Korolev, 1993, 1994) and (Kruglov and Korolev, 1997).

Consider one more variant of inversion of Theorem 4.1.1. Let a distribution function F be arbitrary and a pair of random variables (U, V) belong to the class $\mathcal{K}_0$. Introduce the set $\mathcal{M}_2 = \mathcal{M}_2(F \mid U, V)$ consisting of distribution functions H, whose characteristic functions h allow it to represent the characteristic function f of the distribution function F in the form (1.11) with the fixed pair $(U, V) \in \mathcal{K}_0$:

$$\mathcal{M}_2 = \mathcal{M}_2(F \mid U, V) = \left\{ H : f(t) = \mathsf{E}h^U(t)e^{itV}, \ h(t) = \int_{-\infty}^{\infty} e^{itx} \, dH(x) \right\}.$$

Note that for some distribution functions F and some pairs (U, V) the set $\mathcal{M}_2(F \mid U, V)$ can be empty. Indeed, let, for example, $\mathsf{P}(U = 1) = 1$. Then representation (1.11) should have the form

$$f(t) = h(t)\mathsf{E} \exp \{itV\}. \tag{2.7}$$

If now we take a normal distribution function as F and an arbitrary nondegenerate random variable with distribution function different from normal as V, then by virtue of the Lévy–Cramér theorem on decomposability of normal law only into normal components representation (2.7) becomes impossible for any characteristic function h.

In many situations the set $\mathcal{M}_2$ contains no more than one element. For example, this is so in the following cases:

1) $A(u) = \mathsf{P}(U < u) = 1 - e^{-\lambda u}, u \geq 0$, where λ is a positive constant. Indeed, in this case (1.18) implies that

$$f(t) = \mathsf{E}h^U(t) = \frac{\lambda}{\lambda - \log h(t)}, \tag{2.8}$$

and if (2.8) holds for two functions $h_1(t)$ and $h_2(t)$, then they obviously coincide, see also Sect. 4.6;

2) The random variables $X_{n,k}$ are symmetrical. In this case the characteristic function $h(t)$ is real. But for real $z \in (0, 1)$ the function

$$a(z) = \int_0^{\infty} z^x \, dA(x)$$

is monotone. Therefore in this case the validity of (1.18) for two functions $h_1(t)$ and $h_2(t)$ implies that $h_1(t) = h_2(t)$;

3) The random variables $X_{n,k}$ are nonnegative. In this case the transforms $\chi(s)$ in representation (1.19) are real and the coincidence of the functions $\chi_1(s)$ and $\chi_2(s)$ providing (1.19) with the same $\phi(s)$ and $A(u)$ again follows from the monotonicity of the function $a(z)$ with $z \in (0,1)$;

4) $h_1(t)$ and $h_2(t)$ are analytical infinitely divisible functions. Then the identity

$$\int\limits_0^\infty h_1^u(t)\, dA(u) \equiv \int\limits_0^\infty h_2^u(t)\, dA(u) \tag{2.9}$$

implies the coincidence of $h_1(t)$ and $h_2(t)$ (Szász and Freyer, 1971).

5) The distribution $A(u)$ is arithmetic and

$$\int\limits_0^\infty x\, dA(u) < \infty.$$

Then identity (2.9), where h_1 and h_2 are infinitely divisible characteristic functions, implies the coincidence of $h_1(t)$ and $h_2(t)$ (Szász and Freyer, 1971).

If $P(U = \text{const}) = 1$, then $\mathcal{M}_2$ certainly contains no more than one element if the characteristic function $E\exp\{itV\}$ does not turn into zero anywhere (e.g., is infinitely divisible). Indeed, in this case if h_1 and h_2 are characteristic functions corresponding to the distribution functions H_1 and H_2 from $\mathcal{M}_2(F \mid U, V)$, then

$$f(t) = h_1^U(t)Ee^{itV} = h_2^U(t)Ee^{itV},$$

and therefore for all $t \in \mathbb{R}$

$$Ee^{itV}[h_1^U(t) - h_2^U(t)] = 0,$$

which is possible only if $h_1 = h_2$, since by the condition, $Ee^{itV} \neq 0, t \in \mathbb{R}$.

Possibly, $\mathcal{M}_2$ always contains no more than one element. However, no one has managed either to prove this or to refute this yet.

THEOREM 4.2.2. *Let there exist sequences* $\{k_n\}_{n\geq 1}$, $\{a_n\}_{n\geq 1}$ *and* $\{c_n\}_{n\geq 1}$ *(k_n are natural, a_n and c_n are real) such that (1.9) holds; moreover* $P(U = 0) < 1$. *The convergence of the centered random sums (1.10) takes place if and only if there exists a weakly relatively compact sequence of random variables* $\{Y_n'\}_{n\geq 1}$ *whose distribution functions belong to the set* $\mathcal{M}_2(F \mid U, V)$ *such that*

$$L_1(S_{n,k_n} - a_n, Y_n') \to 0 \qquad (n \to \infty). \tag{2.10}$$

PROOF. Necessity. At first ascertain the weak relative compactness of the sequence $\{S_{n,k_n} - a_n\}_{n\geq 1}$. The condition $P(U = 0) < 1$ implies the existence of $\varepsilon > 0$ and $b > 0$ such that

$$P(N_n \geq bk_n) \geq \varepsilon$$

for all sufficiently large n. As we did in the previous theorem, consider the symmetrized random variables $X_{n,j}^{(s)}$ and their sums $Z_{n,k}$. The sequence $\{Z_{n,k_n}\}_{n\geq 1}$ is weakly relatively compact if and only if so is the sequence $\{Z_{n,[bk_n]}\}_{n\geq 1}$. Indeed, if, for example, $\{Z_{n,k_n}\}_{n\geq 1}$ is not weakly relatively compact, then there exists a sequence $\mathcal{N}$ of natural numbers such that

$$E \exp\{itZ_{n,[bk_n]}\} = |f_n(t)|^{2k} \to g(t) \qquad (n \to \infty, n \in \mathcal{N}),$$

and $g(t)$ is not a continuous function. But then the limit

$$\lim_{\substack{n \to \infty \\ n \in \mathcal{N}}} E \exp\{itZ_{n,[bk_n]}\} = \lim_{\substack{n \to \infty \\ n \in \mathcal{N}}} |f_n(t)|^{2[bk_n]}$$

is also discontinuous, and hence the sequence $\{Z_{n,[bk_n]}\}_{n\geq 1}$ is not weakly relatively compact. Thus if $\{Z_{n,k_n}\}_{n\geq 1}$ and therefore $\{Z_{n,[bk_n]}\}_{n\geq 1}$ are not weakly relatively compact, then there exists an $\varepsilon_0 > 0$ such that for all x

$$\limsup_{n \to \infty} P(|Z_{n,[bk_n]}| > x) > \varepsilon_0.$$

Therefore by the Lévy inequality we will have

$$\limsup_{n \to \infty} P(|Z_{n,[bk_n]}| > x) \geq$$

$$\limsup_{n \to \infty} \sum_{k \geq [bk_n]} P(N_n = k)P(|Z_{n,k}| > x) \geq$$

$$\frac{1}{2} \limsup_{n \to \infty} P(|Z_{n,[bk_n]}| > x)P(N_n \geq [bk_n]) \geq \frac{1}{2}\varepsilon\varepsilon_0$$

for all x, i.e., the sequence $\{Z_{n,N_n}\}_{n\geq 1}$ is not weakly relatively compact. With the help of relation (2.4) we make sure that this contradicts the assumption of the weak convergence of the random variables $S_{n,N_n} - c_n$. Hence we had ascertained that the sequence $\{Z_{n,N_n}\}_{n\geq 1}$ is weakly relatively compact.

According to Lemma 3.1 in (Kruglov, 1984, p. 94-96), the weak relative compactness of the sequence $\{Z_{n,k_n}\}_{n\geq 1}$ implies the shift-compactness of the sequence of random variables $\{S_{n,k_n} - a_n\}_{n\geq 1}$, since Z_{n,k_n} have the same distribution as the difference of two independent copies of the random variable $S_{n,k_n} - a_n$. Thus there exists a sequence $\{d_n\}_{n\geq 1}$ such that

$$\lim_{n \to \infty} \sup_n P(|S_{n,k_n} - a_n - d_n| > x) = 0. \tag{2.11}$$

Show that the sequence $\{d_n\}_{n\geq 1}$ is bounded. It follows from (2.11) that from an arbitrary sequence $\mathcal{N}$ of natural numbers a subsequence $\mathcal{N}_1$ can be chosen so that the random variables $S_{n,k_n} - a_n - d_n$ weakly converge to some random variable Y_1 with $n \to \infty, n \in \mathcal{N}_1$. Applying Theorem 4.1.1 with n running along $\mathcal{N}_1$ and with condition (1.1) written as

$$\left(X_{n,1} - \frac{d_n}{k_n}\right) + \ldots + \left(X_{n,k_n} - \frac{d_n}{k_n}\right) - a_n \Rightarrow Y_1 \quad (n \to \infty, n \in \mathcal{N}_1),$$

we make sure that the sequence of the random variables

$$\left(X_{n,1} - \frac{d_n}{k_n}\right) + \ldots + \left(X_{n,k_n} - \frac{d_n}{k_n}\right) - c_n = (S_{n,N_n} - c_n) - \frac{d_n}{k_n}N_n$$

also converges weakly with $n \to \infty, n \in \mathcal{N}_1$. Hence the arbitrariness of the sequence $\mathcal{N}$ means the weak relative compactness of the sequence $\{S_{n,N_n} - c_n - d_n N_n/k_n\}_{n \geq 1}$. But with regard for (1.10) the sequence $\{d_n N_n/k_n\}_{n \geq 1}$ should be weakly relatively compact. Since by the condition, $N_n/k_n \Rightarrow U$ $(n \to \infty)$ and $P(U = 0) < 1$, the sequence $\{d_n\}_{n \geq 1}$ should be bounded. Thus the weak relative compactness of the sequence $\{S_{n,k_n} - a_n\}_{n \geq 1}$ is established.

Denote

$$\gamma_n = \inf\{L_1(S_{n,k_n} - a_n, Y) : H \in \mathcal{M}_2(F \mid U, V), H(x) = P(Y < x)\}.$$

Prove that $\gamma_n \to 0$ with $n \to \infty$. Assume the contrary. This means that for some $\delta > 0$ and all n from some sequence $\mathcal{N}$ of natural numbers the inequality $\gamma_n \geq \delta$ holds. Choose $\mathcal{N}_1$ so that the sequence of the random variables $\{S_{n,k_n} - a_n\}_{n \in \mathcal{N}_1}$ weakly converges to some random variable Y with $n \to \infty, n \in \mathcal{N}_1$. This and (1.9) by virtue of Theorem 4.1.1 imply that

$$S_{n,N_n} - c_n \Rightarrow \text{ (some) } Z' \quad (n \to \infty, n \in \mathcal{N}_1)$$

But condition (1.10) implies $Z' \overset{d}{=} Z$, i.e., the distribution function of the random variable Y belongs to $\mathcal{M}_2(F \mid U, V)$. By virtue of the choice of $\mathcal{N}_1$ the inequality

$$L_1(S_{n,k_n} - a_n, Y) < \delta$$

holds for all sufficiently large $n \in \mathcal{N}_1$, which contradicts the assumption $\gamma \geq \delta, n \in \mathcal{N}_1$. Hence we proved that $\gamma_n \to 0$ $(n \to \infty)$. For every $n = 1, 2, \ldots$ choose a random variable Y_n' with distribution function belonging to the set $\mathcal{M}_2(F \mid U, V)$ such that

$$L_1(S_{n,k_n} - a_n, Y_n') \leq \gamma_n + \frac{1}{n}.$$

It is obvious that the sequence $\{Y_n'\}$ satisfies condition (2.10). Moreover, its weak relative compactness follows from (2.10) and from the weak relative compactness of the family $\{S_{n,k_n} - a_n\}_{n \geq 1}$ proved above. The necessity is proved.

Sufficiency. Assume that the sequence of the distribution functions

$$F_n(x) = P(S_{n,N_n} - c_n < x)$$

does not weakly converge to F. In this case there exists a sequence $\mathcal{N}$ of natural numbers and $\delta > 0$ such that for all $n \in \mathcal{N}$ the inequality $L_1(F_n, F) \geq \delta$

holds. Choose a subsequence $\mathcal{N}_1 \subseteq \mathcal{N}$ such that $Y'_n \Rightarrow Y$ as $n \to \infty, n \in \mathcal{N}_1$, where Y is some random variable. This can be done according to the assumption. Denoting $h_n(t) = \mathrm{E}\exp\{itY'_n\}$ in the same way we have done in the proof of Theorem 4.1.1 we make sure that for every $t \in \mathbb{R}$,

$$f(t) = \mathrm{E}h_n^U(t)\exp\{itV\} \to \mathrm{E}h^U(t)\exp\{itV\} \qquad (n \to \infty, n \in \mathcal{N}_1),$$

i.e., the limit random variable Y has the distribution function belonging to the set $\mathcal{M}_2(F \mid U, V)$. The inequality

$$L_1(S_{n,k_n} - a, Y) \le L_1(S_{n,k_n} - a_n, Y'_n) + L_1(Y'_n, Y)$$

and condition (2.10) imply that

$$L_1(S_{n,k_n} - a_n, Y) \to 0 \qquad (n \to \infty, n \in \mathcal{N}_1).$$

Apply Theorem 4.1.1 to the sequences of the random variables $\{S_{n,k_n} - a_n\}_{n \in \mathcal{N}_1}$ and the pairs $\{(N_n/k_n, a_n N_n/k_n - c_n)\}_{n \in \mathcal{N}_1}$. As a result we get $L_1(F_n, F) \to 0$ with $n \to \infty, n \in \mathcal{N}_1$, which contradicts the assumption that $L_1(F_n, F) \ge \delta$ for all $n \in \mathcal{N}$. This contradiction completes the proof of the theorem.

REMARK 4.2.1. Pay attention that the condition $\mathrm{P}(U = 0) < 1$ was not used in the proof of sufficiency.

COROLLARY 4.2.1. *If the set $\mathcal{M}_2(F \mid U, V)$ consists of a single element H, then under the conditions of Theorem 4.2.2 weak convergence of the centered random sums (1.10) takes place if and only if (1.1) is fulfilled, where Y is the random variable with the distribution function H.*

Returning to Example 1.1.2 in which we have considered accompanying infinitely divisible laws, we should mention that if the random variables N_n have Poisson distributions with parameters k_n each,

$$\mathrm{P}(N_n = j) = e^{-k_n}\frac{k_n^j}{j!}, \qquad j = 1, 2, \ldots, \tag{2.12}$$

then it is not difficult to make sure that if $k_n \to \infty$ as $n \to \infty$, then

$$\frac{N_n}{k_n} \Rightarrow 1 \qquad (n \to \infty).$$

Therefore setting in Theorem 4.2.2 $a_n = c_n = 0$ we see that with N_n distributed as (2.12), weak convergence of S_{n,k_n} to some infinitely divisible random variable Y takes place if and only if S_{n,N_n} weakly converges to the same random variable. This is a consequence of an obvious fact: if F is an infinitely divisible distribution function, then the set $\mathcal{M}_2(F \mid 1, 0)$ consists of a single element. Thus with the help of Theorem 4.2.2 we have one more proof of the fact that the sums of independent identically distributed random variables converge if and only if the sums of accompanying random variables converge to the same limit (as it was mentioned in Example 1.1.2).

Theorems 4.2.1 and 4.2.2 were proved in (Korolev, 1994). They sharpen and generalize the results (Szász and Freyer, 1971). In the same work (Korolev, 1994) the following two important statements were proved that describe the conditions of convergence of centered random sums of identically distributed summands to a normal law.

THEOREM 4.2.3. *Let sequences $\{a_n\}_{n\geq 1}$ and $\{k_n\}_{n\geq 1}$ provide (1.1) with Y being a nondegenerate infinitely divisible random variable. The convergence*

$$P(S_{n,N_n} - c_n < x) \Rightarrow \Phi_{\alpha,\sigma^2}(x) \qquad (n \to \infty), \tag{2.13}$$

takes place with some $\alpha \in \mathbb{R}$, $\sigma^2 > 0$ and some sequence $\{c_n\}_{n\geq 1}$ if and only if there exist numbers $\alpha_1 \in \mathbb{R}$, $\sigma_1^2 > 0$ and sequences of numbers $\{b_n\}_{n\geq 1}$ and random variables $\{V_n\}_{n\geq 1}$ satisfying the following conditions:

1) $P(Y < x) = \Phi_{\alpha_1,\sigma_1^2}(x), \qquad x \in \mathbb{R},$

2) $0 \leq b_n \leq \dfrac{\sigma^2}{\sigma_1^2},$

3) $P(V_n < x) = \Phi_{\alpha-\alpha_1 b_n,\sigma^2-\sigma_1^2 b_n}(x), \qquad x \in \mathbb{R},$

4) $L_1\left(\dfrac{N_n}{k_n}, b_r\right) \to 0 \qquad (n \to \infty),$

5) $L_1\left(a_n\dfrac{N_n}{k_n} - c_n, V_n\right) \to 0 \qquad (n \to \infty).$

PROOF. Theorem 4.1.1 implies that by virtue of (2.13) either the relation

$$f(t) = \exp\{i\alpha t - \frac{1}{2}\sigma^2 t^2\} = h^\gamma(t)\exp\{itV\}, \tag{2.14}$$

or the relation

$$f(t) = \exp\{i\alpha t - \frac{1}{2}\sigma^2 t^2\} = e^{it\beta}\mathsf{E}h^U(t) \tag{2.15}$$

should hold for some $\gamma > 0$ and $\beta \in \mathbb{R}$. But in case (2.14) according to the Lévy-Cramér theorem on decomposability of the normal law only into normal components both $h^\gamma(t)$ and $\mathsf{E}\exp\{itV\}$ should be normal characteristic functions. In case (2.15) according to the theorem due to V. M. Kruglov and A. N. Titov on the characterization of the normal law in the class of power mixtures of infinitely divisible characteristic functions (Kruglov and Titov, 1986) the random variable U should be degenerate and the characteristic function h should be normal. Thus, first, the random variable Y in (1.1) should have the normal distribution and, second, any pair $(U,V) \in \mathcal{M}_1(\Phi_{\alpha,\sigma^2} \mid \Phi_{\alpha_1,\sigma_1^2})$ should have the form (U_b, V_b), where $P(U_b = b) = 1$,

$$P(V_b < x) = \Phi_{\alpha-\alpha_1 b,\sigma^2-\sigma_1^2 b}(x), \qquad x \in \mathbb{R},$$

for some $b \in \mathbb{R}$. In this connection the weak relative compactness of the sequence of pairs of random variables mentioned in the statement of the theorem follows from the condition $0 \leq b_n \leq \sigma^2/\sigma_1^2$, which in its turn is necessary for the nonnegativeness of variances of the random variables V_b from $\mathcal{M}_1$. Now the desired result immediately follows from Theorem 4.2.1 and Proposition 4.1.1. The proof is completed.

THEOREM 4.2.4. *Let*

$$N_n \overset{P}{\to} \infty \qquad (n \to \infty) \qquad (2.16)$$

and let the sequences $\{a_n\}_{n \geq 1}, \{c_n\}_{n \geq 1}$ and $\{k_n\}_{n \geq 1}$ provide (1.9) with $\mathsf{P}(U = 0) < 1$. *Then (2.13) takes place if and only if there exist numbers $b > 0, \alpha_1 \in \mathbb{R}, 0 < \sigma_1^2 < \sigma^2$ satisfying the following conditions:*

1) $\mathsf{P}(U = b) = 1$,

2) $\mathsf{P}(V < x) = \Phi_{\alpha_1, \sigma_1^2}(x)$,

3) $\mathsf{P}(S_{n,k_n} - a_n < x) \Rightarrow \Phi_{(\alpha - \alpha_1)/b, (\sigma^2 - \sigma_1^2)/b}(x) \qquad (n \to \infty)$.

PROOF. We will show later in the proof of Lemma 4.3.1 that condition (2.16) implies the asymptotic negligibility of the summands as the consequence of convergence (2.13). Therefore the limit random variable for $S_{n,k_n} - a_n$ should be infinitely divisible. Again using Lévy–Cramér and Kruglov-Titov theorems we mention that in this case the random variable U should be degenerate and the random variable V should be normally distributed (possibly degenerate). As this is so, the set $\mathcal{M}_2(\Phi_{\alpha, \sigma^2} \mid U, V)$ consists of a single distribution function

$$H(x) = \Phi_{(\alpha - \alpha_1)/b, (\sigma^2 - \sigma_1^2)/b}(x) \qquad x \in \mathbb{R}.$$

Now this theorem immediately follows from Theorem 4.2.2 and Corollary 4.2.1. The proof is completed.

Similar statements hold for the weak convergence of centered random sums to a Poisson random variable or to the sum of independent normal and Poisson random variables. We should use A. D. Raikov's theorem on the decomposability of a Poisson law only into Poisson components instead of the Lévy-Cramér theorem.

4.3 Necessary and sufficient conditions for the convergence of random sums of independent identically distributed random variables

In this section we will consider necessary and sufficient conditions for the convergence of centered random sums of independent identically distributed random variables without additional assumptions that either normalized indices or nonrandom sums converge under an appropriate choice of centering constants. Similar results for noncentered sums can be obtained as simple corollaries.

Let F be some distribution function and f be its characteristic function. As we have made sure in the previous section, attempts to represent f in the form (1.11) can lead to different representations. To what has been said above, we add, e.g., that exponential characteristic function can be represented in the form (1.11) both as with the help of the characteristic function h degenerate at 1 and the exponentially distributed random variable U as well as with the help of the exponential characteristic function h and the random variable U degenerate at 1 (the random variable V is equal to zero in both cases). Furthermore, this representation is also possible with the help of the characteristic function h degenerate at zero and the exponentially distributed variable V; as this is so, the random variable U is either degenerate or is equal to a linearly transformed variable V.

To each distribution function F we put in correspondence the set $\mathcal{M}_0 = \mathcal{M}_0(F)$ of triples of random variables (Y, U, V) such that f is representable in the form (1.11), where Y is an infinitely divisible random variable with the characteristic function h and the pair (U, V) belongs to the class $\mathcal{K}_0$. The set $\mathcal{M}_0(F)$ is nonempty for any distribution function F since it contains at least one triple (I_0, I_1, Z), where $\mathsf{P}(Z < x) = F(x)$, $x \in \mathbb{R}$, $\mathsf{P}(I_a = a) = 1$, $a \in \mathbb{R}$.

We shall say that a set $\mathcal{M}' \subseteq \mathcal{M}_0(F)$ is weakly relatively compact if from any sequence of triples $\{(Y_n, U_n, V_n)\}_{n \in \mathcal{N}}$ from $\mathcal{M}'$ a subsequence $\{(Y_n, U_n, V_n)\}_{n \in \mathcal{N}_1 \subseteq \mathcal{N}}$ can be chosen so that the sequences of the random variables $\{Y_n\}_{n \in \mathcal{N}_1}$ and the pairs $\{(U_n, V_n)\}_{n \in \mathcal{N}_1}$ weakly converge. Note that if $\mathcal{M}_0(F)$ contains at least one triple (Y, U, V) in which the random variable Y differs from degenerate at zero, then the set $\mathcal{M}_0(F)$ itself is not weakly relatively compact, since along with (Y, U, V) it will contain triples of the form $(Y_\gamma, U/\gamma, V)$ where $\gamma > 0$ and Y_γ is the random variable with the characteristic function $(\mathsf{E} \exp\{itY\})^\gamma$.

To weaken the conditions that were used in the theorems of the previous section and to get complete inverses of Theorems 4.1.1 and 4.1.2 we will consider more explicitly the conditions for the weak relative compactness of centered random sums of independent identically distributed random variables. As before, denote

$$F_n(x) = \mathsf{P}(S_{n,N_n} - c_n < x).$$

In this section we will use centering and normalizing constants of a special form which are called centers and scatters of random variables. These characteristics were introduced by V. M. Zolotarev (Zolotarev, 1986). More explicitly, let G and g, respectively, be the distribution function and characteristic function of some random variable X and let a number $v > 0$ provide $g(t) \neq 0$ for $|t| < v$. Following (Zolotarev, 1986), we will call

$$C(v, X) = C(v, G) = \frac{1}{v} \operatorname{Im} \log g(v)$$

and

$$B(v, X) = B(v, G) = -\frac{6}{v^3} \int_0^v \mathrm{Re} \log g(t) \, dt.$$

the center and the scatter of the random variables X (or the distribution function G), respectively. The properties of these characteristics are described in detail in (Zolotarev, 1986). Here we will mention just some of them. Let $X, X_1, X_2, \ldots$ be independent random variables whose characteristic functions do not turn into zero on the segment $[0, v]$.

PROPERTY 4.3.1. $C(v, -X) = -C(v, X)$, $B(v, -X) = B(v, X)$.

PROPERTY 4.3.2. $C(v, X_1 + \ldots + X_n) = C(v, X_1) + \ldots + C(v, X_n)$, $B(v, X_1, \ldots, X_n) = B(v, X_1) + \ldots + B(v, X_n)$.

PROPERTY 4.3.3. If $X_n \Rightarrow X$ as $n \to \infty$ then $C(v, X_n) \to C(v, X)$ and $B(v, X_n) \to B(v, X)$.

Recall that we denote the greatest lower bound of s-quantiles of the random variable N_n as $l_n(s)$.

In Lemma 4.3.1 given below random variables $\{X_{n,j}\}_{j \geq 1}$ are not assumed to be identically distributed.

LEMMA 4.3.1. (Korolev and Kruglov, 1993) *Assume that the centers* $C(v, X_{n,j})$ *exist for some* $v > 0$ *and all* n *and* j. *The sequence of distribution functions* $\{F_n\}_{n \geq 1}$ *is weakly relatively compact if and only if the sequences of distribution functions*

$$\left\{ P\left(\sum_{j=1}^{l_n(s)} (X_{n,j} - C(v, X_{n,j})) < x \right) \right\}_{n \geq 1} \tag{3.1}$$

and

$$\left\{ P\left(\sum_{j=1}^{N_n} C(v, X_{n,j}) - c_n < x \right) \right\}_{n \geq 1} \tag{3.2}$$

are weakly relatively compact for any $s \in (0, 1)$.

PROOF. "Only if" part. With the help of the Lévy's inequality (Loève, 1963),

$$P\left(\max_{1 \leq k \leq n} \left| \sum_{j=1}^k X_j^{(s)} \right| \geq x \right) \leq 2P\left(\left| \sum_{j=1}^n X_j^{(s)} \right| \geq x \right) \tag{3.3}$$

and the symmetrization inequality (Loève, 1963),

$$P(|X_1^{(s)}| \geq x) \leq 2P\left(|X_1 - a| \geq \frac{x}{2} \right),$$

which hold for any independent random variables $X_1, X_2, \ldots, X_n$ and any $x \geq 0$, $a \in \mathbb{R}$; $X^{(s)}$ denotes the symmetrized variable $X : X^{(s)} = X - X'$, where $X \stackrel{d}{=} X'$ and X and X' are independent, as before, denoting $Z_{n,k} = X_{n,}^{(s)}1 + \ldots + X_{n,k}^{(s)}$ we obtain the chain of inequalities

$$\frac{1}{2}(1-s)P(|Z_{n,l_n(s)}| \geq x) \leq \sum_{k=l_n(s)}^{\infty} P(N_n = k)P(|Z_{n,k}| \geq x) \leq$$

$$\sum_{k=1}^{\infty} P(N_n = k)P(|Z_{n,k}| \geq x) \leq$$

$$2\sum_{k=1}^{\infty} P(N_n = k)P\left(\left|\sum_{j=1}^{k} X_{n,j} - c_n\right| \geq \frac{x}{2}\right) =$$

$$2P(|S_{n,N_n} - c_n| \geq \frac{x}{2}), \tag{3.4}$$

which holds for any $s \in (0,1)$ and $x \geq 0$. As this is so, the weak relative compactness of the sequence of distribution functions $\{F_n\}_{n\geq 1}$ implies that of the sequence of the distribution functions

$$\{P(Z_{n,l_n(s)} < x)\}_{n\geq 1}$$

for every $s \in (0,1)$. Denote

$$Y_{n,k} = S_{n,k} - \sum_{j=1}^{k} C(v, X_{n,j}).$$

Prove that the family of the distribution functions

$$\mathcal{P}_s = \{P(Y_{n,k} < x), k = 1, 2, \ldots, l_n(s); n \geq 1\}$$

is weakly relatively compact for every $s \in (0,1)$. This family is shift-compact since $Z_{n,k} = Y_{n,k} - Y'_{n,k}$ where $Y_{n,k}$ and $Y'_{n,k}$ are independent and identically distributed and $Z_{n,k}$ is contained in $Z_{n,l_n(s)}$ with $1 \leq k \leq l_n(s)$ as an independent summand (see Lemma 3.1 from (Kruglov, 1984) mentioned above). If $\mathcal{P}_s$ is not weakly relatively compact, then for some $\delta > 0$, some sequence of natural numbers $\{r_n\}_{n\geq 1}, 1 \leq r_n \leq l_n(s)$ and some infinitely increasing sequence of real numbers $\{x_n\}_{n\geq 1}$ the inequality $P(|Y_{n,r_n}| \geq x_n) \geq \delta$ holds. Within some choice of real numbers b_n the sequence of the distribution functions $\{P(Y_{n,r_n} - b_n < x)\}_{n\geq 1}$ is weakly relatively compact. Without loss of generality we will assume that this sequence weakly converges to some distribution function G. As this is so, according to the properties of centers, we have

$$-b_n = C(v, Y_{n,r_n} - b_n) \to C(v, G)$$

and therefore $|b_n| \leq M < \infty$, i.e., the sequence of the distribution functions $\{P(Y_{n,r_n} < x)\}_{n\geq 1}$ is weakly relatively compact which contradicts the assumption. Thus the weak relative compactness of the family $\mathcal{P}_s$ is established and implies the weak relative compactness of family (3.1).

According to Theorem 3 from (Gnedenko and Kolmogorov, 1954, p.43), there exists an $r > 0$ such that

$$P(|Y_{n,k}| \geq r) \leq \frac{1}{4}$$

uniformly over the family $\mathcal{P}_s$. Let $\mu_{n,k}$ be the median of the random variable $Y_{n,k}$. In this case according to the definition of a median, $|\mu_{n,k}| < r$. Use one more Lévy's inequality (Loève, 1963):

$$P(\max_{1 \leq k \leq n} |X_1 + \ldots + X_k - \text{med}(X_1 + \ldots + X_k)| \geq x) \leq$$

$$2P(\max_{1 \leq k \leq n} X_1^{(s)} + \ldots + X_k^{(s)} \geq x), \quad (3.5)$$

which holds true for any $x \geq 0$ and any independent random variables $X_1, \ldots, X_n$. With the help of inequalities (3.3) and (3.5) we obtain

$$P(\max_{1 \leq k \leq l_n(s)} |Y_{n,k}| \geq x) \leq P(\max_{1 \leq k \leq l_n(s)} |Y_{n,k} - \mu_{n,k}| \geq x - r) \leq$$

$$2P(\max_{1 \leq k \leq l_n(s)} |Z_{n,k}| \geq x - r) \leq 4P(|Z_{n,l_n(s)}| \geq x - r).$$

Therefore according to the law of total probability we have

$$P\left(\left|\sum_{j=1}^{N_n} (X_{n,j} - C(v, X_{n,j}))\right| \geq x\right) = \sum_{k=1}^{\infty} P(N_n = k)P(|Y_{n,k}| \geq x) \leq$$

$$\sum_{k=1}^{l_n(s)-1} P(N_n = k)P(\max_{1 \leq k \leq l_n(s)} |Y_{n,k}| \geq x) + \sum_{k=l_n(s)}^{\infty} P(N_n = k) \leq$$

$$4sP(|Z_{n,l_n(s)}| \geq x - r) + 1 - s.$$

Thus the weak relative compactness of the sequence of the distribution functions $\{P(Z_{n,l_n(s)} < x)\}_{n \leq 1}$ established before implies

$$\lim_{x \to \infty} \sup_n P\left(\left|S_{n,N_n} - \sum_{j=1}^{N_n} C(v, X_{n,j})\right| \geq x\right) \leq 1 - s.$$

The passage to the limit with $s \uparrow 1$ in this inequality proves the weak relative compactness of the sequence of distribution functions of the random variables $S_{n,N_n} - \sum_{j=1}^{N_n} C(v, X_{n,j})$. But

$$\sum_{j=1}^{N_n} C(v, X_{n,j}) - c_n = (S_{n,N_n} - c_n) - \left(S_{n,N_n} - \sum_{j=1}^{N_n} C(v, X_{n,j})\right). \quad (3.6)$$

Therefore the inequality

$$P\left(\left|\sum_{j=1}^{N_n} C(v, X_{n,j}) - c_n\right| \geq x\right) \leq$$

$$P(|S_{n,N_n} - c_n| \geq \frac{x}{2}) + P\left(\left|S_{n,N_n} - \sum_{j=1}^{N_n} C(v, X_{n,j})\right| \geq \frac{x}{2}\right)$$

according to what has already been proved by virtue of Theorem 3 from (Gnedenko and Kolmogorov, 1954, p.43), implies the weak relative compactness of sequence (3.2). The necessity of conditions of the lemma is proved.

"If" part. The weak relative compactness of the sequence of distribution functions (3.1) for every $s \in (0,1)$ implies the inherence of this property to the distribution functions $\{P(Z_{n,l_n(s)} < x)\}_{n\geq 1}$ which in its turn provides the weak relative compactness of the family of distribution functions $\{P(S_{n,N_n} - \sum_{j=1}^{N_n} C(v, X_{n,j}) < x)\}_{n\geq 1}$ as we have seen while proving the "only if" part. Using the inequality

$$P(|S_{n,N_n} - c_n| \geq x) \leq$$

$$P\left(\left|S_{n,N_n} - \sum_{j=1}^{N_n} C(v, X_{n,j})\right| \geq \frac{x}{2}\right) + P\left(\left|\sum_{j=1}^{N_n} C(v, X_{n,j}) - c_n\right| \geq \frac{x}{2}\right),$$

which takes place by virtue of (2.6), we establish the weak relative compactness of the family of distribution functions $\{F_n\}_{n\geq 1}$. The proof is completed.

COROLLARY 4.3.1. *Assume that for some $v > 0$ and all n and j the centers $C(v, X_{n,j})$ exist. The family of distribution functions $\{F_n\}_{n\geq 1}$ is weakly relatively compact if and only if the sequences of distribution functions (3.1) and*

$$\{P(Y_{n,N_n} < x)\}_{n\geq 1}$$

are weakly relatively compact.

LEMMA 4.3.2. *Assume that condition (2.16) holds. If the sequence of distribution functions $\{F_n\}_{n\geq 1}$ is weakly relatively compact, then $C(v, X_{n,1}) \equiv C_n(v)$ and $B(v, X_{n,1}) \equiv B_n(v)$ exist for some $v > 0$ and all $n \geq 1$; moreover, the sequences of distribution functions*

$$\{P(N_n C_n(v) - c_n < x)\}_{n\geq 1} \quad \text{and} \quad \{P(N_n B_n(v) < x)\}_{n\geq 1}$$

are weakly relatively compact.

PROOF. Let $f_n(t)$ be the characteristic function of the random variable $X_{n,1}$. With the help of inequalities (3.4) we verify the weak relative compactness of the sequence $\{Z_{n,l_n(s)}\}_{n\geq 1}$ for every $s \in (0,1)$. According to Corollary 2 (weak compactness criterion) from (Loève, 1963), for any $\varepsilon > 0$ there exists a $\delta > 0$ such that $1 - |f_n(t)|^{2l_n(s)} < \varepsilon$ for all $t \in [-\delta, \delta]$ and $n \geq 1$. Condition (2.16) implies that $l_n(s) \to \infty$ with $n \to \infty$ for every $s \in (0,1)$. We have

$$1 - |f_n(t)|^2 = \int_{-\infty}^{\infty} (1 - \cos tx)\, dP(X_{n,1}^{(s)} < x) =$$

$$\int_{-\infty}^{\infty} \left(1 - \cos^2 \frac{tx}{2}\right) dP(X_{n,1}^{(s)} < x) \leq 4\left(1 - \left|f_n\left(\frac{t}{2}\right)\right|^2\right).$$

Repeated application of this inequality leads us to the conclusion that $|f_n(t)|^2 \to 1$ with $n \to \infty$ for every $t \in \mathbb{R}$. Hence the sequence of the distribution functions $\{P(X_{n,1}^{(s)} < x)\}_{n\geq 1}$ weakly converges to the distribution function degenerate at zero. But the convergence $|f_n(t)|^2 \to 1$ $(n \to \infty)$ is uniform in every bounded interval. Therefore there exists a $v > 0$ such that

$$\inf_{t \in [-v, v]} |f_n(t)| > 0.$$

Thus $C(v, X_{n,1})$ and $B(v, X_{n,1})$ exist for all $n \geq 1$. As this is so, the weak relative compactness of the sequence $\{P(N_n C_n(v) - c_n < x)\}_{n \geq 1}$ follows from Lemma 4.3.1.

Prove that for every $s \in (0, 1)$

$$\sup_n l_n(s) B_n(v) = \beta(v, s) < \infty. \tag{3.7}$$

Assume the contrary. In this case $l_n(s) B_n(v) \to \infty$ with $n \to \infty, n \in \mathcal{N}$, where $\mathcal{N}$ is some sequence of natural numbers. Without loss of generality we can assume that the sequence of distribution functions $\{P(Z_{n, l_n(s)} < x)\}_{n \geq 1}$ weakly converges to some limit G. According to the second and the third properties of scatters we have

$$B(v, Z_{n, l_n(s)}) = 2l_n(s) B(v, X_{n,1}) \to B(v, G) \quad (n \to \infty, n \in \mathcal{N}).$$

But this contradicts our assumption and therefore (3.7) is proved. Let $x > \beta(v, s)$. By the law of total probability we have

$$P(N_n B_n(v) > x) = \sum_{k=1}^{\infty} P(N_n = k) P(k B_n(v) > x) \leq$$

$$\sum_{k=1}^{l_n(s)-1} P(N_n = k) P(k B_n(v) > x) + \sum_{k=l_n(s)}^{\infty} P(N_n = k) \leq$$

$$\sum_{k=1}^{l_n(s)-1} P(N_n = k) P(l_n(s) B_n(v) > x) + 1 - s \leq 1 - s,$$

and hence

$$\lim_{x \to \infty} \sup_n P(N_n B_n(v) > x) \leq 1 - s.$$

Passing to the limit with $s \uparrow 1$ in this inequality we obtain the weak relative compactness of the sequence of distribution functions $\{P(N_n B_n(v) < x)\}_{n \geq 1}$. The proof is completed.

THEOREM 4.3.1. *Let (2.16) hold and centers $C_n(v) \equiv C(v, X_{n,j})$ and scatters $B_n(v) \equiv B(v, X_{n,j})$ exist for some $v > 0$ and all n. Furthermore, assume that the sequence $\{N_n B_n(v)\}_{n \geq 1}$ contains no subsequences weakly convergent to zero. The convergence of the random sums*

$$S_{n,N_n} - c_n \Rightarrow Z \quad (n \to \infty) \tag{3.8}$$

takes place with some sequence of constants $\{c_n\}_{n \geq 1}$ if and only if there exists a weakly relatively compact sequence of triples $\{(Y'_n, U'_n, V'_n)\}_{n \geq 1}$ from $\mathcal{M}_0(F)$ satisfying the following conditions

$$L_1(S_{n,k_n} - a_n, Y'_n) \to 0 \quad (n \to \infty), \tag{3.9}$$

$$L_2\left(\left(\frac{N_n}{k_n}, a_n\frac{N_n}{k_n} - c_n\right),(U_n',V_n')\right) \to 0 \qquad (n \to \infty), \tag{3.10}$$

where

$$k_n = \left[\frac{1}{B_n(v)}\right], \qquad a_n = \frac{C_n(v)}{B_n(v)}.$$

PROOF. Necessity. Proving Lemma 4.3.2 we have established that condition (3.8) implies

$$\lim_{n\to\infty} k_n B_n(v) = 1 \tag{3.11}$$

by virtue of the fact that $k_n \to \infty$ due to the convergence $f_n(t) \to 1$ with $n \to \infty$, which is uniform on every finite interval. But

$$N_n C_n(v) - c_n = a_n N_n B_n(v) - c_n = a_n\frac{N_n}{k_n} \cdot k_n B_n(v) - c_n,$$

$$N_n B_n(v) = \frac{N_n}{k_n}B_n(v).$$

Therefore (3.11) implies the weak relative compactness of the sequence of the pairs of random variables $\{(N_n/k_n, a_n N_n/k_n - c_n)\}$ with k_n and a_n determined in formulation of the theorem.

To establish the weak relative compactness of the sequence $\{S_{n,k_n} - a_n\}_{n\geq1}$, we choose an arbitrary sequence $\mathcal{N}$ of natural numbers and in its turn choose a subsequence $\mathcal{N}_1 \subseteq \mathcal{N}$ so that the pairs $(N_n/k_n, a_n N_n/k_n - c_n)$ weakly converge as $n \to \infty, n \in \mathcal{N}_1$, to some pair $(U,V) \in \mathcal{K}_0$. According to the assumption, $P(U = 0) < 1$. Applying to the sequence $\mathcal{N}_1$ the reasoning used in the proof of necessity part of Theorem 4.2.2 we verify that the sequence $\{S_{n,k_n} - a_n\}_{n\in\mathcal{N}_1}$ is weakly relatively compact. Therefore a subsequence $\mathcal{N}_2 \subseteq \mathcal{N}_1$ can be chosen such that the random variables $S_{n,k_n} - a_n$ weakly converge with $n \to \infty, n \in \mathcal{N}_2$. The arbitrariness of $\mathcal{N}$ proves the weak relative compactness of the sequence of random variables $\{S_{n,k_n} - a_n\}_{n\geq1}$.

Denote

$$\alpha_n = \inf\{L_1(S_{n,k_n} - a_n, Y)+$$

$$L_2\left(\left(\frac{N_n}{k_n}, a_n\frac{N_n}{k_n} - c_n\right),(U,V)\right) : (Y,U,V) \in \mathcal{M}_0(F)\}.$$

Show that $\alpha_n \to 0$ with $n \to \infty$. Assume the contrary. In this case there exist a sequence $\mathcal{N}_3$ of natural numbers and a $\delta > 0$ such that $\alpha_n \geq \delta$ for all $n \in \mathcal{N}_3$. Choose a subsequence $\mathcal{N}_4 \subseteq \mathcal{N}_3$ so that the sequences of random variables $S_{n,k_n} - a_n$ and of pairs $(N_n/k_n, a_n N_n/k_n - c_n)$ weakly converge to some random variable Y and some pair $(U,V) \in \mathcal{K}_0$, respectively, with $n \to \infty, n \in \mathcal{N}_4$ (this can be done by virtue of the established weak relative compactness of the considered sequences). As this is so, for all sufficiently large $n \in \mathcal{N}_4$ we have

$$L_1(S_{n,k_n} - a_n, Y) + L_2\left(\left(\frac{N_n}{k_n}, a_n\frac{N_n}{k_n} - c_n\right),(U,V)\right) \leq \delta. \tag{3.12}$$

Applying Theorem 4.1.1 to the random variables $S_{n,k_n} - a_n$ and $(N_n/k_n, a_n N_n/k_n - c_n)$ with $n \in \mathcal{N}_4$, by virtue of condition (3.8) we make sure that $(Y, U, V) \in \mathcal{M}_0(F)$. This means that inequality (3.12) contradicts the assumption that $\alpha_n \geq \delta$ for all $n \in \mathcal{N}_4$. Thus we have proved that $\alpha_n \to 0$ with $n \to \infty$. For every $n = 1, 2, \ldots$ choose a triple $(Y'_n, U'_n, V'_n) \in \mathcal{M}_0$ for which the condition

$$L_1(S_{n,k_n} - a_n, Y'_n) + \left(\left(\frac{N_n}{k_n}, a_n \frac{N_n}{k_n} - c_n \right), (U', V') \right) \leq \alpha_n + \frac{1}{n}.$$

is fulfilled. It is obvious that the sequence (Y'_n, U'_n, V'_n) chosen so satisfies (3.9) and (3.10). Its weak relative compactness follows from (3.9) and (3.10) and the inherence of this property to the sequences of random variables $S_{n,k_n} - a_n$ and of the pairs $\{(N_n/k_n, a_n N_n/k_n - c_n)\}_{n \geq 1}$. Necessity is proved.

Sufficiency. Assume that the sequence of the distribution functions $F_n(x) \equiv P(S_{n,N_n} - c_n < x)$ does not weakly converge to $F(x) \equiv P(Z < x)$. In this case for some $\delta > 0$ and all n from some sequence $\mathcal{N}$ of natural numbers the inequality $L_1(F_n, F) \geq \delta$ holds. Since by the condition the sequence $\{(Y'_n, U'_n, V'_n)\}_{n \geq 1}$ is weakly relatively compact, choose a subsequence $\mathcal{N}_1 \subseteq \mathcal{N}$ so that the convergences $Y'_n \Rightarrow Y$ and $(U'_n, V'_n) \Rightarrow (U, V)$ take place as $n \to \infty, n \in \mathcal{N}_1$ for some random variables Y and (U, V). Repeating the reasoning from the proof of Theorem 4.1.1 we make sure that

$$f(t) = \mathsf{E}[h_n^{U'_n}(t) \exp\{itV'_n\}] \to \mathsf{E}[h^U(t) \exp\{itV\}]$$

with $n \to \infty, n \in \mathcal{N}_1$, for any $t \in \mathbb{R}$, where $h_n(t) = \mathsf{E} \exp\{itY'_n\}$. Thus $(Y, U, V) \in \mathcal{M}_0(F)$. The inequality

$$L_1(S_{n,k_n} - a_n, Y) \leq L_1(S_{n,k_n} - a_n, Y'_n) + L_1(Y'_n, Y)$$

and condition (3.9) imply $L_1(S_{n,k_n} - a_n, Y) \to 0$ $(n \to \infty, n \in \mathcal{N}_1)$. Similarly, the inequality

$$L_2\left(\left(\frac{N_n}{k_n}, a_n \frac{N_n}{k_n} - c_n \right), (U, V) \right) \leq$$
$$L_2\left(\left(\frac{N_n}{k_n}, a_n \frac{N_n}{k_n} - c_n \right), (U'_n, V'_n) \right) + L_2((U'_n, V'_n), (U, V))$$

and condition (3.10) imply $L_2((N_n/k_n, a_n N_n/k_n - c_n), (U, V)) \to 0$ $(n \to \infty, n \in \mathcal{N}_1)$. Applying Theorem 4.1.1 to the sequences $\{S_{n,k_n} - a_n\}_{n \in \mathcal{N}_1}$ and $\{(N_n/k_n, a_n N_n/k_n - c_n)\}_{n \in \mathcal{N}_1}$ we verify that $L_1(F_n, F) \to 0$ with $n \to \infty, n \in \mathcal{N}_3$ which contradicts the assumption that $L_1(F_n, F) \geq \delta > 0$ for all $n \in \mathcal{N}$. The proof is completed.

REMARK 4.3.1. In the proof of necessity of conditions of Theorem 4.3.2 the assumption of the existence of centers and scatters is fulfilled automatically by virtue of Lemma 4.3.2, and in the proof of sufficiency the assumption that the sequence $\{N_n/k_n\}_{n \geq 1}$ is separated from zero in the sense of weak convergence was not used.

Theorem 4.3.1 was published in (Korolev, 1993). In a little less general situation when $a_n = c_n = 0$, a complete inversion of Theorem 4.1.2 appears to be possible without additional assumptions concerning the existence of centers and scatters. This complete inversion of Theorem 4.1.2 was proved by V. M. Kruglov (Kruglov and Korolev, 1990). We will present this result without any proof since the steps of the proof of Theorem 4.3.1 are similar to the proof of Kruglov's theorem. An interested reader can get acquainted with it in (Kruglov and Korolev, 1990).

For an arbitrary distribution function F define the set $\mathcal{H}(F)$ of pairs of distribution functions (H, A) such that the characteristic function f corresponding to the distribution function F can be represented in the form (1.18), where $h(t)$ is the infinitely divisible characteristic function corresponding to the distribution function H. We shall say that the sequence of pairs $\{(H_n, A_n)\}_{n \geq 1}$ from $\mathcal{H}(F)$ is weakly relatively compact if each of the sequences $\{H_n\}_{n \geq 1}$ and $\{A_n\}_{n \geq 1}$ is weakly relatively compact.

THEOREM 4.3.2. *Let a distribution function F be given for which the set $\mathcal{H}(F)$ is not empty, and let condition (2.16) be fulfilled. Then*

$$P(S_{n,N_n} < x) \Rightarrow F(x) \qquad (n \to \infty),$$

if and only if there exist a sequence $\{k_n\}_{n \geq 1}$ of natural numbers and a weakly relatively compact sequence of pairs $\{(H_n, A_n)\}_{n \geq 1}$ from $\mathcal{H}(F)$ satisfying the conditions

$$L_1(H_{n,k_n}, H_n) \to 0 \qquad (n \to \infty),$$
$$L_1(A_{n,k_n}, A_n) \to 0 \qquad (n \to \infty),$$

where

$$H_{n,k_n}(x) = P(S_{n,k_n} < x) \qquad x \in \mathbb{R},$$
$$A_{n,k_n}(x) = P(N_n < x k_n) \qquad x \in \mathbb{R}.$$

One elegant result describing necessary and sufficient conditions for the convergence of noncentered random sums to a normal law also belongs to V. M. Kruglov. Formally this result published in 1976 is a consequence of Theorem 4.3.2 and the fact that the set $\mathcal{H}(\Phi)$, where Φ is the standard normal distribution function, contains only pairs of the form $\left(\Phi\left(\frac{x}{\sigma}\right), \mathcal{E}_1(\sigma^2 x)\right), \sigma > 0$, where $\mathcal{E}_1(x)$ is the distribution function having a single jump at the point $x = 1$ (Kruglov, 1976).

COROLLARY 4.3.2 (Kruglov, 1976). *Let (2.16) be fulfilled. Then*

$$P(S_{n,N_n} < x) \Rightarrow \Phi(x) \qquad (n \to \infty),$$

if and only if there exists a sequence $\{k_n\}_{n \geq 1}$ of natural numbers such that

$$P(S_{n,k_n} < x) \Rightarrow \Phi(x) \qquad (n \to \infty),$$
$$P\left(\frac{N_n}{k_n} < x\right) \Rightarrow \mathcal{E}_1(x) \qquad (n \to \infty).$$

4.4 More on some models of financial mathematics

In this section we will apply some results exposed in the preceding sections of this chapter to modify the mathematical model of the distribution of an increment of stock price we have obtained in Sect. 3.5.

When we constructed our model for the distribution of an increment of a stock price process in Sect. 3.5 with the help of limit theorems of random summation in the scheme of "growing" random sums, we assumed that elementary increments of the prelimit process, that is, changes of the price from one contract to another in successive contracts, have zero means. This assumption together with general reasoning typical for the central limit theorem lead us to scale mixtures of normal laws as models of the distribution under consideration. These mixtures are symmetrical and leptokurtic. However, the actually observed distributions are not only leptokurtic, but as a rule, they are skewed as well.

To avoid the necessity of obtaining only symmetrical models which is implied by the reasoning of Sect. 3.5, we generalize our basic model and adjust it to the double array limit scheme. Unlike the situation considered above, at different time intervals the elementary increments of the stock price process may have different systematic drifts (different expectations) since there may occur periods when stock prices rather steadily go up or down. Thus we may assume the identity of distributions of elementary stock price increments within a period of steady change (increase or decrease), but we are not obliged at all to assume this within a long time interval which contains many steady periods.

One more reason for introducing the double array scheme is as follows.

Let n be an auxiliary integer parameter and let t be the length of an interval per which the increments are considered. Let $T = nt$ be the total interval of observations so that n may have the meaning of the sample size if by a sample we mean the set of available observations after the increments per time intervals of the same length t. Let N_n be the number of contracts fixed during the n-th time interval of length t, and let $X_{n,1}, \ldots, X_{n,N_n}$ be the elementary increments of the stock price on the last, n-th, interval. Now let T increase resulting in the increase of n if t remains constant. As this is so, the last, n-th, time interval begins to move rightwards along the time axis. Although we assume that for each $n \geq 1$, $\{X_{n,j}\}_{j\geq1}$ are identically distributed and independent, the distributions of $X_{n,j}$ for different n (i.e., on different time intervals of length t) may not necessarily coincide. Thus we see that the consideration of the double array scheme allows one to construct the models which take into account a possible drift of the distribution of stock price increments.

Now we turn to the formal construction of our models. Our basic model will be written as

$$P_n(t) - P_n(0) = \sum_{j=1}^{N_n} X_{n,j} \qquad (4.1)$$

where $P_n(t)$ and $P_n(0)$ are, respectively, the prices at the end of the n-th interval of length t and at the beginning of this interval; $\{X_{n,j}\}_{j\geq 1}$ are random variables assumed independent and identically distributed for each fixed n, which have the meaning of elementary stock price changes on the n-th time interval of length t; $N_n = N_n(t)$ are natural-valued random variables assumed independent of $\{X_{n,j}\}_{j\geq 1}$ for each $n \geq 1$. They play the roles of the numbers of contracts fixed during the n-th time interval. Note that here various interpretations are possible including those given above as well as those where n has no "physical" sense being just an auxiliary parameter used to construct limit approximations. We can also make n depend on t and study the distributions of the increments of stock prices per time intervals of increasing lengths which makes sense, e.g., for the investigation of the effect of "normalization" of the distributions of the increments of stock prices as the length of the period under study growths. This effect was mentioned in Sect. 3.5.

Our reasoning deals with the absolute increments. But if one is anxious about relative increments, then our additive model (4.1) should be rewritten with respect to the logarithms of the increments involved.

Thus we can see that the double array scheme is flexible enough and admits many ways of interpretation of the involved parameters and random variables.

If k is an integer, then by $S_{n,k}$ we as usual denote the sum $X_{n,1}+\ldots+X_{n,k}$. Concerning the summands $X_{n,j}$, we will make "good" assumptions that they are infinitesimal and satisfy the central limit theorem in the sense that there exist integer k_n such that $k_n \to \infty$ $(n \to \infty)$ and

$$P(S_{n,k_n} < x) \Rightarrow \Phi\left(\frac{x-a}{\sigma}\right) \qquad (n \to \infty) \qquad (4.2)$$

for some $a \in \mathbb{R}$ and $0 < \sigma < \infty$ ($a = 0$ and $\sigma = 1$ included). Note that (4.2) holds in the following rather general situation. In accordance with what has been said in Sect. 3.5, we will assume that $X_{n,j}$ have finite variances. Assume that $X_{n,j}$ are representable as $X_{n,j} = X_{n,j}^* + a_n$ where a_n are nonrandom and describe the systematic drift of an elementary increment $X_{n,j}$ on the n-th time period and $EX_{n,j}^* = 0, DX_{n,j}^* = \sigma_n^2 < \infty$ so that $EX_{n,1} = a_n$ and $DX_{n,1} = \sigma_n^2$. Assume that $a_n k_n \to a$ and $k_n \sigma_n^2 \to \sigma^2$ as $n \to \infty$. Then as a consequence of the well-known result on conditions for convergence to the normal law of the distributions of sums of independent random variables with finite variances in the double array scheme (see, e.g., (Gnedenko and Kolmogorov, 1954)) (4.2) takes place if and only if the Lindeberg condition holds: for any $\varepsilon > 0$

$$\lim_{n\to\infty} k_n E(X_{n,1}^*)^2 I(|X_{n,1}^*| \geq \varepsilon) = 0,$$

that is, the quadratic tails of the distributions of elementary increments should decrease fast enough. By the way, this condition obviously contradicts Mandelbrot's hypothesis of infinite variances but (as we have already seen and will also see in the sequel) also leads to skewed and leptokurtic distributions. Some other reasons for the assumed normality of the limit distribution in (4.2) may be found in Appendix 1.

Now we have everything we need to formulate the main theorem of this section.

THEOREM 4.4.1. *Let $\{X_{n,j}\}$ and N_n satisfy the above conditions. Assume that there exists a sequence $\{k_n\}_{n\geq 1}$ of integers such that (4.2) holds with some $a \in \mathbb{R}$ and $\sigma < \infty$. Then*

$$P_n(t) - P_n(0) - c_n \Rightarrow (\text{some}) \ Z \qquad (n \to \infty) \qquad (4.3)$$

for some $c_n \in \mathbb{R}$ if and only if there exist a weakly relatively compact sequence of random variables $\{U'_n\}_{n\geq 1}$ and a bounded sequence of numbers $\{v'_n\}_{n\geq 1}$ such that

$$\mathsf{E}\exp\{isZ\} = \mathsf{E}\exp\left\{is(U'_n + v'_n) - \frac{1}{2}s^2\sigma^2 U'_n\right\} \qquad (4.4)$$

for each $n \geq 1$, $s \in \mathbb{R}$, and

$$L_1(N_n/k_n, U'_n) \to 0 \qquad (n \to \infty), \qquad (4.5)$$
$$|v'_n + c_n| \to 0 \qquad (n \to \infty).$$

This theorem is a simple consequence of Theorem 4.2.1. Here as usual, L_1 is a metric which metrizes the weak convergence of random variables, e.g., Lévy metric. The limit characteristic function (4.4) corresponds to the distribution function

$$P(Z < x) = \mathsf{E}\Phi\left(\frac{x - b - aU}{\sigma\sqrt{U}}\right) \qquad (4.6)$$

for some $b \in \mathbb{R}$ and nonnegative random variable U. Unfortunately, the family of scale-location mixtures of normal laws of the form (4.6) is not identifiable; therefore there may be several random variables U satisfying (4.6). This is the reason why condition (4.5) of asymptotic "weak rapprochement" of the sequences $\{N_n/k_n\}$ and $\{U'_n\}$ appears in the formulation of Theorem 4.4.1 but not a more interpretable condition $N_n/k_n \Rightarrow U$ (which is sufficient for (4.3) and (4.6) in accordance with Theorem 4.1.1).

There were many attempts to find a convenient parametric form of the distribution of the increments of a stock price process, some of them are recalled in Sect. 3.5. Recently O. Barndorff-Nielsen reported on a very good fit of the so-called Gaussian/inverse Gaussian distribution to empirical data obtained at Danish and German stock exchanges (Barndorff-Nielsen, 1994). The density of a Gaussian/inverse Gaussian distribution is

$$g(x; \alpha, \beta, \mu, \delta) =$$
$$A(\alpha, \beta, \mu, \delta)\left[Q\left(\frac{x-\mu}{\delta}\right)\right]^{-1} K_1\left(\delta\alpha Q\left(\frac{x-\mu}{\delta}\right)\right)e^{\beta x} \qquad (4.7)$$

where K_1 is the modified Bessel function of third order and index 1,

$$A(\alpha, \beta, \mu, \delta) = \frac{\alpha}{\pi}\exp\{\delta\sqrt{\alpha^2 - \beta^2} - \beta\mu\}, \quad Q(x) = \sqrt{1 + x^2}.$$

A good fit of this very distribution may seem surprising since distribution (4.7) is somewhat very special and cumbersome. But actually there is no surprise at all due to Theorem 4.4.1 since Barndorff-Nielsen has demonstrated earlier that the Gaussian/inverse Gaussian distribution is one of generalized hyperbolic distributions, all of which are representable as scale-location mixtures of normal laws (Barndorff-Nielsen, 1977, 1978), see also (Barndorff-Nielsen, Kent and Sørensen, 1982). Namely, the density (4.7) corresponds to the distribution function

$$G(x; \alpha, \beta, \mu, \delta) = \mathsf{E}\Phi\left(\frac{x - \mu - \beta U}{\sqrt{U}}\right) \tag{4.8}$$

where the random variable U has the inverse Gaussian distribution whose density is

$$p(u; \delta, \sqrt{\alpha^2 - \beta^2}) =$$
$$\frac{\delta}{\sqrt{2\pi}} \exp\{\delta\sqrt{\alpha^2 - \beta^2}\} u^{-3/2} \exp\left\{-\frac{1}{2}\left[(\alpha^2 - \beta^2)u + \frac{\delta^2}{u}\right]\right\}. \tag{4.9}$$

This circumstance is the reason why distribution (4.7) is called Gaussian/inverse Gaussian. Note that the right-hand side of (4.8) coincides with that of (4.6) up to notations.

In connection with the paper of Barndorff-Nielsen, the question arises why the mixing distribution (4.9) provides such a good fit with experimental data. One possible and rather simple answer is that the parametric family of inverse Gaussian distributions (4.9) is rich enough so that there always exists a representative of this family which is satisfactorily close to the actual mixing distribution. Another answer may possibly be obtained if someone succeeds in connecting the behavior of the random variables N_n in Theorem 4.4.1 based on the model (4.1) with Barndorff-Nielsen's reasoning which he used when he considered the distributions for the logarithms of particle size and obtained the Gaussian/inverse Gaussian distributions (Barndorff-Nielsen, 1977).

As an example of what Theorem 4.4.1 can give, consider the problem of rational option pricing. Consider a European call option, which is a derivative security giving its owner the right to buy a stock at a certain time t (called exercise or maturity date) for a certain price K (called exercise price).

The classical setting of this problem concerns relative increments of stock prices and assumes that the process of stock prices $P(t)$ follows the geometric Brownian motion,

$$P(t) = P(0) \exp\left\{t\left(\mu_0 - \frac{\sigma_0^2}{2}\right) + \sigma_0 W(t)\right\} \tag{4.10}$$

where $W(t)$ is the standard Wiener process and the parameters μ_0 and σ_0^2 are referred to as the expected return of the stock and the stock price volatility, respectively. The model (4.10) was proposed by P. A. Samuelson (Samuelson, 1965) who also called this model an economical Brownian motion. Hence it follows that the distributions of the increments of this process are lognormal, that is, given $P(0)$,

$$P(\log P(t) - \log P(0) < x) = \Phi\left(\frac{x - t\left(\mu_0 - \frac{\sigma_0^2}{2}\right)}{\sigma_0\sqrt{t}}\right), \quad x \in \mathbb{R}. \qquad (4.11)$$

However, as we have noted in Sect. 3.5, in practice (4.10) does not hold since usually the distribution in the left-hand side of (4.10) is not normal being leptokurtic and skewed. At the same time, the structure of model (4.10) taking account of the proportionality of $dP(t)$ to $P(t)$ with proportionality coefficient linearly dependent on both systematic and stochastic drift in general seems very reasonable (using Ito's lemma it can be shown that the process $P(t)$ given by (4.10) satisfies the stochastic differential equation

$$\frac{dP(t)}{P(t)} = \mu_0 dt + \sigma_0 dW(t)). \qquad (4.12)$$

This means that instead of $\sigma_0 W(t)$ in the exponent in (4.10) some other construction should be used. Unfortunately, the apparatus we developed in this book does not include functional limit theorems and therefore it is not enough to describe possible models of these constructions completely (e.g., to derive equations similar to (4.12)), it can only give us some hints on what these processes may look like. But what we can do is describe one-dimensional distributions of the process $P(t)$. And this is just what we need to evaluate an option.

Using a risk-neutrality valuation argument, the formula for the price of a European call option can be obtained by discounting the expected value of the option at maturity which is $\mathsf{E} \max(0, P(t) - K)$ by the risk-free rate of interest r (assumed constant on the interval $[0, t]$) while simultaneously setting the parameters of the distribution of $P(t)$ so that the expected return on the underlying stock is risk-free (see, e.g., (Hull, 1989)). Namely, in the continuous time case the price C of a European call option is given as

$$C = e^{-rt}\mathsf{E}_0 \max(0, P(t) - K) \equiv e^{-rt}\int \max(0, s - K)d\mathsf{P}_0(P(t) < s) \quad (4.13)$$

with E_0 denoting expectation in a risk-neutral world, that is, with respect to a probability measure P_0 such that, given $P(0)$,

$$\mathsf{E}_0 P(t) \equiv \int s \, d\mathsf{P}_0(P(t) < s) = e^r P(0). \qquad (4.14)$$

For example, for model (4.10) we have

$$\mathsf{P}_0(\log P(t) - \log P(0) < x) = \Phi\left(\frac{x - t\left(r - \frac{\sigma_0^2}{2}\right)}{\sigma_0\sqrt{t}}\right), \quad x \in \mathbb{R}. \qquad (4.15)$$

In accordance with the assertion of Theorem 4.4.1, instead of (4.15) we should have in (4.13)

$$P_0(\log P(t) - \log P(0) < x) = E_0 \Phi \left(\frac{x - b - aU}{\sigma \sqrt{U}} \right), \quad x \in \mathbb{R}, \qquad (4.16)$$

with some real a and b, positive σ and positive random variable U providing (4.14) (of course, (4.15) is a special case of (4.16)). But (4.15) means that, given $P(0)$,

$$P(t) \overset{d}{=} \exp\{\sigma \sqrt{U} X + aU + b + \log P(0)\} \qquad (4.17)$$

where X is the standard normal variable independent of U. Therefore inserting (4.17) into (4.13) we obtain

$$C = e^{-rt} E_0 \max(0, \exp\{\sigma \sqrt{U} X + aU + b + \log P(0)\} - K)$$

and since in accordance with (4.15) X and U are independent with respect to P_0, by the Fubini theorem we finally have

$$C = e^{-rt} \int\limits_0^\infty \int\limits_{-\infty}^\infty \max(0, e^z - K) \, d_z \Phi \left(\frac{z - au - b - \log P(0)}{\sigma \sqrt{u}} \right) dP_0(U < u).$$

REMARK 4.4.1. In order to present a dynamic (t-dependent) model similar to (4.10) instead of the static model (4.17) we can easily see that (4.17) holds, if instead of (4.10) we use the stochastic volatility model

$$P(t) = P(0) \exp\{t(\alpha_1 + \alpha_2 U(t)) + \sqrt{U(t)} W(t)\} \qquad (4.18)$$

with some $\alpha_1, \alpha_2 \in \mathbb{R}$ and positive stochastic process $U(t)$ which may be assumed in some sense bounded. Of course, (4.18) is not a single model satisfying (4.17).

In the conclusion of this section we will touch upon a problem that seems most difficult and amazing from the practical point of view, that is, a problem of statistical determination of the parameters of model (4.5). Note that the distribution of U is also a parameter of (4.5); therefore this problem includes statistical reconstruction of the distribution of U. Due to the fact that the family of mixtures (4.5) is not identifiable, the latter problem admits many solutions. In other words, it is ill-posed.

Indeed, if we assume that the random variable U in (4.5) has a density $q(u)$ and denote $F(x) = P(Z < x)$, then (4.5) is rewritten as

$$F(x) = \int\limits_0^\infty \Phi \left(\frac{x - au + b}{\sigma \sqrt{u}} \right) q(u) \, du, \qquad (4.19)$$

so even with known a, b, σ the problem of reconstruction of the distribution of U reduces to nothing but solving the Fredholm integral equation (4.19) with respect to $q(u)$, and this is a typical example of an ill-posed problem

(Tikhonov, 1977). To regularize the initial problem we replace it with an approximate well-posed one. For this purpose recall that if we restrict ourselves to the search for the desired distribution of U among random variables which can take only a finite number of values (these random variables are called simple or elementary), then we can find no more than one simple random variable U satisfying (4.5), since unlike general scale-location mixtures of normals, finite scale-localtion mixtures of normal laws are identifiable (Teicher, 1963). So instead of one of many possible solutions to the initial problem of reconstruction of the distribution of U, we will look for a simple random variable U which delivers the best possible approximation to $F(x)$ in the set of simple random variables, which in our case will be a so-called pseudosolution.

If U being sought for takes values $u_1, \ldots, u_m$ with probabilities $p_1, \ldots, p_m$, then denoting $\theta_i = \sigma\sqrt{u_i}, c = a\sigma^{-2}$ we come to the problem of determination of the estimates of parameters which provide the best fit of

$$F_m(x) = \sum_{i=1}^{m} p_i \Phi\left(\frac{x - b - c\theta_i^2}{\theta_i}\right) \tag{4.20}$$

to $F(x)$. Here the number of parameters is $2m + 2$. Actually we come to a minimax problem, because although by increasing the number m of possible values of a simple random variable sought for we can provide better approximation, at the same time the amount of required computational resources (time, memory, etc.) increases almost exponentially as m grows. We will not go into further details here since otherwise it will bring us far beyond the limits of this book. The interested reader is referred to numerous works on statistical separation of mixtures e.g., indicated in (Aivazyan, Bukhshtaber, Enyukov and Meshalkin, 1989).

The approximate model (4.19) was obtained on the basis of one of main assumptions in Theorem 4.4.1 which was the identity of distributions of elementary increments within one (steady) period. If we relax this condition and return to the scheme of "growing" random sums of not necessarily identically distributed and not necessarily centered summands, then within the framework of a general approach assuming that the properties of summands guarantee the asymptotic normality of nonrandom sums, we should replace (4.16) with a more general model

$$P(t) \stackrel{d}{=} P(0) \exp\{UX + V\}$$

in which the pair (U, V) is independent of the standard normal variable X resulting in that the approximate model (4.20) should be replaced by the model

$$F_m(x) = \sum_{i=1}^{m} p_i \Phi\left(\frac{x - a_i}{\sigma_i}\right)$$

with $3m$ parameters.

4.5   Limit theorems for supercritical Galton–Watson processes

In Example 1.3.4 we became acquainted with Galton–Watson processes which are mathematical models rich in practical applications, represented as a random sequence formed by recurrent random sums of a special type. This model is interesting for us not only because random sums play a most important role in its definition, but also because random sums appear in a very unexpected way within the investigation of its limit behavior. In this connection the transfer theorems proved in Section 4.1 appear to be very useful. The further presentation is based on the results of the articles (Heyde, 1970), (Heyde, 1971), (Heyde and Brown, 1971).

We are keeping to the notation introduced in Example 1.3.4; therefore it is advisable to look through Example 1.3.4 once more before reading further material.

In this and the next sections we will use the notion of a stable distribution introduced by P. Lévy and A. Ya. Khinchin. In 1925 P. Lévy proved that the class of limit distributions for the sums of independent identically distributed random variables of the form

$$S_n = \frac{U_1 + \ldots + U_n}{B_n}$$

where $B_n > 0$, coincides with the class of distribution functions possessing the property that logarithms of their characteristic functions are defined by the formula

$$\log f(t) = -\left(c_0 + i\frac{t}{|t|}c_1\right)|t|^\alpha, \tag{5.1}$$

where $0 \leq \alpha \leq 2$, $c_0 > 0$, $c_1 \in \mathbb{R}$. If $\alpha = 2$, then $c_1 = 0$. P. Lévy called the laws (5.1) stable. In 1936 A. Ya. Khinchin suggested calling stable the limit laws for the sums of a more general type

$$S_n = \frac{U_1 + \ldots + U_n - A_n}{B_n}, \tag{5.2}$$

where $B_n > 0$, $A_n \in \mathbb{R}$. Such laws have characteristic functions whose logarithms have the form

$$\log f(t) = i\gamma t - c_0|t|^\alpha \left(1 + i\beta\frac{t}{|t|}\omega(t, \alpha)\right), \tag{5.3}$$

where $c_0 > 0$, $0 \leq \alpha \leq 2$, γ and β are real numbers, $-1 \leq \beta \leq 1$,

$$\omega(t, \alpha) = \begin{cases} \text{tg}\dfrac{\pi\alpha}{2}, & \text{if } \alpha \neq 1, \\ \dfrac{2}{\pi}\log|t|, & \text{if } \alpha = 1. \end{cases}$$

The laws (5.1) are now called strictly stable. We will keep to this terminology.

We will say that a distribution F belongs to the domain of attraction of a stable law G if for some constants A_n and B_n the distribution functions of sums (5.2) of independent random variables $U_1, U_2, \ldots$ with the same distribution function F weakly converge to G. It is customary to call the parameter α in representations (5.1) and (5.3) the index of a stable (strictly stable) law.

Further we will consider the situation where the distribution function of the random variable Z_1 from the definition of Galton–Watson process (1.3.1) belongs to the domain of attraction of the stable law with index $\alpha \in (1, 2]$ so that $\mathsf{E} Z_1^a < \infty$ for any $a < \alpha$. Hence it follows that $\mathsf{E} Z_1 \log Z_1 < \infty$ so that by virtue of Theorem 1.3.1 the random variables $X_n = Z_n/m^n$ converge with probability one to a nondegenerate random variable X, where $m = \mathsf{E} Z_1$.

THEOREM 4.5.1. *The distribution function of a random variable X belongs to the domain of attraction of the stable law with index $\alpha \in (1, 2]$ if and only if the distribution function of the random variable Z_1 belongs to the same domain of attraction. Moreover, if the random variables $\{Y_j\}_{j \geq 1}$ are independent and identically distributed and also $Y_j \overset{d}{=} X - 1, j \geq 1$, and the random variables $\{U_j\}_{j \geq 1}$ are also independent and $U_j \overset{d}{=} Z_1, j \geq 1$, then*

$$\frac{1}{c_n} \sum_{j=1}^{n} (U_j - m) \Rightarrow Z \qquad (n \to \infty)$$

if and only if

$$(m^\alpha - m)^{1/\alpha} \frac{1}{c} \sum_{j=1}^{n} Y_j \Rightarrow Z \qquad (n \to \infty),$$

where Z is the random variable with the stable distribution function with index α and $\{c_n\}_{n \geq 1}$ is a monotonically increasing sequence of positive constants.

PROOF. At first mention that relation (1.3.2) considered as an equation with respect to f if ϕ is given, or with respect to ϕ with given f, has a unique solution in both cases. Further the distribution function of the random variable X with mean equal to one belongs to the domain of attraction of the stable law with the index $\alpha \in (1, 2]$ if and only if

$$\phi(s) = 1 - s + s^\alpha L \left(\frac{1}{s} \right) (1 + o(1)) \tag{5.4}$$

with $s \to 0$, where $L(x)$ is a function slowly varying at infinity. Similarly, the distribution function of the random variable Z_1 belongs to the same domain of attraction if and only if

$$1 - f(z) = m(1 - z) - (1 - z)^\alpha M \left(\frac{1}{1 - z} \right) (1 + o(1)) \tag{5.5}$$

with $z \to 1$, where $M(x)$ is also a function slowly varying at infinity. We should verify that (5.4) and (5.5) are equivalent.

At first let (5.4) take place. Then using (1.3.2) and setting $\phi(s) = z$, we have as $s \to 0$

$$1-f(z) = ms - m^\alpha s^\alpha L\left(\frac{1}{sm}\right)(1+o(1)) = ms - m^\alpha s^\alpha L\left(\frac{1}{s}\right)(1+o(1)), \quad (5.6)$$

since L is slowly varying. But as $s \to 0$,

$$z = 1 - s + s^\alpha L\left(\frac{1}{s}\right)(1 + o(1)). \quad (5.7)$$

Therefore using (5.7) from (5.6) we get

$$1 - f(z) = m(1 - z) - (m^\alpha - m)s^\alpha L\left(\frac{1}{s}\right)(1+o(1)). \quad (5.8)$$

However (5.7) implies that

$$s = (1 - z)(1 + o(1))$$

with $z \to 1, s \to 0$. Therefore (5.8) takes the form

$$1 - f(z) = m(1 - z) - (m^\alpha - m)(1 - z)^\alpha L\left(\frac{1}{1-z}\right)(1 + o(1)),$$

which corresponds to (5.5).

Now let

$$1 - f(z) = m(1 - z) - (1 - z)^\alpha M\left(\frac{1}{1-z}\right)(1 + A(1 - z)), \quad (5.9)$$

where $A(1 - z) = o(1)$ with $z \to 1$. Write $\phi(s)$ as

$$\phi(s) = 1 - s + a(s), \quad (5.10)$$

where $a(s) = o(s)$ with $s \to 0$. This is possible since X has unit expectation. In this connection (1.3.2) and (5.9) imply that

$$a(ms) = ma(s) + s^\alpha M\left(\frac{1}{s}\right)(1 + B(s)) \quad (5.11)$$

for some function $B(s)$ depending on $a(s)$ and satisfying the condition $B(s) = o(1)$ with $s \to 0$. Now consider a more general functional equation

$$b(ms) = mb(s) + s^\alpha M\left(\frac{1}{s}\right)(1 + B(s)), \quad (5.12)$$

where $B(s)$ is the same as in (5.11). The homogeneous part of equation (5.12) has general solution $sp(\log s)$, where p is an arbitrary function with period $\log m$. Thus (5.12) has no more than one and therefore the only solution which behaves as $o(s)$ as $s \to 0$ (if there were two solutions, then their difference should have had the form $sp(\log s)$ which is not obviously $o(s)$ as $s \to 0$ if only $p(\log s)$ is not equal to zero identically). It is easy to see that the solution of (5.12) should have the form

$$b(s) = (m^\alpha - m)^{-1} s^\alpha M\left(\frac{1}{s}\right)(1 + o(1)) \tag{5.13}$$

with $s \to 0$, and therefore $a(s)$ has the same form. Now the required form (5.4) follows from (5.10) and (5.13). The reduced arguments imply that the slowly varying functions L and M occurring in (5.4) and (5.5) are connected by the relation

$$L\left(\frac{1}{s}\right) = (m^\alpha - m)^{-1} M\left(\frac{1}{s}\right)(1 + o(1)) \tag{5.14}$$

as $s \to 0$. We exclude from the consideration the case when $\alpha = 2$ and $L\left(\frac{1}{s}\right)$ and $M\left(\frac{1}{s}\right)$ are asymptotically constant since then the variances of Z_1 and X are finite and connected by the relation

$$DX = (m^2 - m)^{-1} DZ_1,$$

and then the assertion of the theorem is obvious. Note also that if $\alpha = 2$ and the variances of X and Z_1 are infinite, then $L\left(\frac{1}{s}\right)$ and $M\left(\frac{1}{s}\right)$ should be unlimited with $s \to 0$. Then since

$$E \exp\{-s Z_1\} = f(e^{-s}),$$

(5.5) implies that

$$E \exp\left\{-\frac{s}{c_n}\sum_{j=1}^{n}(U_j - m)\right\} = \left[f\left(\exp\left\{-\frac{s}{c_n}\right\}\right)\exp\left\{\frac{ms}{c_n}\right\}\right]^n \to e^{cs^\alpha}$$

with $n \to \infty$ for $c > 0$ if and only if

$$\lim_{n \to \infty} \frac{M(c_n)}{c_n^\alpha} = c. \tag{5.15}$$

Using (5.4) and (5.14) we make sure that condition (5.15) is also necessary and sufficient for

$$\lim_{n \to \infty} E \exp\left\{-(m^\alpha - m)^{1/\alpha}\frac{s}{c_n}\sum_{j=1}^{n}Y_j\right\} =$$

$$\lim_{n \to \infty}\left[\phi\left(\frac{s}{c_n}(m^\alpha - m)^{1/\alpha}\right)\exp\left\{\frac{s}{c_n}(m^\alpha - m)^{1/\alpha}\right\}\right]^n = e^{cs^\alpha},$$

which totally completes the proof of the theorem.

Now we have all the information necessary to embark upon the main problem of this section, i.e., the investigation of the convergence rate in Theorem 1.3.1.

The classical central limit theorem for sums of independent random variables can be interpreted as a result in some sense giving an estimate of the convergence rate in the law of large numbers.

Let, for example, random variables $\{Y_j\}_{j\geq 1}$ be independent and identically distributed with $EY_1 = a$, $DY_1 = \sigma^2 < \infty$. Let $S_n = Y_1 + \ldots + Y_n$. Then as is well known,

$$\lim_{n\to\infty} P\left(\frac{S_n - na}{\sigma\sqrt{n}} < x\right) = \Phi(x) \qquad (5.16)$$

uniformly with respect to $x \in \mathbb{R}$. The left-hand side of (5.16) can be written as

$$\lim_{n\to\infty} P\left(\frac{S_n - na}{\sigma\sqrt{n}} < x\right) = \lim_{n\to\infty} P\left(\frac{\sqrt{n}}{\sigma}\left(\frac{1}{n}S_n - a\right) < x\right). \qquad (5.17)$$

This representation can serve as a source of information about the convergence rate in the strong law of large numbers according to which the difference $\frac{1}{n}S_n - a$ aims to zero with probability one under the above assumptions. In fact, to say nonstrictly, with regard for that according to (5.16) and (5.17) the random variable $\frac{\sqrt{n}}{\sigma}\left(\frac{1}{n}S_n - a\right)$ has a nondegenerate proper limit distribution, we can make the conclusion that the difference $\frac{1}{n}S_n - a$ is decreasing as $O\left(\frac{1}{\sqrt{n}}\right)$.

A similar approach will be used with respect to Theorem 1.3.1. Theorem 4.5.1 determines the form of the normalizing constants in the investigation of the asymptotics of the distribution function of the random variable $X - X_n$ with $n \to \infty$. Denote $b_n = b_n(\alpha) = c_n(m^\alpha - m)^{1/\alpha}$, where $\{c_n\}$ is the sequence from Theorem 4.5.1. We will be interested in the asymptotic with $n \to \infty$ distribution of the random variable

$$W_n = \frac{m^n}{b_{m^n}}(X - X_n).$$

Let random variables $\{Y_j\}_{j\geq 1}$ be independent and identically distributed, $Y_j \overset{d}{=} X - 1, j \geq 1$. Assume that for every n the random variables Z_n and $\{Y_j\}_{j\geq 1}$ are independent. Introduce the normalized random sum

$$S_{n,Z_n} = \begin{cases} 0, & \text{if } Z_n = 0, \\ \dfrac{1}{b_{m^n}}(Y_1 + \ldots + Y_{Z_n}), & \text{if } Z_n > 0. \end{cases}$$

All the further reasonings will be based on the following rather unexpected statement.

THEOREM 4.5.2. $W_n \overset{d}{=} S_{n,Z_n}, \quad n \geq 1.$

PROOF. Choose $r > n$ and with the help of the law of total probability consider the characteristic function

$$E\exp\left\{it\frac{m^n}{b_{m^n}}(X_r - X_n)\right\} =$$

$$\sum_{j=0}^{\infty} P(Z_n = j) E\left[\exp\left\{it\frac{m^n}{b_{m^n}}\left(\frac{Z_r}{m^r} - \frac{Z_n}{M^n}\right)\right\} \mid Z_n = j\right] =$$

$$\sum_{j=0}^{\infty} P(Z_n = j) E\left[\exp\left\{it\frac{m^n}{b_{m^n}}\left(\frac{Z_r}{m^r} - \frac{j}{M^n}\right)\right\} \mid Z_n = j\right] =$$

$$\sum_{j=0}^{\infty} P(Z_n = j) \exp\left\{-it\frac{j}{b_{m^n}}\right\} E\left[\exp\left\{it\frac{m^{n-r}}{b_{m^n}}Z_r\right\} \mid Z_n = j\right] =$$

$$\sum_{j=0}^{\infty} P(Z_n = j) \exp\left\{-it\frac{j}{b_{m^n}}\right\} \left[E\exp\left\{it\frac{m^{n-r}}{b_{m^n}}Z_{r-n}\right\}\right]^j =$$

$$E\left[\exp\left\{-it\frac{1}{b_{m^n}}\right\} E\exp\left\{it\frac{m^{n-r}}{b_{m^n}}Z_{r-n}\right\}\right]^{Z_n}. \tag{5.18}$$

Fix n and let $r \to \infty$ in (5.18). For any $\varepsilon > 0$ we can choose an N so that

$$P(Z_n > N) = \sum_{N+1}^{\infty} P(Z_n = j) < \varepsilon. \tag{5.19}$$

Further since $X_{r-n} \Rightarrow X$ as $r \to \infty$, the uniform on every finite interval convergence

$$E\exp\left\{it\frac{m^{n-r}}{b_{M^n}}Z_{r-n}\right\} \to E\exp\left\{it\frac{X}{b_{m^n}}\right\} \quad (r \to \infty)$$

takes place. Therefore we can choose an M such that for all $r > M, t \in \mathbb{R}$ the inequalities

$$\left|E\exp\left\{it\frac{m^{n-r}}{b_{M^n}}Z_{r-n}\right\} - E\exp\left\{it\frac{X}{b_{m^n}}\right\}\right| < \varepsilon. \tag{5.20}$$

are fulfilled. Let $K = \max\{M, N\}$. Then

$$\sum_{j=0}^{\infty} P(Z_n = j) \exp\left\{-it\frac{j}{b_{m^n}}\right\} \left[E\exp\left\{it\frac{m^{n-r}}{b_{m^n}}Z_{r-n}\right\}\right]^j =$$

$$\sum_{j=0}^{\infty} P(Z_n = j) \exp\left\{-it\frac{j}{b_{m^n}}\right\} \left[E\exp\left\{it\frac{X}{b_{m^n}}\right\}\right]^j +$$

$$\sum_{j=0}^{K} P(Z_n = j) \exp\left\{-it\frac{j}{b_{m^n}}\right\} \left(\left[E\exp\left\{it\frac{m^{n-r}}{b_{m^n}}Z_{r-n}\right\}\right]^j - \right.$$

$$\left. \left[E\exp\left\{it\frac{x}{b_{m^n}}\right\}\right]^j\right) +$$

$$\sum_{j=K+1}^{\infty} P(Z_n = j) \exp\left\{-it\frac{j}{b_{m^n}}\right\} \left(\left[E\exp\left\{it\frac{m^{n-r}}{b_{m^n}}Z_{r-n}\right\}\right]^j - \right.$$

$$\left. \left[E\exp\left\{it\frac{x}{b_{m^n}}\right\}\right]^j\right).$$

Denote the last two summands in the right-hand side of this representation as J_1 and J_2, respectively. In this connection (5.19) implies that

$$J_2 \leq 2 \sum_{j=K+1}^{\infty} P(Z_n = j) < 2\varepsilon,$$

at the same time, (5.20) implies

$$|J_1| \leq \sum_{j=0}^{K} P(Z_n = j) \left| \left[E \exp \left\{ it \frac{m^{n-r}}{b_{m^n}} Z_{r-n} \right\} \right]^j - \left[E \exp \left\{ it \frac{X}{b_{m^n}} \right\} \right]^j \right| <$$

$$\varepsilon \sum_{j=0}^{K} j P(Z_n = j) \leq \sum_{j=0}^{\infty} j P(Z_n = j) = \varepsilon m^n,$$

if only $r > k$. However, ε is arbitrary and n is fixed. Therefore we have

$$E \exp \left\{ it \frac{m^n}{b_{m^n}} (X - X_n) \right\} = \lim_{r \to \infty} E \exp \left\{ it \frac{m^n}{b_{m^n}} (X_r - X_n) \right\} =$$

$$\sum_{j=0}^{\infty} P(Z_n = j) \exp \left\{ -it \frac{j}{b_{m^n}} \right\} \left[E \exp \left\{ it \frac{X}{b_{m^n}} \right\} \right]^j =$$

$$E \left[E \exp \left\{ it \frac{X-1}{b_{m^n}} \right\} \right]^{Z_n} = E \left[E \exp \left\{ it \frac{Y_1}{b_{m^n}} \right\} \right]^{Z_n} =$$

$$\sum_{j=0}^{\infty} P(Z_n = j) \left[E \exp \left\{ it \frac{Y_1}{b_{m^n}} \right\} \right]^j =$$

$$\sum_{j=0}^{\infty} P(Z_n = j) E \exp \left\{ it \frac{1}{b_{m_n}} (Y_1 + \ldots + Y_j) \right\} = E \exp \left\{ it S_{n, Z_n} \right\},$$

Q.E.D.

Thus the distribution of the random variable W_n coincides with the distribution of the random sum S_{n, Z_n}, which allows us to investigate easily the asymptotics of the distribution of the random variable W_n with the help of the statements proved earlier in this chapter.

THEOREM 4.5.3. *Let the distribution function of the random variable Z_1 belong to the domain of attraction of the stable law G_α with index $\alpha \in (1, 2]$ and the characteristic function $g_\alpha(t)$. Then for any $t \in \mathbb{R}$,*

$$\lim_{n \to \infty} E \exp \{ it W_n \} = E[g_\alpha(t)]^X = \int_0^{\infty} [g_\alpha(t)]^y \, dP(X < y). \qquad (5.21)$$

PROOF. Theorem 4.5.1 implies that

$$P \left(\frac{m^n}{\sum_{j=1}^{\infty}} X_{n,j} < x \right) \Rightarrow G_\alpha(x) \qquad (n \to \infty), \qquad (5.22)$$

where $X_{n,j} = Y_j/b_{m^n}$. Theorem 1.3.1 implies that

$$\frac{Z_n}{m^n} \Rightarrow X \qquad (n \to \infty). \tag{5.23}$$

Set $N_n = Z_n$, $k_n = m^n$ in Theorem 4.1.2. Then it should follow from (5.22) and (5.23) that

$$\sum_{j=1}^{N_n} X_{n,j} \Rightarrow Z \qquad (n \to \infty),$$

where Z is the random variable with the characteristic function from the right-hand side of (5.21). According to Theorem 4.5.2 in this case we have

$$\sum_{j=1}^{N_n} X_{n,j} \overset{d}{=} W_n,$$

The proof is completed.

Thus having repeated word for word the reasoning preceding Theorem 4.5.2 note that speaking nonstrictly, the random variable $X - X_n$ decreases with $n \to \infty$ as $O(b_{m^n}/m^n) = O(c_{m^n}/[m^n(m^\alpha - m)^{1/\alpha}]) = O(c_{m^n}/m^n)$, if the distribution of Z_1 belongs to the domain of attraction of the stable law with index $\alpha \in (1, 2]$. In particular, if $c_n = n^{1/\alpha}$ which characterizes the so-called domain of normal attraction of the stable law with index α, then $X - X_n$ decreases as $O(m^{n(1/\alpha - 1)})$.

Generally speaking, it is hard to use Theorem 4.5.3 for a more detailed conclusion since the limit law in (5.21) depends on the distribution of the limit variable X itself. However, it can be shown that if we use the normalization by random variables instead of that by constants, then the limit law for the normalized variable $X - X_n$ will be simplified. Namely, under the conditions of Theorem 4.5.3 for any $t \in \mathbb{R}$

$$\lim_{n \to \infty} \mathsf{E}\left[\exp\left\{it\frac{m^n}{b_{Z_n}}(X - X_n)\right\} \mid Z_n > 0\right] = g_\alpha(t),$$

see (Heyde, 1971), whence we can conclude that if Z_1 has the distribution function from the domain of normal attraction of the stable law with index α, then the difference $X - X_n$ behaves as $X_n^{1/\alpha} m^{n(1/\alpha - 1)}$.

4.6 Randomly infinitely divisible distributions

In Chapter 2 investigating random sums of independent identically distributed random variables, in which the number of summands has geometric distribution (geometric random sums), we have paid attention to the fact that there is an exceptionally close analogy between the results for geometric random sums and the results of the classical theory. So there naturally appears a suspicion whether this analogy is universal or not, i.e., can we introduce in the theory of random summation the analogies of such notions of the classical theory as infinitely divisible laws, stable laws, etc. for arbitrary summing indices.

Unfortunately, this suspicion appears to be justified only partially. The randomness of indices brings into the formulations of problems principally new elements as compared to the classical theory. For example, as we have seen in the previous chapters, in contrast to the classical situation, the centering of random sums by constants can lead to the appearance of absolutely arbitrary limit laws, so the information contained in the summands is ignored.

But in a situation which seems more natural from the point of view of the search for the analogies mentioned above, where the random sums of the centered summands are considered, as we will see later, these analogies exist, alas, not always. So the more interesting becomes the problem of describing the classes of situations admitting the construction for random summation of the notion "parallel" to the classical theory.

We begin from an attempt to describe the analogies of the normal law in random summation. Theorem 4.1.2 gives us the reason to try to treat as normal those laws whose characteristic functions have the form

$$f(t) = \int\limits_0^\infty e^{-t^2 y} \, dA(y), \qquad (6.1)$$

where A is the distribution function whose all points of growth are concentrated on the nonnegative semiaxis. In the classical situation a normal law is characterized by the following property of strict stability: if $X_1, X_2, \ldots$ are independent random variables and $P(X_i < x) = \Phi(x)$, $x \in \mathbb{R}$, $i \geq 1$, then for any natural n,

$$X_1 \stackrel{d}{=} n^{-1/2} \sum_{j=1}^n X_j. \qquad (6.2)$$

We shall examine which of random variables with characteristic functions of the form (6.1) can be considered as analogies of Gaussian. At first we should formulate a "random" analog of (6.2). Property (6.2) can be interpreted as the reproducibility of the normal law with respect to the number of summands. It suggests that a "random" analog of (6.2) should describe the reproducibility with respect to a random number of summands from some family. Thus we come to the following definition.

Let $\mathcal{N} = \{N_\theta, \theta \in \Theta\}$ be a family of random variables taking natural values. For convenience further assume that $\Theta \subseteq (0,1)$ and for every $\theta \in \Theta$ the mathematical expectation EN_θ exists; moreover, the parametrization of the family $\mathcal{N}$ is established by expectations in the sense that $EN_\theta = \frac{1}{\theta}$.

DEFINITION 4.6.1. A distribution function F is called Gaussian with respect to the family $\mathcal{N}$ ($\mathcal{N}$-Gaussian) if

$$\int\limits_{-\infty}^\infty x \, dF(x) = 0, \qquad \int\limits_{-\infty}^\infty x^2 \, dF(x) < \infty \qquad (6.3)$$

and for all $\theta \in \Theta$

$$F(x) = \sum_{n=1}^{\infty} P(N_\theta = n) F^{*n}(\theta^{-1/2} x), \qquad x \in \mathbb{R}, \tag{6.4}$$

where as before F^{*n} denotes the n-fold convolution of the distribution function F with itself.

The property of a distribution function to be $\mathcal{N}$-Gaussian can be also defined in terms of random variables, namely, a random variable X is called $\mathcal{N}$-Gaussian if $\mathsf{E} X = 0$, $\mathsf{E} X^2 < \infty$ and for all $\theta \in \Theta$

$$X \overset{d}{=} \theta^{1/2} \sum_{j=1}^{N_\theta} X_j,$$

where $X_1, X_2, \ldots$ are independent random variables and $X \overset{d}{=} X_j$, $j \geq 1$; moreover, the random variables N_θ and $\{X_j\}_{j \geq 1}$ are independent for every $\theta \in \Theta$. However, we will hold to Definition 4.6.1 to pass over the necessity to define the random variables $\{N_\theta, \theta \in \Theta\}$ and $\{X_j, \geq 1\}$ on the same probability space inevitably appearing in this situation.

But not any family of indices $\mathcal{N}$ with the stated properties admits the existence of $\mathcal{N}$-Gaussian distribution function. Consider the problem of description of the families of indices $\mathcal{N}$ which admit $\mathcal{N}$-Gaussian laws. Recall that if $P_1(z)$ and $P_2(z)$ are generating functions of two random variables, then their superposition defined as usual as $P_1 \circ P_2(z) = P_1(P_2(z))$ also is the generating function of some random variable.

Denote the generating function of the random variable N_θ as P_θ, and the semigroup with operation of superposition $\circ$ generated by the family $\{P_\theta, \theta \in \Theta\}$ as $\mathcal{P}$.

THEOREM 4.6.1. *A family $\mathcal{N}$ of natural-valued random variables with the properties stated above admits $\mathcal{N}$-Gaussian law if and only if the semigroup $\mathcal{P}$ is commutative.*

PROOF. Let $f(t)$ be the characteristic function corresponding to an $\mathcal{N}$-Gaussian distribution function F. Then (6.3) is equivalent to

$$f(t) = P_\theta(f(\theta^{1/2} t)) \tag{6.5}$$

for all $t \in \mathbb{R}$ and $\theta \in \Theta$. Consider (6.5) as the equation with respect to f with $t \geq 0$. Set $\phi(t) = f(t^{1/2})$. Since relations (6.3) take place, $f(t)$ is twice continuously differentiable; moreover $f'(0) = 0$, $f''(0) \neq 0$. Therefore $\phi(t)$ is differentiable for $t \geq 0$ and $\phi'(+0) = 0$. If $f(t)$ satisfies (6.5), then $\phi(t)$ satisfies the system

$$\phi(t) = P_\theta(\phi(\theta t)), \qquad \theta \in \Theta \tag{6.6}$$

for $t \geq 0$. Conversely, if $\phi(t)$ satisfies system (6.6), then $f(t) = \phi(t^2)$ satisfies system (6.5). Hence it follows that if $f(t)$ exists, then it is symmetrical: $f(t) = f(-t)$.

Choose an arbitrary $\theta_0 \in \Theta$. For simplicity denote the generating function of the random variable N_{θ_0} as $P(z)$. Relation (6.6) implies that

$$\phi(t) = P(\phi(\theta_0 t)).$$
(6.7)

The functional equation (6.7) is well known. It is usually called the Poincaré equation, after H. Poincaré who investigated the question of existence and uniqueness of analytical solutions of (6.7) (Poincaré, 1890), though similar equations were investigated earlier (Abel, 1881), (Schroeder, 1871), (Koenigs, 1884). We have met this equation in Sections 3.6 and 4.5. It plays an important role in the investigation of the limit behavior of supercritical Galton–Watson processes. In particular, it is known that equation (6.7) has a unique solution with the initial conditions $\phi(0) = 1$, $\phi'(0) = -\alpha$, where $\alpha \geq 0$ is an arbitrary constant (see (Harris, 1963)). This solution is the Laplace-Stieltjes transform of a distribution function $A(x)$ whose all points of growth are concentrated on the nonnegative semiaxis. Thus

$$\phi(t) = \int\limits_0^\infty e^{-tx}\, dA(x),$$
(6.8)

and $\phi(t)$ is determined up to the scale parameter. In this connection if $\phi'(0) = -\alpha \neq 0$, then $A(x)$ is not degenerate at zero.

It is clear that if there exists a solution of the overdetermined system (6.6), then it should satisfy equation (6.7), i.e., coincide with (6.8). Thus system (6.6) has solution if and only if the solution of equation (6.7) does not depend on the choice of $\theta_0 \in \Theta$. In other words, for any fixed $\theta \in \Theta, \theta \neq \theta_0$, equations

$$\phi_\theta(t) = P_\theta(\phi_\theta(\theta t))$$
(6.9)

and (6.7) should have coinciding solutions under initial conditions $\phi(0) = \phi_\theta(0) = 1$, $\phi'(0) = \phi'_\theta(0) = -\alpha$, $\alpha \geq 0$.

Make sure that equations (6.7) and (6.9) have coinciding solutions if and only if the functions P_θ and P commute, i.e.,

$$P_\theta \circ P = P \circ P_\theta.$$
(6.10)

At first assume that (6.10) is fulfilled. Let $\phi(t)$ be the solution of (6.7) with the stated initial conditions. Using (6.7) and (6.10) we can write

$$P_\theta(\phi(t\theta)) = P_\theta(P(\phi(t\theta_0\theta))) = P(P_\theta(\phi(t\theta_0\theta))),$$

i.e., the function $P_\theta(\phi(t\theta))$ satisfies equation (6.7). Moreover, since $\mathbf{E}N_\theta = \frac{1}{\theta}$, we have $P'_\theta(1) = \frac{1}{\theta}$ and hence

$$P_\theta(\phi(t\theta)) = P_\theta(\phi(0)) = P_\theta(1),$$

$$\frac{d}{dt} P_\theta(\phi(t\theta)) = P'_\theta(\phi(t\theta)) \cdot \phi'(t\theta) \cdot \theta|_{t=0} = \frac{1}{\theta}\phi'(0) \cdot \theta = -\alpha.$$

Thus the functions $\phi(t)$ and $P_\theta(\phi(t\theta))$ satisfy equation (6.7) under the same initial conditions. By virtue of the uniqueness of the solution of equation (6.7) satisfying the stated initial conditions we have

$$\phi(t) = P_\theta(\phi(t\theta)), \tag{6.11}$$

i.e., $\phi(t)$ satisfies (6.9). The roles of θ and θ_0 are symmetrical. Therefore we conclude that equations (6.7) and (6.9) have the same solution under condition (6.10).

Now assume that equations (6.7) and (6.9) have one and the same solution and prove (6.10). We have

$$P_\theta(P(\phi(t\theta\theta_0))) = P_\theta(\phi(t\theta)) = \phi(t). \tag{6.12}$$

Here we have used the fact that $\phi(t) = P(\phi(t\theta_0))$, and that $\phi(t) = P_\theta(\phi(t\theta))$. By switching the role of the reasonings we get

$$P(P_\theta(\phi(t\theta\theta_0))) = P(\phi(t\theta_0)) = \phi(t). \tag{6.13}$$

Comparing (6.12) and (6.13) we conclude that

$$P(P_\theta(\phi(t\theta\theta_0))) = P_\theta(P(\phi(t\theta\theta_0))). \tag{6.14}$$

If $\phi'(0) = -\alpha \neq 0$, then as we have already noticed, $\phi(t)$ is the Laplace-Stieltjes transform of a distribution function $A(x)$, which is not degenerate at zero, i.e., the values of the function $\phi(t)$ fill the whole interval $(0, 1]$. In this connection, (6.14) implies that $p(P_\theta(z)) = P_\theta(P(z))$ with $z \in (0, 1]$, and since any generating function is analytical in the unit circle, hence follows equality (6.10).

Return to system (6.5). Relation (6.8) implies that the characteristic function of an $\mathcal{N}$-Gaussian law should have the form (6.1). Moreover system (6.5) is compatible if and only if system (6.6) is compatible, i.e., if and only if for any $\theta, \theta_0 \in \Theta$ (6.10) holds, which is obviously equivalent to the commutativity of the semigroup $\mathcal{P}$. The proof is completed.

REMARK 4.6.1. As we can see from the proof of Theorem 4.6.1, if the semigroup $\mathcal{P}$ is commutative, then the characteristic function of any $\mathcal{N}$-Gaussian law has the form

$$f(t) = \phi(\alpha t^2), \tag{6.15}$$

where $\alpha > 0$ and $\phi(t)$ is the solution of equation (6.7) satisfying the conditions $\phi(0) = -\phi'(0) = 1$.

COROLLARY 4.6.1. *The semigroup $\mathcal{P}$ is commutative if and only if the representation*

$$P_\theta(z) = \phi\left(\frac{1}{\theta}\phi^{-1}(z)\right), \qquad \theta \in \Theta, \tag{6.16}$$

takes place with $z > 0$, where ϕ is the differentiable solution of (6.7), satisfying the conditions $\phi(0) = -\phi'(0) = 1$.

PROOF. The proof of Theorem 4.6.1 implies that the commutativity of the semigroup $\mathcal{P}$ is equivalent to the fact that system (6.6) with the initial conditions $\phi(0) = 1$, $\phi'(0) = -\alpha$ (for definiteness, set $\alpha = 1$) has the solution independent of θ_0, i.e., the commutativity of the semigroup $\mathcal{P}$ is equivalent to the equality

$$\phi(t) = P_\theta(\phi(t\theta))$$

for all θ with ϕ independent of θ. Since ϕ is invertible with $t > 0$ and nonnegative, the latter equation can be written in the form

$$P_\theta(z) = \phi\left(\frac{1}{\theta}\phi'(z)\right), \qquad z > 0.$$

The corollary is proved.

Consider some examples of the families $\mathcal{N}$ of random variables admitting $\mathcal{N}$-Gaussian laws. We should note that an important family of Poisson random variables N_θ with $\mathsf{E}N_\theta = \frac{1}{\theta}$ has no $\mathcal{N}$-Gaussian law, since as is easily seen, in this case the semigroup $\mathcal{P}$ is not commutative.

EXAMPLE 4.6.1 (The classical summation scheme). Let $\mathsf{P}\left(N_\theta = \frac{1}{\theta}\right) = 1$, and $\theta \in \Theta = \{\frac{1}{n}, \, n = 1, 2, \ldots\}$. Then $P_\theta(z) = z^{1/\theta}$. It is obvious that $P_{\theta_1} \circ P_{\theta_2}(z) = z^{1/(\theta_1\theta_2)} = P_{\theta_2} \circ P_{\theta_1}(z)$. According to Theorem 4.6.1 there exists an $\mathcal{N}$-Gaussian distribution which, of course, coincides with the usual normal distribution.

EXAMPLE 4.6.2 (Geometric summation). Let $\mathsf{P}(N_\theta = k) = \theta(1 - \theta)^{k-1}$, $k = 1, 2, \ldots$. Then $P_\theta(z) = \theta z/(1 - (1 - \theta)z)$. It is not difficult to verify that

$$P_{\theta_1} \circ P_{\theta_2}(z) = \frac{\theta_1\theta_2 z}{1 - (1 - \theta_1\theta_2)z} = P_{\theta_2} \circ P_{\theta_1}(z).$$

Therefore an $\mathcal{N}$-Gaussian distribution exists. Equation (6.7) for $\theta_0 \in (0, 1)$ has the form

$$\phi(t) = \frac{\theta_0\phi(\theta_0 t)}{1 - (1 - \theta_0)\phi(\theta_0 t)}. \tag{6.17}$$

We have met similar equations in Chapter 2. The solutions of this equation of the form (6.8) are Laplace-Stieltjes transforms of the exponential laws, i.e., functions of the form $\phi_\alpha(t) = (1 - \alpha t)^{-1}$, $\alpha > 0$. In particular, $\phi(t) = (1 + t)^{-1}$ is a unique solution of equation (6.17) satisfying the conditions $\phi(0) = -\phi'(0) = 1$. Therefore in this case Laplace distributions (double exponential distributions) with the characteristic functions

$$f(t) = \frac{1}{1 + \alpha t^2}, \qquad \alpha > 0. \tag{6.18}$$

are $\mathcal{N}$-Gaussian (geometric Gaussian) laws.

EXAMPLE 4.6.3. Let N be a random variable taking natural values and having mathematical expectation $\mathsf{E}N > 1$. Denote $\theta_0 = 1/\mathsf{E}N$ and let $P(z)$ be the generating function of the random variable N. As $\mathcal{N}$, take the family of random variables whose generating functions are functional iterations (superpositions) of $P(z)$. To introduce in $\mathcal{N}$ the desired parametrization, set

$$P_{\theta_0}(z) = P(z),$$

$$P_{\theta_0^2}(z) = P(P(z)),$$

$$\cdots$$

$$P_{\theta_0^n}(z) = P^{on}(z),$$

where $P^{on}(z)$ denotes the function obtained by n-fold application of the superposition operation o to the same function P, i.e., an n-fold functional iteration of P. Set $\Theta = \{\theta_0^n, n = 1, 2, \ldots\}$. It is clear that in this case the semigroup $\mathcal{P}$ is commutative as the semigroup of powers (in the sense of superposition) of the function $P(z)$. System (6.6) has the form

$$\phi(t) = P(\phi(\theta_0 t)) = P^{o2}(\phi(\theta_0^2 t)) = \ldots = P^{on}(\phi(\theta_0^n t)) = \ldots$$

It coincides with equation (6.7):

$$\phi(t) = P(\phi(\theta_0 t)).$$

We have met this family $\mathcal{N}$ admitting $\mathcal{N}$-Gaussian law in Section 3.6 when we investigated the special model of rarefaction of the renewal processes. The structure of the family $\mathcal{N}$ defined in such a way is obviously closely connected with supercritical Galton–Watson processes.

Theorem 4.6.1 means that the analogies of infinitely divisible, stable laws and other notions of the classical summation theory for random summation can be built only with respect to such families of indices that generate commutative semigroups of generating functions with the operation of superposition. In this connection further we will consider only these families of indices. We will call the solution of equation (6.7) that has form (6.8) and is differentiable on $[0, \infty)$ and satisfies the condition $\phi(0) = -\phi'(0) = 1$, the standard solution.

Describing the conditions of the existence of $\mathcal{N}$-Gaussian laws we have followed the line of presentation of similar results in (Umarov, 1992), where the investigations which started in the works (Klebanov, Manija and Melamed, 1984), (Klebanov, Manija and Melamed, 1985), (Klebanov and Melamed, 1986) are summarized (see also (Kruglov and Korolev, 1990)). The starting point of these investigations was one analytical problem of probability theory proposed by V. M. Zolotarev. This problem is to describe all random variables Y possessing the following property: for any $\theta \in (0, 1)$ there exists a random variable X_θ such that

$$Y \overset{d}{=} X_\theta + Z_\theta Y, \tag{6.19}$$

where the variables X_θ, Z_θ and Y in the right-hand side are independent, $P(Z_\theta = 0) = \theta$, $P(Z_\theta = 1) = 1 - \theta$. If we denote the characteristic functions of the variables Y and Z_θ as f and g_θ, respectively, then relation (6.19) becomes equivalent to the equality

$$f(t) = g_\theta(t)(\theta + (1 - \theta)f(t)), \qquad t \in \mathbb{R}.$$

Expressing $f(t)$ from here and using the formula for the sum of the terms of geometric progression we can write the last relation as

$$f(t) = \sum_{n=1}^{\infty} g_\theta^n(t)\theta(1-\theta)^{n-1} \qquad t \in \mathbb{R}. \tag{6.20}$$

But there is the characteristic function of the random sum $X_1^{(\theta)} + \ldots + X_{N_\theta}^{(\theta)}$ in the right-hand side of (6.20), where $X_j^{(\theta)} \overset{d}{=} X_\theta$, $P(N_\theta - k) = \theta(1-\theta)^{k-1}$, $k = 1, 2, \ldots$ and for every $\theta \in (0,1)$ the random variables $N_\theta, X_1^{(\theta)}, X_2^{(\theta)}, \ldots$ are independent. Thus the original problem of V. M. Zolotarev appeared to be equivalent to the following one: describe all random variables Y for which with every $\theta \in (0,1)$ there exists a sequence of identically distributed random variables $\{X_j^{(\theta)}\}_{j\geq 1}$ such that

$$Y \overset{d}{=} \sum_{j=1}^{N_\theta} X_j^{(\theta)}$$

where N_θ has the above geometric distribution and the variables N_θ, $X_1^{(\theta)}$, $X_2^{(\theta)}, \ldots$ are independent for every θ. In the mentioned works these random variables are called geometrically infinitely divisible. These papers contain a complete description of the properties of geometrically infinitely divisible laws. Also, a more general notion of randomly infinitely divisible distributions was introduced there. Now we turn to the description and investigation of the properties of randomly infinitely divisible laws.

Let $\mathcal{N} = \{N_\theta, \theta \in \Theta\}$ be a family of natural-valued random variables generating a commutative semigroup of generating functions with the operation of superposition.

DEFINITION 4.6.2. A distribution function F is called randomly infinitely divisible by the family $\mathcal{N}$ ($\mathcal{N}$- infinitely divisible) if for any $\theta \in \Theta$ there exists a distribution function F_θ such that

$$F(x) = \sum_{n=1}^{\infty} P(N_\theta = n)F_\theta^{*n}(x) \qquad x \in \mathbb{R}. \tag{6.21}$$

In terms of characteristic functions condition (6.21) is written as

$$f(t) = P_\theta(f_\theta(t)), \qquad t \in \mathbb{R}, \quad \theta \in \Theta, \tag{6.22}$$

where f and f_θ are the characteristic functions corresponding to the distribution functions F and F_θ, P_θ is the generating function of the random variable N_θ.

REMARK 4.6.2. Since the semigroup of generating functions generated by the family $\mathcal{N}$ is commutative, according to Corollary 4.6.1 representation (6.16) takes place for $z \geq 0$. But any generating function is analytical in the unit circle; therefore representation (6.16) takes place also for complex z with

$|z| \leq 1$ if we treat the function $\phi\left(\frac{1}{\theta}\phi^{-1}(z)\right)$ as the analytical continuation of the function $\phi\left(\frac{1}{\theta}\phi^{-1}(z)\right)$, $t \geq 0$. Therefore (6.16) and (6.20) imply that the characteristic function of any $\mathcal{N}$-infinitely divisible law is representable in the form

$$f(t) = \phi\left(\frac{1}{\theta}\phi^{-1}(f_\theta(t))\right), \qquad \theta \in \Theta, \quad t \in \mathbb{R},$$

where ϕ is the standard solution of equation (6.7). Further by $\phi(z)$ and $\phi^{-1}(z)$ for complex z we will mean the analytical continuations of the corresponding functions initially defined for real arguments.

Mention some elementary properties of $\mathcal{N}$-infinitely divisible laws.

PROPERTY 4.6.1. *Characteristic function of any $\mathcal{N}$-infinitely divisible law does not turn into zero on the real axis.*

PROOF. Taking account of Corollary 4.6.1 for every $\theta \in \Theta$, we have the representation

$$f(t) = P_\theta(f_\theta(t)) = \phi\left(\frac{1}{\theta}\phi^{-1}(f_\theta(t))\right), \qquad (6.23)$$

where f and f_θ are the characteristic functions corresponding to the distribution functions F and F_θ in (6.21), ϕ is the standard solution of equation (6.7). The characteristic function f, being continuous, differs from zero in some neighborhood of zero $(-\delta, \delta)$, $\delta > 0$. By virtue of equality (6.23), f_θ differs from zero in the same neighborhood. Moreover, we have $f_\theta(t) \to 1$ with $\theta \to 0$ for $t \in (-\delta, \delta)$. In consequence of the well-known inequality

$$1 - |f_\theta(t)|^2 \leq 4\left(1 - \left|f_\theta\left(\frac{t}{2}\right)\right|^2\right), \qquad (6.24)$$

which is correct for any characteristic function f_θ and any $t \in \mathbb{R}$, the convergence $f_\theta(t) \to 1$ with $\theta \to 0$ takes place on the interval $(-2\delta, 2\delta)$. Repeated application of (6.24) proves the convergence $f_\theta(t) \to 1$ with $\theta \to 0$ on the whole axis and by virtue of (6.23) $f(t)$ differs from zero for any $t \in \mathbb{R}$. The property is proved.

PROPERTY 4.6.2. *A characteristic function which is the limit of the sequence of characteristic functions of $\mathcal{N}$-infinitely divisible laws corresponds to an $\mathcal{N}$-infinitely divisible law.*

PROOF. Let $\{f_n(t)\}$ be a sequence of $\mathcal{N}$-infinitely divisible characteristic functions such that $f_n(t) \to f(t)$ with $n \to \infty$, $t \in \mathbb{R}$, where f is some characteristic function. Then for every $\theta \in \Theta$ by virtue of the continuity of generating functions we have

$$f(t) = \lim_{n \to \infty} f_n(t) = \lim_{n \to \infty} P_\theta(g_\theta^{(n)}(t)) = P_\theta(\lim_{n \to \infty} g_\theta^{(n)}(t)),$$

where $g_\theta^{(n)}(t)$ are some characteristic functions. But $f(t)$ is continuous in zero and P_θ is continuous. Therefore the function $g_\theta(t) = \lim_{n \to \infty} g_\theta^{(n)}$ is continuous in zero and is characteristic being the limit of characteristic functions. The property is proved.

Now we present an analog of the classical de Finetti theorem for $\mathcal{N}$-infinitely divisible laws.

THEOREM 4.6.2. *Let ϕ be the standard solution of equation (6.7). A distribution function F with characteristic function f is $\mathcal{N}$-infinitely divisible if and only if*

$$f(t) = \lim_{n\to\infty} \phi(a_n(1 - g_n(t))), \qquad t \in \mathbb{R} \qquad (6.25)$$

where $a_n > 0$ are some constants and $g_n(t)$ are some characteristic functions.

PROOF. "If" part. At first note that if g is a characteristic function, then the function

$$\psi_\alpha(t) = \phi(\alpha(1 - g(t)), \qquad t \in \mathbb{R} \qquad (6.26)$$

is characteristic for every $\alpha > 0$. Recall that ϕ being the standard solution of equation (6.7), has the form (6.8), i.e. for some non-negative random variable U we have

$$\phi(z) = \mathbf{E}e^{-zU}. \qquad (6.27)$$

Formally substituting $z = \alpha(1 - g(t))$ in (6.27) we get

$$\psi_\alpha(t) = \mathbf{E}(\exp\{\alpha(g(t) - 1)\})^U. \qquad (6.28)$$

The function $\exp\{\alpha(g(t) - 1)\}$ is obviously characteristic. It corresponds to the generalized Poisson law (as we have seen in Chapter 1) and therefore is infinitely divisible. Now we note that $\psi_\alpha(t)$ is the characteristic function of some limit law in transfer Theorem 4.1.2. Moreover, the function $\psi_\alpha(t)$ is $\mathcal{N}$-infinitely divisible. In fact it can be represented as

$$\psi_\alpha(t) = \phi(\alpha(1 - g(t))) =$$
$$\phi\left(\frac{1}{\theta}\phi^{-1}(\phi(\alpha\theta(1 - g(t))))\right) = \phi\left(\frac{1}{\theta}\phi^{-1}(\psi_{\alpha\theta}(t))\right) \qquad (6.29)$$

for every $\theta \in \Theta$. But as we have verified earlier, $\psi_{\alpha\theta}$ is a characteristic function and therefore according to Remark 4.6.2 $\psi_\alpha(t)$ is an $\mathcal{N}$-infinitely divisible characteristic function. Thus every characteristic function under the sign of limit in (6.25) is $\mathcal{N}$-infinitely divisible, and according to Property 4.6.2 the limit characteristic function $f(t)$ is also $\mathcal{N}$-infinitely divisible.

"Only if" part. Let f be an $\mathcal{N}$-infinitely divisible characteristic function. Then according to Remark 4.6.2 for every $\theta \in \Theta$ there exists a characteristic function $f_\theta(t)$ such that

$$f(t) = \phi\left(\frac{1}{\theta}\phi^{-1}(f_\theta(t))\right), \qquad t \in \mathbb{R},$$

whence

$$f_\theta(t) = \phi(\theta\phi^{-1}(f(t))), \qquad t \in \mathbb{R},$$

i.e., $\phi(\theta\phi^{-1}(f(t)))$ is a characteristic function. Setting $\alpha = \frac{1}{\theta}$, $g(t) = \phi(\theta\phi^{-1}(f(t)))$ in the arguments used for the proof of the "if" part, we notice that the function

$$\phi\left(\frac{1}{\theta}(1 - g(t))\right) = \phi\left(\frac{1}{\theta}(1 - \phi(\theta\phi^{-1}(f(t))))\right)$$

is an $\mathcal{N}$-infinitely divisible characteristic function. However, as is easily seen with regard for the properties of the function ϕ,

$$f(t) = \lim_{\theta \to 0} \phi\left(\frac{1}{\theta}(1 - \phi(\theta\phi^{-1}(f(t))))\right).$$

The proof is completed.

The following statement is a simple corollary of Theorem 4.6.2, but its great importance for further reasoning gives us the reason to formulate it as a theorem.

THEOREM 4.6.3. *Let ϕ be the standard solution of equation (6.7). A distribution function F is $\mathcal{N}$-infinitely divisible if and only if the corresponding characteristic function admits the representation*

$$f(t) = \phi(-\log g(t)), \qquad t \in \mathbb{R}, \tag{6.30}$$

where $g(t)$ is some infinitely divisible (in the usual sense) characteristic function.

PROOF. "Only if" part. As we have seen proving the previous theorem, $f(t)$ can be represented in the form

$$f(t) = \lim_{\theta \to 0} \phi\left(\frac{1}{\theta}(1 - f_\theta(t))\right).$$

Continuing this relation with regard for the continuity of ϕ and denoting $g(t) = \lim_{\theta \to 0} \exp\left\{\frac{1}{\theta}(f_\theta(t) - 1)\right\}$, we get

$$\lim_{\theta \to 0} \phi\left(-\log \exp\left\{\frac{1}{\theta}(f_\theta(t) - 1)\right\}\right) =$$

$$\phi\left(-\log \lim_{\theta \to 0} \exp\left\{\frac{1}{\theta}(f_\theta(t) - 1)\right\}\right) =$$

$$\phi(-\log g(t)).$$

For every $\theta \in \Theta$ the function $\exp\left\{\frac{1}{\theta}(f_\theta(t) - 1)\right\}$ is infinitely divisible in the usual sense (as the characteristic function of a generalized Poisson law). Therefore the limit characteristic function g is also infinitely divisible.

"If" part. It is necessary to verify that characteristic function (6.30) can be represented in the form (6.23). It is not difficult to see that

$$\phi(-\log g(t)) = \phi\left(\frac{1}{\theta}\phi^{-1}(\phi(-\theta\log g(t)))\right).$$

Therefore it suffices to show that the function

$$f_\theta(t) = \phi(-\theta\log g(t))$$

is characteristic. Since $g(t)$ is an infinitely divisible characteristic function, the function $g(t)$ will be characteristic for any θ. But then

$$\phi(-\theta \log g(t)) = \int_0^\infty g^{\theta s}(t)\, dA(s) \tag{6.31}$$

for some distribution function A with nonnegative growth points and we deal with a characteristic function of the limit law in the transfer theorem (Theorem 4.1.2). The proof is completed.

Theorem 4.6.3 implies one more property of $\mathcal{N}$-infinitely divisible characteristic functions connecting $\mathcal{N}$-infinite divisibility and usual infinite divisibility.

PROPERTY 4.6.3. *If the standard solution ϕ of equation (6.7) is the Laplace-Stieltjes transform of an infinitely divisible law, then an $\mathcal{N}$-infinitely divisible distribution function is infinitely divisible in the usual sense. Moreover, in this case the module of $\mathcal{N}$-infinitely divisible characteristic function and any of its positive power are infinitely divisible characteristic functions.*

This property follows from representation (6.30) with regard for (6.31) and well-known results about infinite divisibility of power mixtures of infinitely divisible functions (Lukacs, 1970), (Feller, 1971).

As is well known, infinitely divisible laws and only them are the limit laws for sums of independent uniformly asymptotically small random summands in the classical summation theory. Within random summation a similar result holds true for $\mathcal{N}$-infinitely divisible laws if the family $\mathcal{N}$ generates a commutative semigroup of generating functions.

THEOREM 4.6.4. *$\mathcal{N}$-infinitely divisible laws and only them can be the limit laws for random sums of identically distributed random variables with $\theta \to 0$.*

PROOF. The required statement will immediately follow from Theorems 4.1.2 and 4.6.3 if we verify that the condition $\theta \to 0$ guarantees that

$$\theta N_\theta \Rightarrow N, \tag{6.32}$$

where the family $\{N_\theta, \theta \in \Theta\}$ generates a commutative semigroup of generating functions with the operation of superposition and N is a nonnegative random variable whose Laplace-Stieltjes transform coincides with the standard solution ϕ of equation (6.7). It is obvious that the Laplace-Stieltjes transform of the random variable θN_θ is equal to $P_\theta(e^{-s\theta})$, $s \geq 0$. Therefore taking account of Remark 4.6.2 we have

$$\lim_{\theta \to 0} P_\theta(e^{-s\theta}) = \lim_{\theta \to 0} \phi\left(\frac{1}{\theta}\phi^{-1}(e^{-s\theta})\right) = \phi\left(\lim_{\theta \to 0}\frac{1}{\theta}\phi^{-1}(e^{-s\theta})\right) =$$

$$\phi\left(\lim_{\theta \to 0}\frac{\partial}{\partial\theta}\phi^{-1}(e^{-s\theta})\right) = \phi(s), \qquad s \geq 0,$$

since $\phi(0) = -\phi'(0) = 1$. The proof is completed.

The following theorem establishes necessary and sufficient conditions of convergence to $\mathcal{N}$-infinitely divisible laws of random sums in which indices generate a commutative semigroup of generating functions. It connects the components of the analytical representation of characteristic functions of $\mathcal{N}$-infinitely divisible laws with the limit laws for sums with random and non-random indices.

As usual in this section, we assume that the family of natural-valued random variables $\mathcal{N} = \{N_\theta, \theta \in \Theta\}$ generates a commutative semigroup of generating functions. Let random variables $X_{\theta,1}, X_{\theta,2}, \ldots$ be identically distributed for every $\theta \in \Theta$; moreover let $N_\theta, X_{\theta,1}, X_{\theta,2}, \ldots$ be independent. Denote the integer part of the number $1/\theta$ as $m(\theta)$.

THEOREM 4.6.5. *Let $F(x)$ be an $\mathcal{N}$-infinitely divisible distribution function. Then*

$$P\left(\sum_{j=1}^{N_\theta} X_{\theta,j} < x\right) \Rightarrow F(x) \qquad (\theta \to 0), \tag{6.33}$$

if and only if

$$P\left(\sum_{j=1}^{m(\theta)} X_{\theta,j} < x\right) \Rightarrow G(x) \qquad (\theta \to 0), \tag{6.34}$$

where $G(x)$ is the infinitely divisible distribution function corresponding to the characteristic function $g(t)$ in representation (6.30).

PROOF. The sufficiency of condition (6.34) follows from Theorem 4.1.2 with regard for the obvious relation $\theta m(\theta) \to 1$ with $\theta \to 0$ and the condition $\theta N_\theta \Rightarrow N$ proved in Theorem 4.6.4.

Show the necessity of condition (6.34). Let $f_\theta(t)$ be the characteristic function of the random variable $X_{\theta,1}$. Then statement (6.33) written in terms of characteristic functions takes the following form: for every $t \in \mathbb{R}$,

$$\lim_{\theta \to 0} P_\theta(f_\theta(t)) = f(t) = \phi(-\log g(t)).$$

But taking account of Remark 4.6.2 and continuity of ϕ we get from here that

$$\lim_{\theta \to 0} \frac{1}{\theta} \phi^{-1}(f_\theta(t)) = -\log g(t), \qquad t \in \mathbb{R},$$

whence as we have seen above it follows that

$$\lim_{\theta \to 0} g_\theta(t) = 1, \qquad t \in \mathbb{R}. \tag{6.35}$$

Just as in the proof of Theorem 4.6.2, we can make sure that the function $\phi\left(\frac{1}{\theta}(1 - f_\theta(t))\right)$ is an $\mathcal{N}$-infinitely divisible characteristic function for every $\theta \in \Theta$. But then according to Property 4.6.2, $\lim_{\theta \to 0} \phi\left(\frac{1}{\theta}(1 - f_\theta(t))\right)$ is an $\mathcal{N}$-infinitely divisible characteristic function. Hence according to Theorem

4.6.3 we conclude that there exists an infinitely divisible (in the usual sense) characteristic function h such that

$$\lim_{\theta \to 0} \phi \left(\frac{1}{\theta}(1 - f_\theta(t)) \right) = \phi(-\log h(t)), \qquad t \in \mathbb{R},$$

i.e.,

$$\lim_{\theta \to 0} \frac{1}{\theta}(1 - f_\theta(t)) = -\log h(t), \qquad t \in \mathbb{R}. \tag{6.36}$$

Further we have

$$\log f_\theta^{m(\theta)}(t) = m(\theta) \log(1 - (1 - f_\theta(t))) = \\ m(\theta)(f_\theta(t) - 1) + \kappa m(\theta)|f_\theta(t) - 1|^2, \tag{6.37}$$

where $|\kappa| \leq 1$. With regard for (6.35) and (6.36) it follows from (6.37) that

$$\lim_{\theta \to 0} f_\theta^{m(\theta)}(t) = h(t), \qquad t \in \mathbb{R}. \tag{6.38}$$

Applying Theorem 4.1.2 with regard for conditions (6.38) and (6.32) we make sure that

$$\lim_{\theta \to 0} \mathsf{E} \exp \left\{ it \sum_{j=1}^{N_\theta} X_{\theta,j} \right\} = \phi(-\log h(t)), \qquad t \in \mathbb{R},$$

whence due to (6.33) it follows that we can assume $h(t) \equiv g(t)$, i.e., (6.34) holds true with G having the required properties. The proof is completed.

Theorems 4.6.3 and 4.6.5 give us very simple prescriptions for constructing analogs of such notions as stable laws and domains of their attraction for random sums with indices generating commutative semigroups of generating functions.

We formulate an analog of the well-known definition of a domain of attraction for random summation. We have already met some particular cases of these analogs in Sections 2.5 and 3.6. Let $N_\theta, X_1, X_2, \ldots$ be independent random variables; moreover let $N_\theta \in \mathcal{N}$ where $\mathcal{N}$ is a family generating a commutative semigroup of generating functions. Let the variables $\{X_j\}_{j \geq 1}$ be identically distributed with distribution function F.

DEFINITION 4.6.3. If there exist some constants $A(\theta)$ and $B(\theta)$ such that the distribution functions of the random sums

$$S_{N_\theta} = \frac{1}{B(\theta)} \sum_{j=1}^{N_\theta} (X_j - A(\theta))$$

weakly converge to some distribution function H as $\theta \to 0$, then we shall say that F is attracted by the family $\mathcal{N}$ to H (in brief: F is $\mathcal{N}$-attracted to H). The population of all distribution functions that are $\mathcal{N}$-attracted to H is called the domain of $\mathcal{N}$-attraction of the distribution function H.

THEOREM 4.6.6. *A distribution function H has a nonempty domain of $\mathcal{N}$-attraction if and only if its characteristic function h admits the representation*

$$h(t) = \phi(-\log g(t)), \tag{6.39}$$

where $g(t)$ is a stable characteristic function,

$$\log g(t) = i\gamma t - c|t|^\alpha \left[1 + i\beta \frac{t}{|t|} \omega(t, \alpha)\right]. \tag{6.40}$$

Here $\gamma \in \mathbb{R}$, $-1 \le \beta \le 1$, $0 < \alpha \le 2$, $c \ge 0$ and

$$\omega(t, \alpha) = \begin{cases} \mathrm{tg} \dfrac{\pi\alpha}{2}, & \text{if } \alpha \ne 1, \\ \dfrac{2}{\pi} \log |t|, & \text{if } \alpha = 1. \end{cases} \tag{6.41}$$

The proof of this result appears to be a combination of Theorems 4.6.3, 4.6.5 and the classical results (Gnedenko and Kolmogorov, 1954) and therefore is omitted.

DEFINITION 4.6.4. *A distribution function H whose characteristic function admits representation (6.39) where ϕ is the standard solution of equation (6.7) and $g(t)$ is defined by relations (6.40) and (6.41), is called $\mathcal{N}$-stable.*

Combining Theorems 4.6.3, 4.6.5 and 4.6.6 with the classical theorems from (Gnedenko and Kolmogorov, 1954, Sect. 35) we obtain the description of domains of $\mathcal{N}$-attraction of $\mathcal{N}$-stable laws.

THEOREM 4.6.7. *Let the family $\mathcal{N} = \{N_\theta, \theta \in \Theta\}$ of natural-valued random variables generate a commutative semigroup of generating functions.*

I. A distribution function F belongs to the domain of $\mathcal{N}$-attraction of an $\mathcal{N}$-stable distribution function with characteristic parameter $\alpha \in (0, 2)$ if and only if

$$\lim_{x \to \infty} \frac{F(-x)}{1 - F(x)} = \frac{c_1}{c_2},$$

$$\lim_{x \to \infty} \frac{1 - F(x) + F(-x)}{1 - F(kx) + F(-kx)} = k^\alpha$$

for every $k > 0$, where c_1 and c_2 are connected with the parameters of stable characteristic function (6.40) by the relations

$$\beta = \frac{c_1 - c_2}{c_1 + c_2}, \qquad 0 < \alpha < 2,$$

$$c = \begin{cases} \Gamma(1 - \alpha)(c_1 + c_2) \cos \dfrac{\pi\alpha}{2}, & 0 < \alpha < 1, \\ (c_1 + c_2) \dfrac{\pi}{2}, & \alpha = 1, \\ \dfrac{1}{1 - \alpha} \Gamma(2 - \alpha)(c_1 + c_2) \cos \dfrac{\pi\alpha}{2}, & 1 < \alpha < 2. \end{cases}$$

Here $\Gamma(\cdot)$ is Euler's gamma function.

II. *A distribution function F belongs to the domain of $\mathcal{N}$-attraction of the $\mathcal{N}$-Gaussian law if and only if*

$$\lim_{t \to \infty} \frac{t^2 \displaystyle\int_{|x|>t} dF(x)}{\displaystyle\int_{|x|<t} x^2\, dF(x)} = 0.$$

Chapter 5

Mathematical theory of reliability growth. A Bayesian approach

5.1 Bayesian reliability growth models

This chapter is devoted to the application of the methods of asymptotic theory of random summation to the solution of an important problem of reliability prediction of complex systems during testing or adjustment when changes are brought into the system being tested.

Any aggregate made for the first time, e.g., a new software, an airplane or a managing system, as usual does not possess the required reliability. Let us agree to deal with a complicated system. Before the system operation begins, some adjusting tests are performed to detect and correct its defects and therefore to increase its reliability. A great many questions appear in this connection. For example, what is the reliability of a system on a certain stage of its testing, and how long should the system be tested and modified to achieve the required reliability?

A perculiarity of this problem is that after every modification the system changes and therefore the data for its statistical reliability analysis, i.e., time intervals between the system failures (or between modifications), cannot be interpreted as identically distributed random variables or which is the same, as a homogeneous sample which is characteristic for the classical problems of reliability theory and quality control. This circumstance even had stimulated some authoritative specialists to make a hasty conclusion that the methods of probability theory and mathematical statistics are not applicable for getting the answers to the questions stated above (e.g., see (Henley and Kumamoto, 1981)). However, as we will see below these methods make it possible to form adequate conclusions concerning this problem.

First works describing mathematical reliability growth models for repeatedly changing systems appeared in the middle of 1950-s, and there are a lot of them at present. According to the communication (Van Pul, 1990) more than 400 works are devoted only to software reliability growth models. At the same time we should note that there is no universal method for solving this problem. In this connection the problem of suitable classification of the existing reliability growth models is urgent. However, up to now the attempts to construct such classifications were purely descriptive and practically useless. Until now, there was no unified mathematical approach to the construction of reliability growth models. Possibly, that is why reliability growth models are bypassed in the canonical textbooks and monographs on mathematical reliability theory.

We will formulate some main assumptions (axioms) formalizing common perceptions about the process of error removal while testing software or technical systems. On the one hand, these assumptions allow us to describe correctly the object of the further mathematical investigation and on the other hand, they are rather easy to verify in practice with regard to real conditions. According to our assumptions, the main object of the investigation is not the system itself but the space of its input values which is represented as a union of two disjoint sets. The first one, called defective below, is characterized by the property that any of its elements being given to the input of the system implies its malfunction, while the system reacts correctly on the elements from the second set. This representation naturally leads to Bayesian models.

As is mentioned in the review (Shanthikumar, 1983) many known methods of reliability growth analysis of complex systems (and practically all methods of software reliability analysis) are intended for the estimation of the number of remaining defects or use information about the number of corrected defects as initial data. However, the orientation of methods on this very indicator is not quite justified. In fact, some defects become apparent only in a certain combination. So their numbering becomes rather conditional. If we consider software, then it is also very difficult to estimate the number of errors corrected after detection of a software failure.

As we have already mentioned, in our opinion, the set of the system input values should be considered as the main object of our investigation, but not the system itself, especially in the development of probability methods of software reliability analysis. The structure of a system becomes practically unessential within this approach. The last circumstance is especially important for reliability analysis of complex systems.

Let $\mathcal{X}$ be the set of system input values, $\mathcal{Y}$ be the set of its output values, $f : \mathcal{X} \to \mathcal{Y}$ be a transform realized by the system and $\mathfrak{C}$ be some system of subsets of the set $\mathcal{Y}$.

ASSUMPTION 5.1.1. A reference function $I : \mathcal{X} \to \mathfrak{C}$ is defined with the help of which we can determine, which reaction of the system to an input value is correct and which is not. If $f(x) \in I(x)$, then the system reacts correctly to the input x; if $f(x) \notin I(x)$, then the system reacts incorrectly to the input x.

In the latter case we will say that a failure takes place. The function $I(x)$ is defined by the specification. If this function is defined for all $x \in \mathcal{X}$ as we assume here, then the specification is complete. Otherwise the specification is incomplete.

A particular structure of the set $\mathcal{X}$ is unessential for our construction. Furthermore, attempts to consider it in more detail can lead to considerable difficulties. For example, by virtue of discreteness of the representation of information in a computer, in the investigation of software reliability, $\mathcal{X}$ can be considered as a finite set. However, for some complicated programs the cardinality of $\mathcal{X}$ is approaching 10^{100} (Palchun, 1989) which makes the application of the methods traditional for finite sets unacceptable. (Note that the number of atoms of hydrogen, which is the most widespread element, in the observable part of the universe is approximately equal to 10^{48} (Parnov, 1967).)

Our assumptions about the structure of the set $\mathcal{X}$ are reduced to the following.

ASSUMPTION 5.1.2. A σ-algebra $\mathfrak{A}$ of subsets of the set $\mathcal{X}$ is given with a nonnegative measure V defined on it.

In other words, we consider the set of the system input values to be a measurable space with a measure $(\mathcal{X}, \mathfrak{A}, V)$. We will consider all subsets of the set $\mathcal{X}$ mentioned below to be $\mathfrak{A}$-measurable without stating it additionally.

The use of the Bayesian ideology for solving the problems of reliability prediction turns out to be very fruitful and has quite a reasonable basis, since within testing complex systems we can distinguish two sources of randomness in the functioning of the system, which is deterministic itself in the sense that if the system remains unchanged, then it reacts identically each time it is fed with the same input. First, the randomness appears as a result of unpredictable hitherto (before the testing begins) changes brought in the system and therefore changing the transformation performed by the system. Second, the randomness appears as a result of the transformation carried out by the system of a random input. In our further constructions the randomness of the first of the two mentioned types will be formalized with the help of random parameters of Bayesian models.

It is not an exaggeration to say that the Bayesian approach, being rather simple and natural, can play the same role in the mathematical theory of reliability growth, as the scheme of summation of random variables plays in classical statistics explaining the normality of errors by virtue of the central limit theorem, or as the scheme of maximums (minimums) of random variables plays in classical reliability theory explaining the forms of lifetime distributions by virtue of limit theorems for extreme order statistics.

Among the authors whose ideas served as a basis for our reasoning which resulted in the mathematical theory of reliability growth presented here, we mention E. Nelson (Nelson, 1973), who suggested taking into account cardinalities of subsets of the set of input values to determine software reliability, and B. Littlewood (Littlewood, 1981), who was the first to consider a Bayesian reliability growth model.

To formalize accurately what has been said above, let us introduce some notions that may seem rather abstract from the practical point of view.

Let $(\Omega, \mathcal{F}, \mathsf{P})$ be an original probability space. All random variables and random elements introduced below will be assumed to be defined on this probability space.

DEFINITION 5.1.1. A mapping $f : \mathcal{X} \to \mathcal{Y}$ is called specified with respect to the pair $(I, \mathfrak{A})$ (or $(I, \mathfrak{A})$-specified) if

$$\{x \in \mathcal{X} : f(x) \notin I(x)\} \in \mathfrak{A}$$

It is clear that we can get an equivalent definition if we require that σ-algebra $\mathfrak{A}$ should contain the set $\{x \in \mathcal{X} : f(x) \in I(x)\}$. Denote the set of all specified mappings f operating from $\mathcal{X}$ to $\mathcal{Y}$ as $\mathfrak{F}$. Let $\mathfrak{B}$ be the Borel σ-algebra of subsets of the real axis.

DEFINITION 5.1.2. A random set (random subset of the set $\mathcal{X}$) is a mapping $D(\omega) : \Omega \to \mathfrak{A}$ such that for any $B \in \mathfrak{B}$,

$$\{\omega \in \Omega : \mathsf{V}(D(\omega)) \in B\} \in \mathfrak{F}.$$

In other words, the measure of a random set should be a random variable.

DEFINITION 5.1.3. A mapping $f(\omega, \cdot) : \Omega \to \mathfrak{F}$ is called a stochastic transformation if the set

$$D(\omega) = \{x \in \mathcal{X} : f(\omega, x) \notin I(x)\}$$

is random.

Every realization of a stochastic transformation is determinate, i.e., for any fixed $\omega \in \Omega$, whatever number of times the same value x is supplied to the system input, the same value $y = f(\omega, x)$ will be at the output. Further to make it brief, we will omit the argument ω for random sets and stochastic transformations and write simply D and $f(x)$ instead of $D(\omega)$ and $f(\omega, x)$, respectively.

Consider a sequence $\{f_i\}_{i \geq 0}$ of stochastic transformations, which will be interpreted in the following way. The process of testing is characterized by some changes being brought in the system during this process and as a result the function f performed by the system is changed. Assume that $f = f_0$ before the testing begins, and $f = f_i, i \geq 1$ after the i-th change. As this is so, the stochastics of the transformations can be clarified by the following example. Let two absolutely identical systems be tested independently of each other (for instance, by different teams). Then the sequences of the detected defects in the systems will probably be different. If new defects are not brought in the system while correcting the detected ones, then these sequences will differ most likely only by rearrangement of the number of defects. If new defects can be brought while correcting, then the detected defects themselves for these two systems may be different. In this example the testing process of each system corresponds to its own trajectory, i.e., to its own $\omega \in \Omega$.

Consider the sequence of $\mathcal{X}$-valued random elements $\{\xi_j\}_{j\geq 1}$, which we will interpret as the sequence of the input values, used to test the system. In this connection we will call the random elements $\{\xi_j\}_{j\geq 1}$ tests or test values.

In the investigation of the variation of the system reliability characteristics during its debugging, the process of bringing in changes proves to be determining, but not the process of the failures, as in the classical reliability theory (these two processes are undoubtedly connected with each other, but not in a one-to-one manner: the changes influencing reliability can be brought in both after failures and after successful tests).

Let $\{M_i\}_{i\geq 0}$ be the sequence of the numbers of the tests after which the changes were brought in the system. Namely, let $M_0 = 0$ and for $i \geq 1$

$$M_i = \min\{j \geq M_{i-1} : f(\xi_j) = f_{i-1}(\xi_j), f(\xi_{j+1}) = f_i(\xi_{j+1})\},$$

i.e. M_i is the number of the test after which the transformation performed by the system is switched from f_{i-1} to f_i. With the help of the sequence $\{M_i\}_{i\geq 0}$ we define the sequence $\{K_i\}_{i\geq 1}$ by setting $K_i = M_i - M_{i-1}, i \geq 1$. Thus K_i is the number of tests between the $(i-1)$-th and the i-th changes of the system. Generally speaking, both M_i and K_i are random variables, $i \geq 1$.

We are not aware of the works where a rigorous definition of reliability growth model is given. Usually, as this model, a method of computing the reliability characteristics varying in time in a certain way is meant. Instead we formulate

DEFINITION 5.1.4. A discrete reliability growth model is the family of finite-dimensional distributions of the random sequence $\{K_i\}_{i\geq 1}$.

Since the sequences $\{M_i\}_{i\geq 0}$ and $\{K_i\}_{i\geq 1}$ are connected bijectively, if we replace the sequence $\{K_i\}$ by the sequence $\{M_i\}$, we will obtain an equivalent definition.

Discrete reliability growth models are suitable in those cases where time unit is, for instance, a time of the system reaction to an input value (e.g., a run of the program), which is advisable for analyzing reliability of interactive program systems. This approach allows one to ignore time delays that are not connected directly with the work of programs, but caused, for example, by pauses in the work of an operator (user).

To define a reliability growth model in a general case consider the sequence of the series $\{\tau_{i,j}\}_{j\geq 1}, i = 1,2,\dots$ of nonnegative random variables. Set $n(i,j) = M_{j-1} + j$. Assume that $\tau_{i,j}$ is the time of the reaction of the system to the test $\xi_{n(i,j)}$. Introduce random variables

$$X_i = \tau_{i,1} + \dots + \tau_{i,k}, \quad Y_i = X_1 + \dots + X_i, \quad i \geq 1.$$

X_i is the length of the time interval between the $(i-1)$-th and the i-th changes, and if we assume that the changes are brought in instantly, then Y_i is the time of the i-th change.

DEFINITION 5.1.5. A reliability growth model is the family of finite-dimensional distributions of the random sequence $\{X_i\}_{i\geq 1}$.

As in the case of a discrete reliability growth model, since the sequences $\{X_i\}$ and $\{Y_i\}$ are connected bijectively, we can formulate an equivalent definition of reliability growth model as the family of finite-dimensional distributions of the sequence $\{Y_i\}$.

DEFINITION 5.1.6. The sets

$$D_i = \{x \in \mathcal{X} : f_i(x) \notin I(x)\}, \quad i = 0, 1, \ldots$$

are called defective.

Denote

$$V = \mathsf{V}(\mathcal{X}), \quad S_i = \mathsf{V}(D_i), \quad i \geq 0.$$

Suppose that $\infty > V \gg 1$ (the meaning of the relation $V \gg 1$ will be defined more exactly below).

DEFINITION 5.1.7. A reliability growth model is called Bayesian, if for every $n \geq 1$ there is a function

$$H_n(x_1, \ldots, x_n; s_0, \ldots, s_{n-1}) : \mathbb{R}^n \times \mathbb{R}_+^n \to [0, 1],$$

possessing the following properties:

1) For every fixed collection $(s_0, \ldots, s_{n-1}) \in \mathbb{R}_+^n$, H_n is an n-dimensional distribution function as a function of the variables $x_1, \ldots, x_n$;

2) For every fixed collection $(x_1, \ldots, x_n) \in \mathbb{R}^n$ H is measurable as a function of the variables $s_0, \ldots, s_{n-1}$;

3) For any $x_1, \ldots, x_n$ $(x_i \in \mathbb{R}^1, i = 1, \ldots, n)$,

$$P(X_1 < x_1, \ldots, X_n < x_n) = \mathsf{E}H_n(x_1, \ldots, x_n; S_0, \ldots, S_{n-1}). \quad (1.1)$$

In Bayesian reliability growth models the changes of the system are formalized with the help of one-dimensional random variables $S_0, S_1, \ldots$ which are random parameters. Thus (1.1) implies that in Bayesian models the random variables $S_0, S_1, \ldots$ accumulate all information concerning the changes of the stochastic transforms performed by the system as a result of its modifications.

Our further aim is to describe rather wide classes of Bayesian reliability growth models which have convenient analytical properties.

5.2   Conditionally geometric models

Further on we will suppose that the following assumptions are correct.

ASSUMPTION 5.2.1. The random elements $\{\xi_j\}_{j \geq 1}$ are independent and have identical distribution, which is uniform on $\mathcal{X}$ with respect to the measure V: for any set $A \in \mathfrak{A}$,

$$P(\xi_j \in A) = V^{-1}\mathsf{V}(A), \quad j \geq 1.$$

Thus the measure V characterizes not the "physical" volume of the set $A \subseteq \mathcal{X}$, but the probability that an element from a set A will be supplied to the input of the system.

ASSUMPTION 5.2.2. The random variables $\{\tau_{i,j}\}$ are jointly independent and row-wise (for every fixed $i \geq 1$) identically distributed; moreover, for each $i \geq 1$ the random variables K_i and $\{\tau_{i,j}\}$ are independent.

Imposing these restrictions upon the random variables $\tau_{i,j}$, we, first, take into account that duration of the system reaction to the same input values may vary after its modification, since the distributions of the random variables $\tau_{i,j}$ may not coincide for different i; second, keep to the requirement introduced in Assumption 5.2.1 of the independence of tests; and third, pass over difficulties that can appear in the detailing of the dependence of duration of the system reaction on the tests $\{\xi_j\}$ and the transformations $\{f_m\}$.

ASSUMPTION 5.2.3. Conditional distribution of the random variables K_i under fixed S_{i-1} is geometrical with some parameter $p_i = p_i(S_{i-1})$:

$$P(K_i = k \mid S_{i-1}) = p_i^{k-1}(1 - p_i), \qquad k = 1, 2, \ldots; i \geq 1. \qquad (2.1)$$

Before we give some examples of situations when assumption 5.2.3 is fulfilled, we shall formulate

DEFINITION 5.2.1. Reliability growth models corresponding to the testing process, when the changes are brought in the system after and only after a failure takes place, are called regular.

The overwhelming majority of the known reliability growth models of software are regular.

EXAMPLE 5.2.1. For regular reliability growth models, Assumption 5.2.3 is true automatically as an implication of Assumption 5.2.1. In this case $p_i = 1 - V^{-1}S_{i-1}$.

Now consider two less trivial examples.

EXAMPLE 5.2.2. Let changes be brought in the system necessarily after each failure. But besides assume that the changes may be also brought in after successful tests. In this connection, if $i - 1$ changes have already been brought in, then assume that after a successful test the changes are brought in with probability r_i and tests of the nonmodified system continue with probability $1 - r_i$. Then obviously $K_i = \min(K_i^{(1)}, K_i^{(2)})$, where

$$P(K_i^{(1)} = k \mid S_{i-1}) = (1 - V^{-1}S_{i-1})^{k-1}V^{-1}S_{i-1}, \quad k = 1, 2, \ldots; i \geq 1,$$

$$P(K_i^{(2)} = k \mid S_{i-1}) = (1 - r_i)^{k-1}r_i, \quad k = 1, 2, \ldots; i \geq 1;$$

moreover, the random variables $K_i^{(1)}$ and $K_i^{(2)}$ can be considered conditionally (under fixed S_{i-1}) independent. As is known, in this case the conditional distribution of the random variable K_i under fixed S_{i-1} is geometric with parameter $p_i = p_i(S_{i-1}) = (1 - V^{-1}S_{i-1})(1 - r_i)$.

Consider a more general example of the situation when Assumption 5.2.3 is correct.

EXAMPLE 5.2.3. Let the possibility of not bringing changes in the system after some failures be admitted along with the possibility of bringing changes in after successful tests. Namely, if $i - 1$ changes have already been brought in, then assume after a failure the changes are brought in with probability u_i and tests of the nonmodified system continue with probability $1 - u_i$. Then with the help of the law of total probability we observe that the conditional distribution of the random variable K_i under fixed value of S_{i-1} is again geometric, but this time with the parameter

$$p_i = (1 - V^{-1}S_{i-1})(1 - r_i) + V^{-1}S_{i-1}(1 - u_i). \tag{2.2}$$

Pay attention that Example 5.2.2 is a particular case of Example 5.2.3, corresponding to the value $u_i = 1$. Models similar to the ones described in Examples 5.2.1–5.2.3, but with nonrandom S_i, $i \geq 0$, and a special functional dependence of r_i, u_i on S_{i-1}, were considered in the works (Volkov, 1981), (Volkov and Shishkevich, 1975) and (Kabak and Rappoport, 1987).

DEFINITION 5.2.2. Bayesian reliability growth models satisfying Assumptions 5.2.1–5.2.3 are called conditionally geometric.

By virtue of Assumption 5.2.1 with a fixed sequence $S_0, S_1, \ldots$, the random variables $K_1, K_2, \ldots$ are conditionally independent. This circumstance allows us to render concrete the form of the distribution function H_n in (1.1) for discrete conditionally geometric models. Denote the random variable $p_i(S_{i-1})$ as Q_i.

THEOREM 5.2.1. *For any discrete conditionally geometric reliability growth model the relation*

$$P(K_1 = k_1, \ldots, K_n = k_n) = E \prod_{i=1}^{n} (Q_i^{k_i - 1} - Q_i^{k_i})$$

takes place for any natural $n, k_1, \ldots, k_n$.

Later in Section 5.4, we will consider continuous reliability growth models, in which the random variables $S_0, S_1, \ldots$ are independent, which are called renewing. For discrete renewing models the following statement holds true.

COROLLARY 5.2.1. *For any discrete renewing conditionally geometric reliability growth model the relation*

$$P(K_1 = k_1, \ldots, K_n = k_n) = \prod_{i=1}^{n} (EQ_i^{k_i - 1} - EQ_i^{k_i})$$

takes place for any natural $n, k_1, \ldots, k_n$.

As an example of a discrete renewing conditionally geometric reliability growth model we examine a model which can be considered as an irregular generalization of the discrete variant of the classical Jelinski-Moranda model (Jelinski and Moranda, 1972).

EXAMPLE 5.2.4. Assume that there are n defects in the system before the tests begin. Associate each defect with the set $E_i \subseteq \mathcal{X}$, whose any element being supplied to the system input activates that part of the system that contains one and the same defect. Assume that $E_i \cap E_j = \emptyset$ for $i \neq j$ and $V(E_i) = \alpha$, $i = 1, \ldots, n$. Every modification of the system is reduced to the correction of exactly one defect, so that $S_0 = n\alpha, S_1 = (n - 1)\alpha, \ldots, S_{i-1} = (n - i + 1)\alpha, \ldots, S_{n-1} = \alpha$. Further assume that the tests are conducted according to the scheme described in Example 5.2.3, where the conditional probability of the system modification after a failure is equal to u, and the conditional probability of the modification after a successful test is equal to r (both these probabilities do not depend on the number of a test or on the number of a modification). Then

$$
p_i = \begin{cases} (1 - r)(1 - \alpha V^{-1}(n - i + 1)) + (1 - u)\alpha V^{-1}(n - i + 1), \\ \qquad\qquad\qquad\qquad\qquad\qquad\qquad i = 1, 2, \ldots, n; \\ 1, \qquad\qquad\qquad\qquad\qquad\qquad\quad i > n. \end{cases}
$$

In particular, hence it follows that

$$
\mathsf{E}K_i = \begin{cases} [(1 - r)(1 - \alpha V^{-1}(n - i + 1)) + (1 - u)\alpha V^{-1}(n - i + 1)]^{-1}, \\ \qquad\qquad\qquad\qquad\qquad\qquad\qquad i = 1, 2, \ldots, n; \\ \infty, \qquad\qquad\qquad\qquad\qquad\qquad\quad i = n + 1. \end{cases}
$$

Since the variables S_i are constant with probability one for every i, they are independent. Therefore according to the classification suggested in Section 5.4, the considered model is a globally parametrized renewing model whose parameters are $r, u, \delta = V^{-1}\alpha$ and n. Statistical analysis of the model under consideration can be conducted by means of the maximum likelihood method. Let a sample $K_1, \ldots, K_m$ be given, where K_i is the number of tests between the $(i - 1)$-th and the i-th modifications of the system. Then the logarithm of the likelihood function has the form

$$
\log L(r, u, \delta, n; K_1, \ldots, K_m) =
$$
$$
\sum_{i=1}^{m} [(K_i - 1) \log\{\delta(n - i + 1)(2 - r - u) + 1 - r\} + \qquad (2.3)
$$
$$
\log\{r - \delta(n - i + 1)(2 - r - u)\}].
$$

The values of r, u, δ and n providing the maximum of function (2.3) can be found in the following way. At first, for every $n \geq 1$ $r(n), u(n)$ and $\delta(n)$ are found as solutions of the system of equations

$$
\begin{cases}
\displaystyle\sum_{i=1}^{m}\left[\frac{(K_i-1)(n-i+1)(2-r-u)}{\delta(n-i+1)(2-r-u)+1-r}-\frac{(n-i+1)(2-r-u)}{r-\delta(n-i+1)(2-r-u)}\right]=0 \\[3mm]
\displaystyle\sum_{i=1}^{m}\left[\frac{(K_i-1)(1+\delta(n-i+1))}{\delta(n-i+1)(2-r-u)+1-r}-\frac{1+\delta(n-i+1)}{r-\delta(n-i+1)(2-r-u)}\right]=0 \\[3mm]
\displaystyle\sum_{i=1}^{m}\left[\frac{(K_i-1)\delta(n-i+1)}{\delta(n-i+1)(2-r-u)+1-r}-\frac{\delta(n-i+1)}{r-\delta(n-i+1)(2-r-u)}\right]=0.
\end{cases}
$$

Further

$$\hat{n}=\arg\max\log L(r(n),u(n),\delta(n),n;K_1,\ldots,K_m)$$

is determined, and $\hat{n}, r(\hat{n}), u(\hat{n})$ and $\delta(\hat{n})$ are taken as the maximum likelihood estimates of the parameters n, r, u and δ correspondingly.

However, note that the rigorous proof of the convergence of the described algorithm to the maximum likelihood estimators of the parameters of the considered model and the investigation of the computing efficiency should be subjected to a special study.

It is obvious that from the formal point of view discrete conditionally geometric reliability growth models are particular cases of general conditionally geometric models, corresponding to $\tau_{i,j}\equiv 1$.

For the fixed volume S_{i-1} of the defect domain, the random variable X_i in conditionally geometric models is a random sum of independent identically distributed random variables where the number K_i of summands has geometric distribution and does not depend on the summands. Such objects are usually called geometric random sums. They possess a number of favorable analytical properties, e.g., see (Kruglov and Korolev, 1990, Chapter 8) and Chapter 2 of this book. In particular, with the help of the Wald identity it is easy to get

$$E(X_i\mid S_{i-1})=(1-Q_i)^{-1}E\tau_{i,j}.$$

To describe general conditionally geometric models more explicitly we need additional notations. Denote the distribution function and Laplace-Stieltjes transform of the random variable $\tau_{i,j}$ as $T_j(x)$ and $v(s)$, respectively. Using the notation $p_j(S_{j-1})=Q_j$ introduced above, according to Assumptions 5.2.2 and 5.2.3, by the law of total probability we can write

$$P(X_j<x\mid S_{j-1})=(1-Q)\sum_{k=1}^{\infty}Q_j^{k-1}T_j^{*k}(x),\qquad(2.4)$$

where T_j^{*k} denotes the k-fold convolution of the distribution function T_j with itself. Using representation (2.4) and conditional independence of the random variables $X_1,\ldots,X_n$ under a fixed sequence $S_0,\ldots,S_{n-1}$, it is easy to obtain the unconditional distribution of the random variables $X_1,\ldots,X_n$:

$$P(X_1<x_1,\ldots,X_n<x_n)=E\prod_{j=1}^{n}(1-Q_j)\sum_{k=1}^{\infty}Q_j^{k-1}T_j^{*k}(x_j)\qquad(2.5)$$

for any $n \geq 1$ and any $x_j \in \mathbb{R}^1, j = 1, \ldots, n$.

In contrast to multivariate distribution functions, multivariate Laplace-Stieltjes transforms corresponding to conditionally geometric reliability growth models can be written out in a finite form.

THEOREM 5.2.2. *For any conditionally geometric reliability growth model the representation*

$$E \exp \left\{ -\sum_{j=1}^{n} s_j X_j \right\} = E \prod_{j=1}^{n} \frac{(1 - Q_j) v_j(s_j)}{1 - Q_j v_j(s_j)} \qquad (2.6)$$

takes place for any $n \geq 1$ and $s_j \geq 0, j = 1, \ldots, n$.

PROOF. Assumptions 5.2.2 and 5.2.3 and representation (2.4) imply that

$$E(\exp\{-s X_j\} \mid S_{j-1}) = \frac{(1 - Q_j) v_j(s)}{1 - Q_j v_j(s)}, \quad s \geq 0. \qquad (2.7)$$

Conditional independence of the random variables $X_1, \ldots, X_n$ under fixed $S_0, \ldots, S_{n-1}$ and the fact, that the distribution of the random variables X_j depends only on S_{j-1}, with the account of (2.7) imply that

$$E \exp \left\{ -\sum_{j=1}^{n} s_j X_j \right\} = EE \left(\exp \left\{ -\sum_{j=1}^{n} s_j X_j \right\} \Big| S_0, \ldots, S_{n-1} \right) =$$

$$E \prod_{j=1}^{n} E(\exp\{-s_j X_j\} | S_0, \ldots, S_{n-1}) =$$

$$E \prod_{j=1}^{n} E(\exp\{-s_j X_j\} | S_{j-1}) = E \prod_{j=1}^{n} \frac{(1 - Q_j) v_j(s_j)}{1 - Q_j v_j(s_j)},$$

Q.E.D.

COROLLARY 5.2.2. *For any renewing conditionally geometric reliability growth model the representation*

$$E \exp \left\{ -\sum_{j=1}^{n} s_j X_j \right\} = \prod_{j=1}^{n} E \frac{(1 - Q_j) v_j(s_j)}{1 - Q_j v_j(s_j)}, \qquad (2.8)$$

takes place for any $n \geq 1$ and $s_j \geq 0, j = 1, \ldots, n$.

In spite of the fact that relation (2.8) can be used for statistical analysis of renewing globally parametrized reliability growth models using, for instance, a modified method of moments for empirical Laplace-Stieltjes transforms, still both analytical and statistical investigation of general conditionally geometric reliability growth models with the help of representation (2.6) is rather difficult. The use of exponential approximations, which we will consider next, turns out to be an essential support here.

5.3   Conditionally exponential models

DEFINITION 5.3.1. A Bayesian reliability growth model such that for every $n \geq 1$ and any $x_1 \geq 0, \ldots, x_n \geq 0$ the representation

$$P(X_1 \geq x_1, \ldots, X_n \geq x_n) = \mathsf{E} \exp \left\{ -\sum_{j=1}^{n} W_j x_j \right\}, \qquad (3.1)$$

takes place, where $W_1, \ldots, W_n$ are some random variables, $W_j = W_j(S_{j-1})$, $j = 1, \ldots, n$, is called conditionally exponential.

By using the notation $W(S)$ here and above, where W and S are random variables, we mean that the random variable W is measurable with respect to the σ-algebra generated by the random variable S.

Conditionally exponential models possess a big amount of favorable properties since the right-hand side of (3.1) is, on the one hand, a multivariate Laplace-Stieltjes transform, and on the other hand, is a special form of mixture of exponential laws. Later in Section 5.4 we will show that the class of conditionally exponential models includes almost all known reliability growth models (they are counted in dozens). At the same time in many cases (for instance, at the final stages of testing or adjustment) these models can be used as approximations for conditionally geometric models, introduced with the help of a rather general and natural assumptions that can be easily verified in practice. Further we will observe the interconnection between conditionally geometric and conditionally exponential models more explicitly. Theorem 3.3.5 proven in Chapter 3 is the main connecting link here. For convenience, we give its formulation once more.

Let $Z_1, Z_2, \ldots$ be independent identically distributed random variables with $\mathsf{E} Z_j = \alpha > 0$, $\{N_n\}_{n \geq 1}$ be natural-valued random variables; moreover, let the random variables $N_n, Z_1, Z_2, \ldots$ be independent for every n and $N_n \to \infty$ in probability as $n \to \infty$. Then

$$\frac{1}{n\alpha} \sum_{j=1}^{N_n} Z_j \Rightarrow \text{(some) } Z \qquad (n \to \infty) \qquad (3.2)$$

if and only if

$$N_n/n \Rightarrow Z \qquad (n \to \infty). \qquad (3.3)$$

This statement sharpens and generalizes the well-known Rényi theorem where sufficient conditions for convergence of distributions of geometric random sums to exponential law are established. In fact, if the random variable N_n in theorem 3.3.5 has the geometric distribution with parameter p_n and $n(1 - p_n) \to 1$ with $n \to \infty$, then as is known, convergence (3.3) takes place, where the random variable Z has the standard exponential distribution, and thus (3.2) holds true. Therefore for Q_j sufficiently close to one, Theorem 3.3.5

allows us to approximate geometric convolutions in (2.4) by means of exponential laws. Namely, for Q_j sufficiently close to one, for any conditionally geometric reliability growth model we have

$$P(X_1 > x_1, \ldots, X_n > x_n) \approx E \prod_{j=1}^n \exp\left\{-\frac{(1-Q_j)x_j}{E\tau_{i,j}}\right\} =$$
$$E \exp\left\{-\sum_{j=1}^n \frac{(1-Q_j)x_j}{E\tau_{i,j}}\right\} \qquad (3.4)$$

If in the last expression we denote $W_j = (1-Q_j)/E\tau_{i,j}$, then we exactly obtain the right-hand side of (3.1).

To obtain estimates of the accuracy of approximation (3.4) of conditionally geometric models by means of conditionally exponential ones, we need two lemmas.

LEMMA 5.3.1. *Let* $a_1, \ldots, a_n, b_1, \ldots, b_n$ *be complex numbers such that* $|a_i| \leq 1$ *and* $|b_i| \leq 1$ *for* $i = 1, \ldots, n$. *Then*

$$\left|\prod_{i=1}^n a_i - \prod_{i=1}^n b_i\right| \leq \sum_{i=1}^n |a_i - b_i|. \qquad (3.5)$$

PROOF can be conducted by induction which is a simple exercise.

LEMMA 5.3.2. *Let* $Z_1, Z_2, \ldots$ *be nonnegative identically distributed random variables with* $EZ_1^s < \infty, 1 \leq s \leq 2, EZ_1 = 1$ *and let* N *be a geometrically distributed random variable with parameter* p; *moreover, let the random variables* $N, Z_1, Z_2, \ldots$ *be independent. Then there exists an absolute constant* $C_0 > 0$ *such that*

$$\sup_x \left|P\left((1-p)\sum_{j=1}^N Z_j < x\right) - 1 + e^{-x}\right| \leq C_0(1-p)^{s-1}EZ_1^s.$$

This statement describing the rate of convergence in the Renyi theorem was proved in (Kalashnikov and Vsekhsvyatskii, 1985) by the method of probability metrics, discussion of which may lead us far beyond the limits of this book.

THEOREM 5.3.1. *The following estimate of the accuracy of approximation of a conditionally geometric reliability growth model with* $E\tau_{i,1}^2 < \infty$ *by the corresponding conditionally exponential model is valid: for any* $n \geq 1$,

$$\sup_{x_1, \ldots, x_n} \left|P(X_1 \geq x_1, \ldots, X_n \geq x_n) - E\exp\left\{-\sum_{j=1}^n \frac{(1-Q_j)x_j}{E\tau_{j,1}}\right\}\right| \leq$$
$$C_0 \sum_{j=1}^n E(1-Q_j)E\tau_{j,1}^2 \leq nC_0 \max_{1 \leq j \leq n} E(1-Q_j)E\tau_{j,1}^2,$$

where C_0 is the constant from Lemma 5.3.2.

PROOF. Taking account of (2.5) we have

$$P(X_1 \geq x_1, \ldots, X_n \geq x_n) = \mathsf{E} \prod_{j=1}^{n} \mathsf{P}\left(\sum_{i=1}^{K_j} \tau_{j,i} \geq x_j\right).$$

Therefore

$$\sup_{x_1, \ldots, x_n} \left| P(X_1 \geq x_1, \ldots, X_n \geq x_n) - \mathsf{E}\exp\left\{-\sum_{j=1}^{n} \frac{(1-Q_j)x_j}{\mathsf{E}\tau_{j,1}}\right\} \right| \leq$$

$$\sup_{x_1, \ldots, x_n} \mathsf{E}\left| \prod_{j=1}^{n} \mathsf{P}\left(\sum_{j=1}^{K_j} \tau_{j,i} \geq x_j\right) - \prod_{j=1}^{n}\exp\left\{-\frac{(1-Q_j)x_j}{\mathsf{E}\tau_{j,1}}\right\} \right|.$$

Now we only have to apply successively Lemmas 5.3.1 and 5.3.2 taking account of the fact that the supremum of a sum does not exceed the sum of supremums. The proof is completed.

An approximation of type (3.4) can be used for discrete conditionally geometric reliability growth models too. In this case the absolute constant contained in a similar estimate can be written explicitly.

THEOREM 5.3.2. *The following estimate of the accuracy of approximation of an arbitrary discrete conditionally geometric reliability growth model by the corresponding conditionally exponential model is valid: for any $n \geq 1$,*

$$\sup_{x_1, \ldots, x_n} \left| P(K_1 > x_1, \ldots, K_n > x_n) - \mathsf{E}\exp\left\{-\sum_{j=1}^{n}(1-Q_j)x_j\right\} \right| \leq$$

$$n \max_{1 \leq j \leq n} \mathsf{E}[(1-Q_j)/Q_j].$$

PROOF of this theorem is similar to the proof of Theorem 5.3.1. We only have to apply the following result instead of Lemma 5.3.2. Let K be a geometrically distributed random variable with parameter p. Denote $q = 1-p$, let $I(A)$ be the indicator of a set A, put

$$\Delta(x) = |\mathsf{P}(qK \leq x) - 1 + e^{-x}|, \quad \Delta = \sup_{x} \Delta(x).$$

LEMMA 5.3.3. *For any $x \geq 0, q \in (0,1)$ we have*

$$\Delta(x) \leq xI(x < q) + \frac{q}{1-q}e^{-x}\left[1 + \frac{x}{2(1-q)}\right] I(x \geq q), \qquad (3.6)$$

$$\Delta \leq \frac{q}{1-q}. \qquad (3.7)$$

PROOF. Let $æ$ be the integer part of the number x/q. Then

$$P(qK > x) = \sum_{k=æ+1}^{\infty} q(1-q)^{k-1} = q\sum_{k=æ}^{\infty}(1-q)^k = (1-q)^{æ}.$$

If $x < q$, then $æ = 0$ and therefore for $x \in (0, q)$

$$\Delta(x) \leq 1 - e^{-x} \leq x. \tag{3.8}$$

For $x \geq q$ we have

$$\frac{x}{q} - 1 \leq æ \leq \frac{x}{q},$$

and hence

$$\exp\left\{\frac{x}{q}\log(1-q)\right\} \leq (1-q)^{æ} =$$

$$\exp\{æ\log(1-q)\} \leq \exp\left\{\left(\frac{x}{q}-1\right)\log(1-q)\right\}.$$

Therefore for x and q specified above we have

$$\Delta(x) \leq \max\{\Delta^{(1)}(x), \Delta^{(2)}(x)\}, \tag{3.9}$$

where

$$\Delta^{(1)}(x) = \left|e^{-x} - \exp\left\{\frac{x}{q}\log(1-q)\right\}\right|,$$

$$\Delta^{(2)}(x) = \left|e^{-x} - \exp\left\{\left(\frac{x}{q}-1\right)\log(1-q)\right\}\right|.$$

At first consider $\Delta^{(1)}(x)$. We have

$$\log(1+x) = \sum_{k=1}^{\infty}(-1)^{k+1}\frac{x^k}{k}, \qquad -1 \leq x \leq 1.$$

Hence,

$$\exp\left\{\frac{x}{q}\log(1-q)\right\} = \exp\left\{-\frac{x}{q}\sum_{k=1}^{\infty}\frac{q^k}{k}\right\} = \exp\left\{-x\sum_{k=1}^{\infty}\frac{q^{k-1}}{k}\right\} =$$

$$\exp\left\{-x\left(1+\sum_{k=1}^{\infty}\frac{q^k}{k+1}\right)\right\}.$$

Therefore

$$\Delta^{(1)}(x) = \left|e^{-x} - e^{-x}\exp\left\{-x\sum_{k=1}^{\infty}\frac{q^k}{k+1}\right\}\right| =$$

$$e^{-x}\left(1 - \exp\left\{-x\sum_{k=1}^{\infty}\frac{q^k}{k+1}\right\}\right). \tag{3.10}$$

But

$$\sum_{k=1}^{\infty}\frac{q^k}{k+1} \leq \frac{1}{2}\sum_{k=1}^{\infty}q^k = \frac{q}{2(1-q)}.$$

Hence continuing (3.10) we obtain

$$\Delta^{(1)}(x) \leq e^{-x}\left(1 - \exp\left\{-\frac{xq}{2(1-q)}\right\}\right) \leq e^{-x}\frac{xq}{2(1-q)}. \tag{3.11}$$

Now consider $\Delta^{(2)}(x)$. We have

$$\Delta^{(2)}(x) = \left| e^{-x} - \exp\left\{\frac{x}{q}\log(1-q)\right\}\exp\left\{\log\frac{1}{1-q}\right\} \right| =$$

$$= \left| e^{-x} - \frac{1}{1-q}\exp\left\{\frac{x}{q}\log(1-q)\right\} \right| = \tag{3.12}$$

$$\frac{1}{1-q}\left| e^{-x} - \exp\left\{\frac{x}{q}\log(1-q)\right\} - qe^{-x} \right| \leq \frac{1}{1-q}\Delta^{(1)}(x) + \frac{qe^{-x}}{1-q}.$$

Using estimate (3.11), from (3.12) we obtain

$$\Delta^{(2)}(x) \leq \frac{q}{1-q}e^{-x}\left[1 + \frac{x}{2(1-q)}\right]. \tag{3.13}$$

Comparing the right-hand sides of (3.11) and (3.13), by virtue of (3.9) we have

$$\Delta(x) \leq \frac{q}{1-q}e^{-x}\left[1 + \frac{x}{2(1-q)}\right] \tag{3.14}$$

for $x \geq q$. By unifying (3.14) and (3.8) we obtain (3.6).

To prove (3.7) note that (3.6) implies the estimate

$$\Delta(x) \leq \frac{q}{1-q}\left[\mathbf{I}(x < q) + e^{-x}\left(1 + \frac{x}{2(1-q)}\right)\mathbf{I}(x \geq q)\right],$$

and it follows from here that

$$\sup_x \Delta(x) \leq \frac{q}{1-q}\max\left\{1, \sup_{x \geq q}e^{-x}\left(1 + \frac{x}{2(1-q)}\right)\right\}. \tag{3.15}$$

The derivative of the function $f(x) = e^{-x}\left(1 + \frac{x}{2(1-q)}\right)$ is equal to $e^{-x}\frac{2q-1-x}{2(1-q)}$, and if $q < \frac{1}{2}$, then this derivative is negative for all $x \geq 0$. But this means that

$$\sup_{x \geq q}e^{-x}\left(1 + \frac{x}{2(1-q)}\right) \leq \sup_{x \geq 0}e^{-x}\left(1 + \frac{x}{2(1-q)}\right) = 1,$$

and therefore (3.15) implies (3.7) for $q < \frac{1}{2}$. As is easy to see, for $q \geq \frac{1}{2}$ estimate (3.7) is trivial. The proof of the lemma and hence of Theorem 5.3.2 also, is completed.

COROLLARY 5.3.1. *For all $x \geq 0$ and $q < \frac{1}{2}$,*

$$\Delta(x) \leq \frac{q}{1-q}\exp\{-[x - \log(x+1)]\}.$$

Practically all known conditionally exponential reliability growth models are regular (an example of irregular conditionally exponential model will be given in Section 5.4). For regular models which have the form

$$P(X_1 \geq x_1, \ldots, X_n \geq x_n) =$$

$$E\exp\left\{-\frac{1}{V}\sum_{j=1}^{n}\frac{S_{j-1}x_j}{ET_{j,1}}\right\}, \qquad n \geq 1, x_j \geq 0, j = 1, \ldots, n, \tag{3.16}$$

in view of Definition 5.2.1 and representation (3.4), we can assume without loss of generality that $E\tau_{j,1} = 1$ for all $j \geq 1$. In fact, otherwise it would have been possible to consider the random variables $S'_k = S_k/E\tau_{k+1,1}$ instead of the random variables $\{S_k\}_{k \geq 0}$ playing the key role in conditionally exponential models, by the same token taking account of time characteristics of the system performance. Below, unless otherwise indicated, we will consider regular models. In such models on account of the assumption $E\tau_{j,1} = 1, j \geq 1$, the representation

$$P(X_k \geq x) = \int_0^\infty \exp\left\{-\frac{sx}{V}\right\} dG_{k-1}(s), \qquad k \geq 1,$$

takes place, where G_{k-1} is the distribution function of the random variable S_{k-1}. Rigorously speaking, according to the definition of Bayesian reliability growth models $G_j(s) = 1$ for all $s \geq V, j \geq 0$. However, if G is an arbitrary distribution function whose growth points lie on the positive semiaxis, then

$$\left| \int_0^V e^{-sxV^{-1}} dG(s) - \int_0^\infty e^{-sxV^{-1}} dG(s) \right| \leq e^{-x}(1 - G(V+0)), \qquad (3.17)$$

and for proper distribution functions G the right-hand side of (3.17) tends to zero uniformly in x as $V \to \infty$. Further we will assume that the volume V of the set of the input values $\mathcal{X}$ is sufficiently large in order that the right-hand side of (3.17) is negligibly small after replacing G by $G_j, j \geq 0$. This considerably simplifies arguments under a negligible loss of accuracy and allows us to consider arbitrary distribution functions of nonnegative random variables as G_j, essentially expanding the class of conditionally exponential reliability growth models. This is the sense in which we should understand the condition $V \gg 1$ introduced in Section 5.1.

5.4 Renewing models

DEFINITION 5.4.1. Bayesian reliability growth models in which $S_0, S_1, \ldots$ are independent random variables, are called renewing.

Definition 5.3.1 implies that for renewing conditionally exponential models for any $n \geq 1$ we have

$$P(X_1 \geq x_1, \ldots, X_n \geq x_n) = \prod_{k=1}^n \int_0^\infty e^{-sx_k} dF_{k-1}(s) = \prod_{k=1}^n P(X_k \geq x_k), \quad (4.1)$$

where $F_{k-1}(s)$ are the distribution functions of the random variables $W_k = W_k(S_{k-1})$, i.e., the random variables $X_1, \ldots, X_n, \ldots$ are also independent.

Renewing Bayesian models are characterized by the testing beginning as if anew after each change of the system, since generally speaking, the defective sets D_i and D_{i+1} are not related in any way (the distributions of the sets volumes turn out to be connected, not the sets themselves).

It follows from (4.1) that every renewing conditionally exponential reliability growth model is completely determined by the family of the distribution functions

$$\left\{ H_j(x) = 1 - \int\limits_0^\infty e^{-szV^{-1}}\, dF_{j-1}(s), \quad x \geq 0 \right\}_{j\geq 0} \qquad (4.2)$$

where H_j is the distribution function of the random variable X_j.

DEFINITION 5.4.2. We will say that a renewing conditionally exponential reliability growth model is parametric if a parametrized family of the distribution functions $\mathcal{H} = \{H(\cdot, \theta), \theta \in \Theta\}$ where Θ is some set, and a sequence of parameters $\{\theta_j\}_{j\geq 0}$, $\theta \in \Theta$, $j \geq 0$, are given such that $H_j(x) = H(x, \theta_j)$, $j \geq 0$.

Thus in parametric renewing reliability growth models, variation of reliability is provided by means of variation of parameters. Every parametric renewing conditionally exponential reliability growth model is completely determined by the family $\mathcal{H}$ and the sequence $\{\theta_j\}_{j\geq 0}$. It turns out that the overwhelming majority of reliability growth models that are mentioned in the review (Shanthikumar, 1983) as "analytical", are parametric renewing conditionally exponential models. As this is so, as a rule the sequence $\{\theta_j\}_{j\geq 0}$ is given in the form

$$\theta_j = \theta(j, n, \alpha_1, \ldots, \alpha_r),$$

where $r \geq 1$, and $\alpha_1, \ldots, \alpha_r$ are some global parameters, identical for all $j \geq 1$, and n is an integer-valued parameter playing the role of the total number of defects in the system before the tests begin, θ is a fixed function, $\theta : \mathbb{R}^{r+2} \to \Theta$. We will call these models globally parametrized.

The class of distributions representable in the form of the mixture of exponential laws (4.2) is rather wide. Now we will give some examples of its representatives and corresponding parametric renewing conditionally exponential reliability growth models.

1) At first, there is, obviously, exponential distribution itself that corresponds to a degenerate distribution function F_{j-1}. Reliability growth models of complex systems with exponential distribution of time intervals between failures are rather popular. Historically the first model of this type was the Jelinski-Moranda model (Jelinski and Moranda, 1972). In this model $H_j(x) = 1 - \exp\{-\theta_j x\}$, and $\theta_j = \alpha(n - j + 1)$, where α is some global parameter and n is the total number of defects in the system before the tests begin. Exponential distribution is also used in Moranda (Moranda, 1975), Goel-Okumoto (Goel and Okumoto, 1979), Shooman (Shooman, 1972) and Musa (Musa, 1980) models, in the works (Kopylov, 1990) and (Volkov and Shishkevich, 1975). Mentioned models differ by their parametrization.

2) Hyperexponential distribution – a discrete mixture of exponential laws – corresponds to a discrete distribution function F_{j-1}. Hyperexponential distribution characterizes the models considered in (Kabak and Rappoport, 1987), (Yamada and Osaki, 1984), (Yamada and Osaki, 1986).

3) Pareto distribution defined by the density

$$h_j(x) = (b_j - 1)c_j^{b_j-1}(x + c_j)^{-b_j}, \quad c_j > 0, \quad b_j > 0, \quad x \geq 0.$$

This distribution appears when the mixing law F_{j-1} is a gamma distribution. Reliability growth models of this type were considered in the works of B. Littlewood and J. Verrall (Littlewood and Verrall, 1973), (Littlewood, 1981), where it was assumed, in particular, that $b_j = b = \text{const}$, c_j is a polynomial in j. By the consideration of parametrizations of more general forms we can essentially extend the family of renewing models of the Pareto type.

4) Using the properties of completely monotone functions (e.g. see (Feller, 1971)) we can verify that the Weibull distribution with the density

$$h_j(x) = c_j b_j^{-c_j} x^{c_j-1} \exp\{-(x/b_j)^{c_j}\}, \quad b_j > 0, \quad 0 < c_j \leq 2, \quad x \geq 0.$$

can be represented in form (4.2). This distribution is used in the reliability growth model for software suggested by Wagoner (Wagoner, 1973).

5) A special case of the Weibull distribution corresponding to the value $c = 2$, the Rayleigh distribution, is used in the Shick-Wolverton model (Shick and Wolverton, 1973).

One can get acquainted more explicitly with criterions for distributions to belong to the class of laws representable in form (4.2), by the work (Heckmann, Robb and Walker, 1990).

There is no need to describe the abovementioned parametric renewing conditionally exponential reliability growth models in detail. An interested reader can refer to the review (Shanthikumar, 1983).

All mentioned renewing models are regular (see Definition 5.2.1). Now we will present an example of irregular renewing model which may be considered as an irregular generalization of the classical Jelinski-Moranda model (Jelinski and Moranda, 1972) or the continuous analog of the model considered in Example 5.2.4.

EXAMPLE 5.4.1. Let the testing be carried out according to the scheme described in Example 5.2.4 with the parameter u equal to one (i.e., after every failure the system is necessarily modified), and let the model be characterized by the relation

$$P(X_1 \geq x_1, \ldots, X_n \geq x_n) =$$

$$\exp\left\{-\sum_{j=1}^{n} X_j[1 - (1 - r)(1 - \delta(n - j + 1))]\right\}. \tag{4.3}$$

Here the parameters n, r and δ have the same sense as in Example 5.2.4. Thus, model (4.3) may be considered as a continuous approximation of the model from Example 5.2.4 with $u = 1$. The logarithm of the likelihood function constructed from the sample $X_1, \ldots, X_m$ has the form

$$\log L(r, \delta, n; X_1, \ldots, X_m) =$$

$$\sum_{j=1}^{m} \log[1 - (1 - r)(1 - \delta(n - j + 1))] - \sum_{j=1}^{m} X_j[1 - (1 - r)(1 - \delta(n - j + 1))]$$

To find the maximum likelihood estimates of the parameters n, δ and r, the same prescription as in Example 5.2.4 may be proposed: first, for every $n \geq 1$ find $r(n)$ and $\delta(n)$ as the solutions of the system

$$
\begin{cases}
\displaystyle\sum_{j=1}^{m} \left[\frac{1 - \delta(n - j + 1)}{1 - (1 - r)(1 - \delta(n - j + 1))} - X_j[1 - \delta(n - j + 1)] \right] = 0 \\
\displaystyle\sum_{j=1}^{m} \left[\frac{(n - j + 1)(1 - r)}{1 - (1 - r)(1 - \delta(n - j + 1))} - X_j(n - j + 1)(1 - r) \right] = 0,
\end{cases}
$$

and afterwards determine

$$
\hat{n} = \arg\max_{n} \log L(r(n), \delta(n), n; X_1, \ldots, X_m)
$$

and take $\hat{n}, r(\hat{n})$ and $\delta(\hat{n})$ as the desired estimates.

Renewing models admit an increase of the volume of the defective domain with a low-quality correction of a current defect. At the same time the stochastic independence of the random variables $\{S_j\}_{j \geq 0}$ means that under detecting the next trouble, the whole system is constructed anew, which happens very seldom in reality (and in fact seems to be impossible in the design and development of complex systems). This circumstance explains mainly those difficulties that appear while fitting renewing models to real data.

The statistical analysis of reliability growth of complicated systems with the help of renewing models reduces to applying the maximum likelihood method (Shanthikumar, 1983) or the combination of the methods of trend analysis of time series and of the method of moments, which will be described further.

5.5   Models with independent decrements of volumes of defective sets

In this and the next two sections we will consider regular models only.

Reliability growth models in which only a part of the defective set is modified during a modification of the system after a failure is detected, but not the whole defective set, will be called partially renewing. In this section and in the following one we will reduce a few examples of partially renewing models. In these models the random variables $S_0, S_1, \ldots$ are not independent, except for the situation when they are nonrandom.

DEFINITION 5.5.1. Bayesian reliability growth models in which for some $n \geq 1$, $S_0 = \zeta_1 + \ldots + \zeta_n$, and the decrements of volumes of the defective sets $\zeta_1 = S_0 - S_1, \zeta_2 = S_1 - S_2, \ldots, \zeta_n = S_{n-1} - S_n$ are independent random variables, are called models with independent decrements of volumes of defective sets (briefly: models with independent decrements).

Denote $G_i(s) = P(\zeta_i < s)$, $i = 1, \ldots, n$.

THEOREM 5.5.1. *For partially renewing conditionally exponential relia-bility growth model with independent decrements we have*

$$P(X_1 \geq x_1, \ldots, X_n \geq x_n) = \prod_{k=1}^{n} \int \exp\{-\frac{s}{V}(x_1 + \ldots + x_n)\} \, dG_k(s) \quad (5.1)$$

for any $x_i \geq 0$, $i = 1, \ldots, n$.

PROOF. From Definition 5.3.1 it follows that

$$P(X_1 \geq x_1, \ldots, X_n \geq x_n) =$$

$$\int \cdots \int \exp\left\{-\frac{1}{V} \sum_{k=1}^{n} x_k u_k\right\} dP(S_0 < u_1, \ldots, S_{n-1} < u_n). \quad (5.2)$$

But $S_0 = \zeta_1 + \ldots + \zeta_n$, $S_1 = \zeta_2 + \ldots + \zeta_n, \ldots, S_{n-1} = \zeta_n$. Therefore the integral in the right-hand side of (5.2) can be written as

$$\int \cdots \int \exp\left\{-\frac{1}{V} \sum_{k=1}^{n} x_k(s_k + \ldots + s_n)\right\} dP(\zeta_1 < s_1, \ldots, \zeta_n < s_n).$$

Whence, regrouping the summands in the exponent and using the indepen-dence of the random variables $\zeta_1, \ldots, \zeta_n$, we finally obtain:

$$P(X_1 \geq x_1, \ldots, X_n \geq x_n) =$$

$$\int \cdots \int \exp\left\{-\frac{1}{V} \sum_{k=1}^{n} s_k(x_1 + \ldots + x_k)\right\} dG(s_1) \ldots dG_n(s_n) =$$

$$\int \cdots \int \prod_{k=1}^{n} \exp\left\{-\frac{1}{V} s_k(x_1 + \ldots + x_k)\right\} dG_k(s_k) =$$

$$\prod_{k=1}^{n} \int \exp\{-\frac{s}{V}(x_1 + \ldots + x_k)\} \, dG_k(s),$$

Q.E.D.

COROLLARY 5.5.1. *For a Bayesian reliability growth model with indepen-dent decrements, for any* $m = 1, \ldots, n$ *and* $x \geq 0$ *we have*

$$P(X_m \geq x) = P(\min_{m \leq k \leq n} Z_k \geq x) = P\left(\frac{VX}{\zeta_m + \ldots + \zeta_n} \geq x\right), \quad (5.3)$$

where the random variables $Z_1, \ldots, Z_n$ *are independent and*

$$P(Z_k \geq x) = \int \exp(-sx/V) \, dG_k(s), \qquad k = 1, \ldots, n,$$

and X *is a random variable with the standard exponential distribution, which is independent of the random variables* $\zeta_1, \ldots, \zeta_n$.

PROOF. The left-hand equality (5.3) is almost obvious. To prove the right-hand one it is sufficient to use the fact that the Laplace-Stieltjes transform of a sum of independent random variables is equal to the product of Laplace-Stieltjes transforms of the summands, and therefore

$$P(X_m \geq x) = \prod_{k=m}^{n} \int \exp(-sx/V)\, dG_k(s) =$$

$$\int \exp(-sx/V)\, dP(\zeta_m + \ldots + \zeta < s) =$$

$$P\left(\frac{VX}{\zeta_m + \ldots + \zeta_n} > x\right).$$

The corollary is proved.

The right-hand equality (5.3) in its turn implies:

COROLLARY 5.5.2. Let $E|\zeta_k| < \infty$, $k = 1, \ldots, n$. Then for any $m = 1, \ldots, n$ we have

$$EX_m \geq V\left[\sum_{k=m}^{n} E\zeta_k\right]^{-1}. \tag{5.4}$$

Futhermore, the equality is attained if $\zeta_1, \ldots, \zeta_n$ are degenerate.

PROOF. The right-hand equality (5.3) means that the distributions of the random variables X_m and the random variables $VX/(\zeta_m + \ldots + \zeta_n)$, coincide, where X is a standard exponential random variable independent of $\zeta_m, \ldots, \zeta_n$. Therefore due to Jensen's inequality we have

$$EX_m = EX \cdot E\frac{V}{\zeta_m + \ldots + \zeta_n} = VE\frac{1}{\zeta_m + \ldots + \zeta_n} \geq$$

$$V\left[E\sum_{k=m}^{n} \zeta_k\right]^{-1} = V\left[\sum_{k=n}^{n} E\zeta_k\right]^{-1}.$$

The corollary is proved.

DEFINITION 5.5.2. A Bayesian reliability growth model with independent decrements whose changes of volumes of defective sets $\zeta_1, \ldots, \zeta_n$ are identically distributed is called homogeneous.

Consider a homogeneous conditionally exponential reliability growth model with independent decrements. Denote $E\zeta_1 = \alpha$. Then it follows from (5.4) that

$$EX_m \geq \frac{V}{\alpha(n - m + 1)}, \tag{5.5}$$

i.e., for these models the mean time between successive failures as a function of m, which is the number of a failure, increases very fast: faster than a hyperbola.

As this is so, if $P(\zeta_1 = \alpha) = 1$, then inequality (5.5) turns into equality and the model itself turns into the classical Jelinski-Moranda model.

Unfortunately, the statistical analysis of reliability growth with the help of conditionally exponential models with independent decrements is rather difficult due to inconvenience of the expression for the joint density of the random variables $X_1, \ldots, X_n$ even in a homogeneous case.

In practice, independence of variations of volumes of defective sets can take place if modifications are brought into the system many times and these variations are small.

A reliability growth model similar to the one just considered, namely, a model of a random walk with absorption type has a wider field of application. In this model $S_0, S_1, \ldots$ are coordinates of a point on the straight line observed in the moments $t = 0, 1, 2, \ldots$. Transitions of the point are random and independent on all steps. The walk starts from the random point $S_0 > 0$ and has an absorbing screen in zero. Unfortunately, not only statistical, but also analytical investigation of these models appears to be very complicated, so now the only method of their study is simulation.

The models described in this section admit (unlike renewing models) a partial change of the system during the removal of defects. They also admit an increase of volumes of defective sets during low-quality error correction.

5.6 Order-statistics-type (mosaic) reliability growth models

In reliability growth models with independent decrements, the volume S_0 of the defective domain before testing begins, is represented as a sum $S_0 = \zeta_1 + \ldots + \zeta_n$ of independent random variables, whose numbers correspond to the order numbers of modifications of the system. In other words, every time the defect is detected, a random experiment is performed, whose result is the random variable ζ_i. Moreover, the trials are independent.

In this section we will also consider regular reliability growth models assuming the representation of S_0 as the sum $S_0 = \zeta_1 + \ldots + \zeta_n$ of independent random variables. However, now we consider that the defects are numbered not in the order of their detection, but beforehand. Namely, assume that the defective set D_0 is a union of disjoint sets $D_0 = E_1 + \ldots + E_n$. Denote $\zeta_i = V(E_i)$, $i = 1, \ldots, n$. Let us assume that if a current test ξ_j falls into one of the sets $E_1, \ldots, E_n$, then this subset is entirely excluded from the defective set. This construction can be interpreted as follows: the system initially contains n defects. To each defect the set of input values is put in correspondence such that if the system is fed with a value from this set, then the element of the system is initiated which contains the given defect.

Assume that the random variables ζ_i, $i = 1, \ldots, n$ are independent. This may be so if the sets $E_1, \ldots, E_n$ are situated sufficiently far from each other.

We will call these reliability growth models order-statistics-type models. These models are also called mosaic. This name is explained if we consider $\mathcal{X}$ as a part of a plane and paint the defective subsets in different colors. Then during the simulation of the testing process described above we may observe pictures which resemble a mosaic.

In Section 5.1 we introduced the random variables Y_j, $j \geq 1$, where $Y_j = X_1 + \ldots + X_j$, i.e., Y_j is the time (integer-valued optionally), when the j-th failure is detected, if we assume that the changes are brought into the system instantly. The following statement is a more accurate variant of a theorem proved in (Galtsov and Solov'ev, 1991).

THEOREM 5.6.1. *Let in addition to Assumptions 5.2.1 and 5.2.2, $\mathsf{E}\tau_{i,j} = 1$, $i \geq 1, j \geq 1$. Assume that the testing is performed according to the scheme described above in this section, and the random variables $\zeta_1, \ldots, \zeta_n$ are independent and identically distributed with a common distribution function G. Then*

$$\lim_{V \to \infty} \mathsf{P}(Y_1 \leq y_1 V, \ldots, Y_n \leq y_n V) = \mathsf{P}(Z_{(1)} \leq y_1, \ldots, Z_{(n)} \leq y_n)$$

for any $y_i \in \mathbb{R}$, $i = 1, \ldots, n$, where $Z_{(1)}, \ldots, Z_{(n)}$ are order statistics, constructed from the independent homogeneous sample $Z_1, \ldots, Z_n$ from the distribution function $H(x)$,

$$H(x) = \int_0^\infty (1 - e^{-sx}) \, dG(s), \qquad x \geq 0. \tag{6.1}$$

PROOF. It is easy to see that in the situation under consideration, some permutation $(i_1, \ldots, i_n)$ of the set $1, \ldots, n$ one-to-one corresponds to each sequence $S_0, S_1, \ldots, S_n$ ($i_1, \ldots, i_n$ are the numbers of the defective sets that the tests fall into in the moments $Y_1, \ldots, Y_n$, respectively); moreover all permutations are equiprobable due to the symmetry of the problem. Therefore, without loss of generality we may consider that in the moment Y_1 the test has entered into the subset E_1, in the moment Y_2 into the subset E_2, etc. Then taking account of the definition of the random variables $\tau_{i,j}$ (see Section 5.1) we have

$$\mathsf{P}(X_1 \geq x_1, \ldots, X_n \geq x_n) =$$

$$n! \int_{\mathbb{R}_+^n} \prod_{i=1}^n \left[\mathsf{P}\left(\sum_{j=1}^{N_i} \tau_{i,j} \geq x_i \right) \frac{s_i}{s_1 + \ldots + s_n} \, dG(s_i) \right]$$

for any $x_i \in \mathbb{R}$, $i = 1, \ldots, n$, where by virtue of the regularity of the considered model by definition, the random variable N_i has the geometric distribution with parameter $1 - V^{-1}(s_1 + \ldots + s_n)$, $i = 1, \ldots, n$. Taking account of the condition $\mathsf{E}\tau_{i,j} = 1$, according to Theorem 3.3.5 (to be more exact, according to the Rényi theorem implied by Theorem 3.3.5) we have

$$\lim_{V \to \infty} \mathsf{P}\left(\sum_{i=1}^{N_i} \tau_{i,j} \geq x_i V \right) =$$
$$\exp\{-(s_1 + \ldots + s_n)x_i\}, \qquad x_i \geq 0, \quad i = 1, \ldots, n.$$

Therefore according to the dominated convergence theorem,

$$\lim_{V\to\infty} P(X_1 \geq x_1 V, \ldots, X_n \geq x_n V) =$$

$$\lim_{V\to\infty} n! \int_{\mathbf{R}_+^n} \prod_{i=1}^{n} \left[P\left(\sum_{j=1}^{N_i} \tau_{i,j} \geq x_i V\right) \frac{s_i}{s_1 + \ldots + s_n} \, dG(s_i) \right] = \quad (6.2)$$

$$n! \int_{\mathbf{R}_+^n} \prod_{i=1}^{n} \left[\exp\{-(s_1 + \ldots + s_n)x_i\} \frac{s_i}{s_1 + \ldots + s_n} \, dG(s_i) \right].$$

Denote the right-hand side of (6.2) as $\mathcal{U}(x_1, \ldots, x_n)$. By differentiating this function we obtain

$$\frac{\partial^n \mathcal{U}(x_1, \ldots, x_n)}{\partial x_1 \cdots \partial x_n} =$$

$$(-1)^n n! \int_{\mathbf{R}_+^n} \prod_{i=1}^{n} [s_i \exp\{-(s_1 + \ldots + s_n)x_i\} \, dG(s_i)] =$$

$$(-1)^n n! \int_{\mathbf{R}_+^n} \prod_{i=1}^{n} [s_i \exp\{-s_i(x_1, \ldots, x_i) \, dG(s_i)] =$$

$$(-1)^n n! h(x_1) h(x_1 + x_2) \cdots h(x_1 + \ldots + x_n),$$

where h is the density corresponding to the distribution function H from (6.1). Hence, as a limit, the random vector $(X_1, \ldots, X_n)$ has the density

$$p(x_1, \ldots, x_n) = n! h(x_1) h(x_1 + x_2) \cdots h(x_1 + \ldots + x_n).$$

But since $Y_1 = X_1$, $Y_2 = X_1 + X_2, \ldots, Y_n = X_1, \ldots, X_n$, the random vector $(Y_1, \ldots, Y_n)$ also has the limit density

$$g(y_1, \ldots, y_n) = n! h(y_1) h(y_2) \cdots h(y_n),$$

concentrated on the set $0 < y_1 < y_2 < \ldots < \infty$. But, as is known, the vector of order statistics constructed from an independent homogeneous sample of volume n from the distribution function (6.1) has the very same density. The proof is completed.

Note that from the practical point of view the assumption that the random variables $\zeta_1, \ldots, \zeta_n$ are identically distributed is not unnecessarily restrictive, since the tested systems are as a rule unique, and therefore in practice we may substitute the theoretical distribution of the random variables ζ_i with its statistical image.

Evidently, for the first time the order-statistics-type reliability growth model was suggested in (Miller, 1984). In (Galtsov, 1992) and (Korolev, 1992) with the help of the well-known properties of order statistics and corresponding spacings some general results for order-statistics-type reliability growth models were obtained. In particular, in (Korolev, 1992) the expression for the conditional density $p_{j+1}(x|t_0)$ of the random variables X_{j+1} under the condition $Y_j = t_0$, $j = 1, \ldots, n-1$, was presented. If h is the density corresponding to the distribution function H from (6.1), then

$$p_{j+1}(x|t_0) = (n-j)h(x+t_0)[1 - H(x+t_0)]^{n-j-1}[1 - H(t_0)]^{j-n}. \quad (6.3)$$

Consider the conditional distribution function $R_t^{(m+1)}(x)$ of the random variable X_{m+1} under the condition that the last failure occurred at time t_0 and afterwards the system has worked successfully until time t. This distribution function is of special interest for practical investigation of reliability growth of complex systems during their testing, since, as a rule, the testing is not terminated after some failure, but continues for some time more, and not all the defects are necessarily detected in this case. It turns out that the distribution function $R_t^{(m+1)}$ does not depend on t_0. Indeed, using (6.3) we obtain the following representation:

$$R_t^{(m+1)}(x) = \frac{\displaystyle\int_{t-t_0}^{t-t_0+x} h(y+t_0)[1 - H(y+t_0)]^{n-m-1}dy}{\displaystyle\int_{t-t_0}^{t-t_0+x} h(y+t_0)[1 - h(y+t_0)]^{n-m-1}dy} =$$

$$\frac{\displaystyle\int_{t}^{t+x} d[1 - (1 - H(y))^{n-m}]}{\displaystyle\int_{t}^{\infty} d[1 - (1 - H(y))^{n-m}]} = 1 - \left[1 - \frac{1 - H(t+x)}{1 - H(t)}\right]^{n-m}.$$

Now consider the situation when the volumes $\zeta_1, \ldots, \zeta_n$ of elementary defective subsets are nonrandom. This situation may be regarded as an approximation to the real situation in the case when the theoretical distribution G is substituted by its statistical image we have talked about.

In this case the probability of the chain $\{S_0, S_1, \ldots, S_n\}$ is determined only by the order of the defective subsets that successive tests get into, and therefore this probability is equal to

$$\left(1 - \frac{S_1}{S_0}\right)\left(1 - \frac{S_2}{S_1}\right) \cdots \left(1 - \frac{S_n}{S_{n-1}}\right).$$

So Definition 5.3.1 immediately implies:

THEOREM 5.6.2. Let the assumptions of theorem 5.6.1 hold, but the variables $\zeta_1, \ldots, \zeta_n$ be nonrandom. Then

$$\lim_{V \to \infty} P(X_1 \geq x_1 V, \ldots, X_n \geq x_n V) =$$

$$\sum_{\{(i_1, \ldots, i_n)\}} \exp\left\{-\sum_{j=1}^{n} x_j(\zeta_{i_j} + \ldots + \zeta_{i_n})\right\} \prod_{j=1}^{n} \frac{\zeta_{i_j}}{\zeta_{i_j} + \ldots + \zeta_{i_n}}, \quad (6.4)$$

where the indices in the outer sum run over the set of all permutations of the numbers $\{1, 2, \ldots, n\}$.

COROLLARY 5.6.1. *The asymptotic distribution of the random variables X_m ($m = 1, 2, \ldots, n$) in the considered situation is hyperexponential:*

$$\lim_{V \to \infty} P(X_m \geq xV) = \sum_k q_k^{(m)} \exp\{-t_k^{(m)} x\}, \quad x \geq 0,$$

where the points $t_k^{(m)}$ have the form $t_k^{(m)} = \zeta_{i_m} + \ldots + \zeta_{i_n}$, and the probabilities $q_k^{(m)}$ can be represented as

$$q_k^{(m)} = \sum_{I_{mk}} \prod_{j=1}^{n-1} \frac{\zeta_{i_j}}{\zeta_{i_j} + \ldots + \zeta_{i_n}},$$

where $I_{mk} = \{(i_1, \ldots, i_n) : \zeta_{i_m} + \ldots + \zeta_{i_n} = t_k^{(m)}\}$. As this is so, the number of different points $t_1^{(m)}, t_2^{(m)}, \ldots$ does not exceed C_n^{n-m+1}.

It follows from relation (6.4) that the limit as $V \to \infty$ of the joint distribution of the random variables X_1/V, $\ldots$, X_m/V has the density $p_{X_1, \ldots, X_m}(x_1, \ldots, x_m)$ which for any $m = 1, 2, \ldots, n$ has the form

$$p_{X_1, \ldots, X_m}(x_1, \ldots, x_m) =$$

$$\sum \exp\left\{-\sum_{j=1}^{m} x_j(\zeta_{i_j} + \ldots + \zeta_{i_n})\right\} \prod_{j=1}^{m} \zeta_{i_j}, \quad x_j \geq 0, \tag{6.5}$$

where the outer summation, as before, is conducted over the set of all permutations $\{(i_1, \ldots, i_m)\}$ of the integer numbers from 1 to n. In particular, (6.5) implies that for large V the conditional distribution of the random variable X_m/V under the condition $X_1/V = x_1, \ldots, X_{m-1}/V = x_{m-1}$ may be approximated by an absolutely continuous law with the density

$$p_m(x|x_1, \ldots, x_{m-1}) =$$

$$\frac{p_{X_1, \ldots, X_m}(x_1, \ldots, x_{m-1}, x)}{p_{X_1, \ldots, X_{m-1}}(x_1, \ldots, x_{m-1})} =$$

$$\frac{\sum \exp\left\{-\sum_{j=1}^{m-1} x_j(\zeta_{i_j} + \ldots + \zeta_{i_n})\right\} \exp\{-x(\zeta_{i_m} + \ldots + \zeta_{i_n})\} \zeta_{i_1} \cdots \zeta_{i_m}}{\sum \exp\left\{-\sum_{j=1}^{m-1} x_j(\zeta_{i_j} + \ldots + \zeta_{i_n})\right\} \zeta_{i_1} \cdots \zeta_{i_{m-1}}}.$$

Using the fact that the random variable X_m has a hyperexponential limit distribution as $V \to \infty$, it is not difficult to verify that

$$V^{-1}\mathsf{E}X_m \approx \sum_k \frac{q_k^{(m)}}{t_k^{(m)}},$$

$$V^{-2}\mathsf{D}X_m \approx 2\sum_k \frac{q_k^{(m)}}{(t_k^{(m)})^2} - \left(\sum_k \frac{q_k^{(m)}}{t_k^{(m)}}\right)^2.$$

Mosaic reliability growth models appear to be rather convenient for statistical analysis (see Sect. 5.9). Perhaps, their only deficiency is that they do not admit the possibility of an increase of the volume of defective domain when the system is corrected falsely.

5.7 Generalized conditionally exponential models

Consider the following generalization of the regular Jelinski-Moranda model. Assume that the defective set D_0 consists of K "elementary" defective subsets $E_1, \ldots, E_K$, which are randomly scattered over $\mathcal{X}$. A subset $\mathcal{X}_0$ is defined in the set $\mathcal{X}$ and tests with values in $\mathcal{X}_0$ are used for the system adjustment, which can be formalized as $\mathsf{V}(\mathcal{X} \setminus \mathcal{X}_0) = 0$. It is obvious that due to a random displacement of the sets $E_1, \ldots, E_K$ on $\mathcal{X}$, the number N of those elementary defective sets which have nonempty intersection with $\mathcal{X}_0$, is random. However, note that such a testing procedure gives adequate results if the system is not only tested, but also operates only on input values from $\mathcal{X}_0$.

Assume that during the adjustment m failures have been detected and the corresponding changes have been brought into the system. This means that a realized value of the random variable N is not less than m. Then it makes sense to consider the conditional distributions

$$\mathsf{P}(X_1 \geq x_1, \ldots, X_m \geq x_m, X_{m+1} \geq x_{m+1} | N \geq k),$$

where $k \geq m$. Let us assume that the elementary defective subsets $E_1, E_2, \ldots$ are pairwise disjoint and have identical volumes in the sense that $\mathsf{V}(E_i \cap \mathcal{X}_0) = s$, $i \geq 1$. Then with the help of Definition 5.3.1 we obtain

$$\mathsf{P}(X_1 \geq x_1, \ldots, X_m \geq x_m, X_{m+1} \geq x_{m+1} | N \geq k) =$$

$$\frac{1}{\mathsf{P}(N \geq k)} \sum_{n \geq k} \mathsf{P}(N = k) \exp\left\{-\frac{s}{V} \sum_{j=1}^{m+1} (n - j + 1)x_j\right\}. \tag{7.1}$$

Whence it follows that the distribution of the random variable X_{m+1} is formally hyperexponential; moreover, when $k = m$, this distribution is improper with the defect $\mathsf{P}(N = m)/\mathsf{P}(N \geq m)$ (in this case the exponent does not depend on x_{m+1}). Therefore if $\mathsf{P}(N = m) > 0$, then $\mathsf{E}(X_{m+1} | N \geq m)$ is infinite, and when $k > m$ the following estimate holds.

THEOREM 5.7.1. *For the generalized Jelinski-Moranda model, when $k > m$ and $P(N \geq k) > 0$, we have*

$$E(X_{m+1}|N \geq k) \geq \frac{V}{s}\left[\frac{E N I(N \geq k)}{P(N \geq k)} - m\right]^{-1},$$

where $I(A)$ is the indicator function of a set A.

PROOF. Let $Z_{m,k}$ be a random variable taking values $t_j = sV^{-1}(k+j-m)$ with probabilities $p_j = P(N = k+j)/P(N \geq k)$, $j = 0, 1, 2, \ldots$. Then as is easy to see,

$$P(X_{m+1} \geq x|N \geq k) = \sum_{j=0}^{\infty} p_j \exp\{-t_j x\} = P(X Z_{m,k}^{-1} \geq x),$$

where X is the standard exponential random variable independent of $Z_{m,k}$. In other words, the conditional distribution of the random variable X_{m+1} under the condition $N \geq k$, coincides with the unconditional distribution of the random variable $X Z_{m,k}^{-1}$. Therefore due to Jensen's inequality we have

$$E(X_{m+1}|N \geq k) = E X Z_{m,k}^{-1} = E X \cdot E Z_{m,k}^{-1} \geq (E Z_{m,k})^{-1} =$$

$$\left[\frac{s}{V}\sum_{j=0}^{\infty}(k+j-m)\frac{P(N=k+j)}{P(N \geq k)}\right]^{-1} = \frac{s}{V}\left[\frac{E N I(N \geq k)}{P(N \geq k)} - m\right]^{-1},$$

Q.E.D.

Unfortunately, attempts to write an explicit form of distribution (7.1) for particular distributions of the random variable N lead to very cumbersome expressions, unsuitable for the direct analysis. Let, for instance, $\mathcal{X}$ be a plane and the elementary defective subsets $E_1, E_2, \ldots$ be circles with the same radius, whose centers form a stationary Poisson point process on a plane, which corresponds to a "purely stochastic" chaotic arrangement of points, as is mentioned in (Ambartsumyan, Mecke and Stoyan, 1989). Let the parameter λ of the Poisson process and the radius of the circles guarantee that the probability of intersections of circles be negligibly small. Then (7.1) implies that for $k \geq m$

$$P(X_{m+1} \geq x|N \geq k) = \frac{e^{smx/V}\left(\exp\{\lambda e^{-sx/V}\} - \sum_{n<k}\left(\lambda e^{-sx/V}\right)^n / n!\right)}{e^{\lambda} - \sum_{n<k}\frac{\lambda^n}{n!}}.$$

Though this formula can be used for computations, it is too cumbersome for qualitative analysis. This illustrates that it is reasonable to use approximations we are passing to.

Let N_k be a random variable taking the values $t_i = isV^{-1}$ with probabilities $p_i = P(N = k+i)/P(N \geq k)$, $i = 0, 1, 2, \ldots$.

THEOREM 5.7.2. *Assume that the distribution of the random variable N_k depends on some parameter $\lambda \in (0, \infty)$ and for some positive infinitely increasing function $c(\lambda)$ (possibly, dependent on k), the random variable $N_k/c(\lambda)$ weakly converges as $\lambda \to \infty$ to some random variable ν with distribution function Q. Then for any integer $k \geq l$, $l \geq 1$ and nonnegative finite x_j, $1 \geq j \geq l$, in the generalized Jelinski-Moranda model we have*

$$\lim_{\lambda \to \infty} P(c(\lambda)X_1 \geq x_1, \ldots, c(\lambda)X_l \geq x_l | N \geq k) = \int_0^\infty \exp\left\{ -t \sum_{j=1}^l x_j \right\} dQ(t).$$

PROOF. Relation (7.1) implies that

$$P(c(\lambda)X_1 \geq x_1, \ldots, c(\lambda)X_l \geq x_l | N \geq k) =$$

$$\sum_{n \geq k} \frac{P(N = n)}{P(N \geq k)} \exp\left\{ -\frac{s}{Vc(\lambda)} \sum_{j=1}^l (n - j + 1)x_j \right\} =$$

$$\sum_{i=0}^\infty p_i \left\{ -\frac{s}{Vc(\lambda)} \sum_{j=1}^l (k + i - j + 1)x_j \right\} =$$

$$\exp\left\{ -\frac{s}{Vc(\lambda)} \sum_{j=1}^l (k - j + 1)x_j \right\} \sum_{i=0}^\infty p_i \exp\left\{ -\frac{t_i}{c(\lambda)} \sum_{j=1}^l x_j \right\}. \qquad (7.2)$$

Let $Q_k^{(\lambda)}$ be the distribution function of the random variable N_k. Weak convergence of the random variable $N_k/c(\lambda)$ to ν as $\lambda \to \infty$ means that for any continuous bounded function $\phi(t)$

$$\int_0^\infty \phi(t) \, dQ_k^{(\lambda)}(tc(\lambda)) \to \int_0^\infty \phi(t) \, dQ_k^{(\lambda)}(t) \qquad (\lambda \to \infty).$$

It holds also for $\phi(t) = \exp\left\{ -t \sum_{j=1}^l \right\}$ (with fixed finite positive $x_1, \ldots, x_l$). Therefore

$$\sum_{i=1}^\infty p_i \exp\left\{ -\frac{t_i}{c(\lambda)} \sum_{j=1}^l x_j \right\} = \int_0^\infty \exp\left\{ -\frac{t}{c(\lambda)} \sum_{j=1}^l \right\} dQ_k^{(\lambda)}(t) =$$

$$\int_0^\infty \exp\left\{ -t \sum_{j=1}^l x_j \right\} dQ_k^{(\lambda)}(tc(\lambda)) \to \int_0^\infty \exp\left\{ -t \sum_{j=1}^l x_j \right\} dQ_k^{(\lambda)}(t) \qquad (7.3)$$

as $\lambda \to \infty$. At the same time, as $\lambda \to \infty$,

$$\exp\left\{ -\frac{s}{Vc(\lambda)} \sum_{j=1}^l (k + 1 - j)x_j \right\} \to 1. \qquad (7.4)$$

Relations (7.2), (7.3) and (7.4) prove the theorem.

Pay attention to a rather unexpected fact: the limit distribution turns out to be constant on hyperplanes $x_1 + \ldots + x_l = \text{const}$.

Estimates of the convergence rate in Theorem 5.7.2, when the limit law Q is degenerate, may be obtained from the estimates of stability of an exponential distribution to a random perturbation of the parameter, which is contained in the following statement. Let, as before, X be a random variable with the standard exponential distribution, Y be an arbitrary positive random variable independent of X. Denote

$$\Delta_1(x) = |1 - e^{-x} - P(XY^{-1} < x)|, \qquad \Delta_1 = \sup_x \Delta_1(x),$$

$g_Y(x)$ is the Laplace-Stieltjes transform of the random variable Y. By the symbol $\alpha \vee \beta$ we will denote the greatest of the numbers α and β. Our aim is to estimate Δ_1 and $\Delta_1(x)$ in terms of moment characteristics which describe the departure of the random variable Y from a degenerate one.

THEOREM 5.7.3. If $E(Y \vee Y^{-1}) < \infty$, then

$$\Delta_1 \leq 0.3679 E(Y \vee Y^{-1} - 1). \tag{7.5}$$

If $E(Y \vee Y^{-1})^2 < \infty$ and $EY = 1$, then

$$\Delta_1 \leq 0.2707 E(Y \vee Y^{-1} - 1)^2. \tag{7.6}$$

If $EY^2 < \infty$, then for any $x \geq 0$

$$\Delta_1(x) \leq [x(e^{-x} + g_Y^{1/2}(2x))][E(Y - 1)^2]^{1/2}. \tag{7.7}$$

If $EY^4 < \infty$ and $EY = 1$, then for any $x \geq 0$

$$\Delta_1(x) \leq \frac{1}{2}[x^2(e^{-x} + g_Y^{1/2}(2x))][E(Y - 1)^4]^{1/2}. \tag{7.8}$$

PROOF. Prove estimate (7.5). Let $G(s)$ be the distribution function of the random variable Y. According to the Lagrange formula we have

$$e^{-x} - e^{-sx} = x(s - 1)\exp\{-x[s + \theta_s(1 - s)]\},$$

where $0 \leq \theta_s \leq 1$. Therefore for $x \geq 0$

$$\Delta_1(x) = \left| \int_0^\infty (e^{-x} - e^{-sx}) \, dG(s) \right| \leq$$

$$\int_0^\infty |s - 1| x \exp\{-x[s + \theta_s(1 - s)]\} \, dG(s) =$$

$$\int\limits_0^1 |s-1|x\exp\{-x[s+\theta_s(1-s)]\}\,dG(s)+$$

$$\int\limits_1^\infty |s-1|x\exp\{-x[s+\theta_s(1-s)]\}\,dG(s) \le$$

$$\int\limits_0^1 (1-s)xe^{-sx}\,dG(s) + \int\limits_1^\infty (s-1)xe^{-x}\,dG(s) \equiv I_1 + I_2. \qquad (7.9)$$

Note that $\sup_{x\ge 0} xe^{-sx} = (ce)^{-1}$ for $c > 0$. Therefore, continuing (7.9), we obtain

$$\Delta_1(x) \le \frac{1}{e}\int\limits_1^0 (s^{-1}-1)\,dG(s) + \frac{1}{e}\int\limits_1^\infty (s-1)\,dG(s) =$$

$$0.3679 E(Y \vee Y^{-1} - 1).$$

To prove (7.6), note that by the Taylor formula

$$e^{-x} - e^{-sx} = x(s-1)e^{-x} + \frac{1}{2}x^2(1-s)^2 \exp\{-x[s+\theta_s(1-s)]\}$$

where $0 \le \theta_s \le 1$. Therefore, taking account of the condition $EY = 1$, we have

$$\Delta_1(x) = \left| xe^{-x}\int\limits_0^\infty (s-1)\,dG(s)+ \right.$$

$$\left. \frac{1}{2}x^2\int\limits_0^\infty (1-s)^2 \exp\{-x[s+\theta_s(1-s)]\}\,dG(s) \right| =$$

$$\frac{1}{2}x^2\left| \int\limits_0^\infty (1-s)^2 \exp\{-x[s+\theta_s(1-s)]\}\,dG(s) \right|. \qquad (7.10)$$

Further arguments are similar to those used to prove (7.5). Prove (7.7). Use inequality (7.9). According to the Cauchy-Schwarz inequality we have

$$I_1 \le x\left[\int\limits_0^1 (1-s)^2\,dG(s)\right]^{1/2}\left[\int\limits_0^1 e^{-2xs}\,dG(s)\right]^{1/2} \le$$

$$x\left[\int\limits_0^\infty (1-s)^2\,dG(s)\right]^{1/2}\left[\int\limits_0^\infty e^{-2xs}\,dG(s)\right]^{1/2} =$$

$$x[E(Y-1)^2 g_Y(2x)]^{1/2}. \qquad (7.11)$$

Using Lyapunov's inequality we obtain

$$I_2 \le xe^{-x}E|Y-1| \le xe^{-x}[E(Y-1)^2]^{1/2}. \qquad (7.12)$$

By unifying (7.11) and (7.12) we obtain (7.7). Estimate (7.8) is proved by similar reasoning applied to relation (7.10). The proof is completed.

Since for the example described above with the Poisson point process $N_k/\lambda \Rightarrow 1$ as $\lambda \to \infty$, for large λ by virtue of Theorem 5.7.2 we may use the relation

$$P(X_{m+1} \geq x | N \geq k) \approx \exp\{-\frac{s}{V}(\lambda - m)x\}. \tag{7.13}$$

The estimates contained in the above theorem, will give that the error of approximation (7.13) is $O(\lambda^{-1/2})$. To obtain an error of order $O(\lambda^{-1})$ we should use the approximation

$$P(X_{m+1} \leq x | N \leq k) \approx \exp\{-\frac{s}{V}[\mathbf{E}NI(N \leq k) - m]x\}.$$

Now turn to more general generalized homogeneous reliability growth models with independent decrements of volumes of defective sets. Let $S_0 = \xi_1 + \ldots + \xi_N$, where $\{\xi_j\}_{j \geq 1}$ are independent identically distributed random variables and N is a natural-valued random variable independent of $\{\xi_j\}_{j \geq 1}$. Remember that here ξ_j has the sense of the change of the volume of the defective set, after the j-th modification of the system.

For such a model, as is easily seen,

$$P(X_{m+1} \geq xV | N > m) = \sum_{n < m} \frac{P(N = n)}{P(N > m)} \int e^{-sx} dP(\xi_{m+1} + \ldots + \xi_n < s). \tag{7.14}$$

Let N_n be a random variable taking the value k with probability $q_k = P(N = m + k)/P(N > m)$, $k = 1, 2, \ldots$. Then, since the random variables $\{\xi_j\}_{j \geq 1}$ are independent and identically distributed, having continued (7.14) we obtain

$$P(X_{m+1} \leq xV | N > m) = \int e^{-sx} d\left[\sum_{k=1}^{\infty} q_k P(\xi_1 + \ldots + \xi_k < s)\right],$$

i.e., it is the Laplace-Stieltjes transform of a random sum of independent identically distributed random variables. Some properties of these objects were described in Chapters 3 and 4. They are convenient for the construction of approximating distributions. For convenience, consider the scheme of series and deal as if with a sequence of systems with the volumes of defective sets $S_{1,0}, \ldots, S_{n,0}, \ldots$. Let each of the variables $S_{n,0}$ be represented in the form

$$S_{n,0} = \xi_{n,1} + \ldots + \xi_{n,N_n}, \qquad n \geq 1,$$

where the random variables $\{\xi_{n,j}\}_{j \geq 1}$ are independent and identically distributed for every n, and the integer-valued positive random variables N_n are independent of the sequence $\{\xi_{n,j}\}_{j \geq 1}$ for every n. Let $X_{n,m}$ be the time between the $(m - 1)$-th and the m-th failures during the testing of the n-th system.

Let $N_{n,m}$ be the random variable taking the value k with probability $q_{n,m}(k) = P(N_n = m + k)/P(N_n > m)$, $k = 1, 2, \ldots$, and independent of the sequence $\{\xi_{n,j}\}_{j \geq 1}$.

THEOREM 5.7.4. *Let, in addition to the assumptions made above, there exist an infinitely increasing sequence of natural numbers $\{k_n\}_{n\geq 1}$ (possibly, also dependent on m) and random variables Z and N such that*

$$\xi_{n,1} + \ldots + \xi_{n,k_n} \Rightarrow Z$$

and

$$\frac{N_{n,m}}{k_n} \Rightarrow N$$

as $n \to \infty$. Then for any $x \geq 0$

$$\lim_{n\to\infty} P(X_{n,m+1} \geq xV|N_n > m) = \int\limits_0^\infty g^s(x)\, dW(s),$$

where $g(x)$ is the Laplace-Stieltjes transform of the random variable Z, $g(x) = Ee^{-Zx}$ and W is the distribution function of N.

In other words, approximations for the conditional reliability function $P(X_{m+1} \geq xV|N > m)$ in the generalized conditionally exponential reliability growth model with independent decrements of volumes of defective sets should be sought for among exponential mixtures of Laplace-Stieltjes transforms of infinitely divisible laws.

The proof of this statement differs from the proof of Theorem 4.1.1 only in notation.

Based on the result of Corollary 5.6.1, one can note that Theorem 5.7.4 is a kind of a connecting link between limit theorems for random sums and limit theorems for order statistics, constructed from samples with random sizes (Gnedenko, 1983) in the scheme of series. Note that in (Korolev, 1992) a similar result was formulated for the distribution of the variable X_1 in order-statistics-type reliability growth models, which may be interpreted as a characteristic of reliability of the system when the information on times of its preceding failures is missing. The necessity to investigate the distribution function of the random variable X_1 may also appear in the process of designing a system.

5.8 Statistical prediction of reliability by renewing models

In this section as well as in the next one we will consider the problem, that is in some sense converse to the one considered above: from the observed values of times of failures $Y_1, \ldots, Y_m$ in continuous models, or values $M_1, \ldots, M_m$ of numbers of tests that have led to failures in discrete models (or, which is equivalent, from the lengths of time intervals $X_1, \ldots, X_m$ between failures in continuous models, or numbers $K_1, \ldots, K_m$ of successive successful tests between failures in discrete models) determine probability characteristics of the random variable X_{m+1} (K_{m+1} in the discrete case), to predict reliability characteristics of the tested system. We will mainly consider the methods of statistical analysis of continuous models. Their modification for discrete case is obvious. We begin with the consideration of prediction of reliability by renewing models.

Consider a globally parametrized renewing model. For this model one can easily write the joint density of the random variables $X_1, \ldots, X_m$ with the help of relation (4.1), and hence, the likelihood function

$$L(\theta_1, \ldots, \theta_m; X_1, \ldots, X_m) =$$

$$\prod_{j=1}^{m} h_j(X_j; \theta_j) = \prod_{j=1}^{m} h(X_j; n, j, \alpha_1, \ldots, \alpha_r),$$

where $h_j(x, \theta_j)$ is the density of the random variable X_j, that has the form $h(x; n, j, \alpha_1, \ldots, \alpha_r)$ for a globally parametrized model (h is a fixed density, $n, \alpha_1, \ldots, \alpha_r$ are global parameters). To find the estimates $\hat{n}, \hat{\alpha}_1, \ldots, \hat{\alpha}_r$ of the global parameters the maximum likelihood method is traditionally used. Then, setting the density of the random variable X_{m+1} equal to

$$h(x; \hat{n}, m + 1, \hat{\alpha}_1, \ldots, \hat{\alpha}_r),$$

we can obtain the required reliability characteristics of the system. The corresponding examples may be found in (Shanthikumar, 1983).

Favorable properties of the maximum likelihood method are well known. However, it has one essential deficiency in the considered situation: if a chosen parametrization is wrong, then the prediction will not be exact. In other words, the maximum likelihood method seriously depends on the choice of the density h and of the function $\theta(n, j, \alpha_1, \ldots, \alpha_r)$. This is well-illustrated by the examples from the paper (Brocklehurst et al., 1990). A review of methods allowing improving the estimators obtained with the help of the maximum likelihood method is presented in this work.

One more way to increase stability of procedures of reliability growth statistical analysis by renewing models consists of applying the method of least squares, which in our case is as follows. For a globally parametrized model denote

$$\mathsf{E}X_j = \mu(n, j, \alpha_1, \ldots, \alpha_r), \quad \mathsf{D}X_j = \sigma^2(n, j, \alpha_1, \ldots, \alpha_r).$$

The least squares estimators $\tilde{n}, \tilde{\alpha}_1, \ldots, \tilde{\alpha}_r$ minimize the form

$$\sum_{j=1}^{m} [X_j - \mu(n, j, \alpha_1, \ldots, \alpha_r)]^2.$$

However, the nonlinearity of the problem is essentially complicated by its heteroscedasticity, since as a rule, $\mathsf{D}X_j$ essentially depends on j.

The heteroscedasticity may be compensated by the following iterative procedure of weighted least squares of an EM-algorithm type. We will take

$$(\tilde{n}^{(0)}, \tilde{\alpha}_1^{(0)}, \ldots, \tilde{\alpha}_r^{(0)}) = \arg \min_{n, \alpha_1, \ldots, \alpha_r} \sum_{j=1}^{m} [X_j - \mu(n, j, \alpha_1, \ldots, \alpha_r)]^2.$$

as an initial approximation (ordinary least squares estimates). Estimators of
the global parameters on the k-th iteration are determined as

$$(\widetilde{n}^{(k)}, \widetilde{\alpha}_1^{(k)}, \ldots, \widetilde{\alpha}_r^{(k)}) = \arg \min_{n, \alpha_1, \ldots, \alpha_r} \sum_{j=1}^{m} \left(\frac{X_j - \mu(n, j, \alpha_1, \ldots, \alpha_r)}{\sigma(\widetilde{n}^{(k-1)}, j, \widetilde{\alpha}_1^{(k-1)}, \ldots, \widetilde{\alpha}_r^{(k-1)})} \right)^2.$$

Simulation shows that this iterative procedure leads to essentially more accurate estimates than an ordinary method of least squares.

Note that the least squares estimators depend not on the distribution H,
but only on its moments of the first (ordinary) or the first two (iterative)
orders which determine the tendency of the changes of the random variables
$X_1, \ldots, X_m$ (or are determined by this tendency), which may be almost identical for different models. This is why we consider the method of least squares
to be more stable.

Unfortunately, as a rule, both the maximum likelihood method and the
method of least squares (especially the iterative one) require a large amount of
computations, which can cause a considerable computing error. The method
of moments described below is less accurate from the statistical point of view,
but it may appear to be a less laborious one or may have smaller computing
error, which compensates the statistical error. In our situation the method of
moments is performed in two steps. The first step is to find in any way (for
instance, by means of procedures of trend analysis or smoothing of time series
with the help of autoregressive moving average models) the trend $T(j)$, $j =
1, \ldots, m$, which is an estimate of the mean value of the random sequence
$X_1, \ldots, X_m$. The second step is to find the solution $\widetilde{n}, \widetilde{\alpha}_1, \ldots, \widetilde{\alpha}_r$ of the system

$$\begin{cases} T(m-r) & = \mu(n, m-r, \alpha_1, \ldots, \alpha_r) \\ T(m-r+1) & = \mu(n, m-r+1, \alpha_1, \ldots, \alpha_r) \\ \quad \cdots \\ T(m) & = \mu(n, m, \alpha_1, \ldots, \alpha_r). \end{cases}$$

It is clear that this method may be applied only if $r \leq m$. In its described
variant we take the trend values (and the corresponding expressions for EX_j)
in the last r points. It is done so since the last stages of testing determine
the tendency of the system reliability variation to a greater extent than the
initial ones.

The methods described above are oriented to globally parametrized reliability growth models. One more method which may be regarded as a modification of the method of moments, is described in (Korolev, 1992).

The base of this method is the assumption that the distribution of the
random variables X_j, $j \geq 1$ (lengths of time intervals between successive
failures) is exponential. The following reasons may be suggested in favor of
this choice of the distribution. First, volumes of changes of defective domains
on the final stages of testing are small as compared to the total volume V of
the space of input values $\mathcal{X}$. This means that mixing laws in representation
(4.2) are close to degenerate. Second, it is well known that spacings, which are
models of lengths of time intervals between failures in mosaic (order-statistics-
type) models, are asymptotically independent and exponential as the sample

size increases (e.g., see (Sweeting, 1986)). Finally, using the connection of an exponential law with a geometric one (see Lemma 5.3.3), it is very convenient to pass from discrete time to continuous and vice versa. The described method is equally applicable for both continuous and discrete models.

When solving the problem by this method, the tendency of variation of mean values of the random variables X_i, $i \geq 1$ (the trend $T(i)$) is determined first. The method of least squares is used for this purpose. Generally speaking, the type of the trend may be arbitrary and is determined according to some additional arguments. For instance, relation (5.4) implies that the trend should be sought for as the hyperbola

$$T(i) = a(b-i)^{-1}, \quad 1 \leq i \leq [b], \quad a < 0, \quad b \geq 1.$$

However, note that in this case the heteroscedasticity of the least squares problem, caused by the difference in distributions of the random variables X_i, $i \geq 1$, is complicated by its nonlinearity. At the same time, tests of the system may be conducted according to schemes different from the suggested one. For example, defects may be removed not after every failure. By this reason, it is advisable to foresee the possibility of searching $T(i)$ as a polynomial of an arbitrary degree, which simplifies computations essentially, or as an exponent which characterizes the trend in, say, the Moranda model (Moranda, 1975). As a rough approximation for EX_{m+1} we take

$$EX_{m+1} \approx \hat{X}_{m+1} = T(m+1).$$

We may try to make the predictor $\hat{X}_{m+1}$ more accurate. Since the random variables $X_1, \ldots, X_m$ may not be independent in real tests, to the series $X_i' = (T(i))^{-1} X_i$ obtained from the original series in order to stabilize the variance, we can adjust an autoregressive moving average model, with the help of which we can construct the prediction $\hat{X}_{m+1}'$ and take $\hat{X}_{m+1} = T(m+1)\hat{X}_{m+1}'$ as the predicted value of the original series.

One more way to improve a predictor is to use the so-called U-plot. This method goes back to the works (Dawid, 1984) and (Brocklehurst et al., 1990). Its essence is as follows. Let $\hat{W}_i(x)$ be the predicted distribution of the random variable X_i, while its real distribution function is $W_i(x)$. Let Q_i be the transform connecting W_i and $\hat{W}_i : W_i(x) = Q_i(\hat{W}_i(x))$. If the transform Q_i was known, one would have been able to obtain the real distribution W_i from the approximate $\hat{W}_i$. According to (Brocklehurst et al., 1990), the sequence $\{Q_i\}$ is almost stationary in many real situations, i.e., Q_i does not practically change as i changes. This allows one to substitute the transform Q_i with its estimate Q_i^* constructed from the preceding observations. Note that Q_i is the distribution function of the random variable $U_i = \hat{W}_i(X_i)$. Therefore it is suggested to take as Q_i^* the estimate of the distribution function of the random variable U_i constructed from the sample $X_{i-k}, X_{i-k+1}, \ldots, X_i$, where $k \leq i - 1$. For example, Q_i^* can be a smoothed empirical distribution function constructed from the mentioned sample. Having constructed Q_{m+1}^*, one should take $W_{m+1}^* = Q_{m+1}^*(1 - \exp\{-T(m+1))^{-1}x\})$ as an estimate of the distribution function of the random variable X_{m+1}, and the mean value computed with respect to this distribution function should be taken as $\hat{X}_{m+1}$.

When the pointwise prediction $\hat{X}_{m+1}$ is obtained, at the second step of solving the reliability prediction problem with the help of the method of moments we find an estimator $\hat{p}_{m+1}$ of p_{m+1}, which is the probability of a correct reaction of the system to a next input after the debugging is finished. Due to approximation (3.4) for the conditional distribution of the random variable X_{m+1}, $\hat{p}_{m+1} = 1 - \hat{X}_{m+1}^{-1}$ is taken as an estimate of p_{m+1}. To construct the confidence interval $[\underline{p}_{m+1}, \overline{p}_{m+1}]$ for p_{m+1}, we again use the exponential approximation and for $\gamma \in (0,1)$ set

$$\underline{p}_{m+1} = 1 - \frac{1}{\hat{X}_{m+1}} \log \frac{2}{1-\gamma}, \qquad \overline{p}_{m+1} = 1 - \frac{1}{\hat{X}_{m+1}} \log \frac{2}{1+\gamma}.$$

As this is so, $\mathsf{P}(\underline{p}_{m+1} \leq p_{m+1} \leq \overline{p}_{m+1}) \approx \gamma$. Under such a choice of the bounds $\underline{p}_{m+1}$ and $\overline{p}_{m+1}$, the hypothetically observed value of X_{m+1} is substituted with its predictor $\hat{X}_{m+1}$, and the quantiles of orders $\frac{1-\gamma}{2}$ and $\frac{1+\gamma}{2}$ of the geometric distribution are approximated by the same quantiles of the exponential law. To characterize the accuracy of this approximation, denote by $l(\alpha)$ an arbitrary α-quantile of the random variable $(1-p)X$, where X has the geometric distribution with parameter p, $\alpha \in (0,1)$. Then

$$|l(\alpha) - \log(1-\alpha)^{-1}| \leq 2(1-p)(2-\alpha)/(1-\alpha),$$

provided $p > 2/(3-\alpha)$. This inequality can be proved directly using a piecewise linear approximation of the distribution function of the random variable $(1-p)X$ and Lemma 5.3.3.

In order not to lose information about successful runs of the program after the last failure, the interval estimate may be revised using the standard techniques for estimating the parameter of a binomial law.

The precision of the prediction method can be characterized by the magnitude

$$\delta = \frac{|p_{m+1} - \hat{p}_{m+1}|}{|1 - \hat{p}_{m+1}|}.$$

With the help of the mosaic simulation model pseudotests were generated to which the described above approximate method of software reliability prediction was applied, and the trend was sought as a polynomial of degree not higher than three. The mean value of δ by one hundred pseudosamples appeared to be equal to 0.35, and the mean square deviation of δ was 0.39. This means that on the average, the described method correctly specifies the number of nines after the decimal point in the decimal representation of p_{m+1}.

This "nonparametric" method as well as the method of moments described before are based on the analysis of a trend. Along with traditional reliability characteristics, this feature gives us a possibility to predict such important characteristics as the expected number of tests remaining to achieve the required reliability (the probability of trouble-free performance in one test), or the expected number of tests remaining to achieve the required mean time between failures.

In fact, let $T_d(i)$ be the trend of the random sequence $\{K_i\}_{i \geq 1}$, where K_i is the number of tests between the $(i-1)$-th and the i-th failures and let p_0

be the required reliability of the system, that is the probability of successful operation of the system in one test. Put

$$N_d = N_d(p_0) = \min\left(i : T_d(i) > \frac{1}{1-p_0}\right).$$

Then as an estimate of the expected number of tests remaining to achieve the required reliability p_0 we can suggest

$$n_d = n_d(p_0) = \left[\sum_{j=m+1}^{N_d} T_d(j)\right].$$

Analogously, if $T_c(i)$ is the trend of the random sequence $\{X_i\}_{i\geq 1}$, where X_i is the time interval between the $(i-1)$-th and the i-th failures and τ_0 is the required mean time between failures (mean time of trouble-free performance), we put

$$N_c = N_c(\tau_0) = \min(i : T_c(i) > \tau_0).$$

Then as an estimate of the expected number of tests remaining to achieve the required mean time between failures (mean time of trouble-free performance) we can suggest

$$n_c = n_c(\tau_0) = \left[\sum_{j=m+1}^{N_c} T_c(j)\right].$$

5.9 Statistical prediction of reliability by order-statistics-type models

Theorem 5.6.1 allows us to use well-known properties of order statistics and corresponding spacings for the analysis of the failures times $Y_1,\ldots,Y_m$ in mosaic models. However, statistical analysis of reliability by mosaic models has one peculiarity that leads to unusual statistical problems. Namely, we know, as a rule, only the times $Y_1,\ldots,Y_m$ of the first m failures, where $m \leq n$ and n, which is the size of the complete sample and has the sense of the total number of defects in the program (system) before testing or adjustment, is unknown.

Due to Theorem 5.6.1 it is possible to interpret the failures times $Y_1,\ldots,Y_m$ as a set of order statistics, constructed from the sample of size m from a reduced (truncated) distribution. However, unlike traditional statistical problems connected with truncated distributions (e.g., see (Schneider, 1986), we cannot restrict ourselves to the estimation of the distribution. It is also necessary to determine (for instance, to estimate) the number of missing observations that were as if rejected during the truncation, which plays the role of the number of the remaining defects. This problem is not considered in any of the known works on processing of data with gaps (e.g., see references in (Little and Rubin, 1987)).

By analogy with different types of censoring, we will consider different types of truncation (as is mentioned in (Schneider, 1986), censoring with an unknown size of a complete sample is equivalent to truncation) depending on the interpretation of the initial data.

We will say that the sample $Y_1, \ldots, Y_m$ is truncated by the first type, if tests are held until a fixed time t. In this case the number of failures observed by the time t is random.

We will say that the sample $Y_1, \ldots, Y_m$ is truncated by the second type, if the number of observed failures m is assumed nonrandom.

We begin with the second situation. Its analysis will be performed in two stages. As we have already mentioned, according to Theorem 5.6.1 we consider $Y_1, \ldots, Y_m$ as a sample of size m from the distribution function

$$F(x) = \begin{cases} 0, & x < 0, \\ H(x)/H(t), & 0 \le x \le t, \\ 1, & x > t, \end{cases}$$

where t is the time when testing stopped and the distribution function $H(x)$ is defined in (6.1).

On the first stage, the distribution function $F(x)$ is identified. For instance, the distribution function H may be given in a parametric form: $H(x) = H(x; \theta)$. In Section 5.4 we gave examples of distributions representable in the form (6.1). This class of distributions contains exponential and hyperexponential distributions, as well as Weibull, Rayleigh and Pareto distributions. Thus the class of laws (6.1) is rather wide, and the parametrization may be set by choosing a finite number of types of laws differing, say, by the behavior of tails, and then we can introduce a natural parametrization within each type. Therefore, $F(x) = F(x; \theta)$ and the problem of identification is reduced to finding an estimator $\hat{\theta}$ of the parameter θ of the distribution function $F(x; \theta)$ (and hence, of $H(x; \theta)$), for example, with the help of the maximum likelihood method followed by a goodness-of-fit testing for the correspondence of the experimental data $Y_1, \ldots, Y_m$ to the chosen model (completely determined by the distribution function F).

The considered problem may be formalized by the statistical structure $(\mathbb{R}^m, \mathcal{B}_m, \mathcal{P}_m)$, where $\mathcal{B}_m$ is the Borel σ-algebra of subsets of $\mathbb{R}^m$, and the family of probability measures $\mathcal{P}_m$, defined on $\mathcal{B}_m$, is parametric:

$$\mathcal{P}_m = \{ \mathrm{P}^*_{n,\theta} : (n, \theta) \in \mathbf{Z}_m \times \Theta \},$$

where $\mathbf{Z}_m$ is the set of integer numbers greater or equal to m, and Θ is the set of possible values of the parameter θ. (Further, concerning some asymptotic problems, we will assume that we deal with the family of statistical structures $\{ (\mathbb{R}^m, \mathcal{B}_m, \mathcal{P}) \}_{m \ge 1}$.

Since the parameter n (the total number of elementary defective subsets) plays an important role in prediction of reliability characteristics of the system, but was not determined on the first stage, its estimate is sought for on the second stage. Mention an important feature of the problem under consideration: on the first stage, the distribution function H can be identified independently of the value of n, and as we will see below the methods of estimating n on the second stage do not depend on t, which is the time when testing stopped.

We are coming to the construction of the parameter n according to the following reasoning. First assume that the distribution function H is known. From the formal point of view this means that we consider a subfamily $\mathcal{P}_m^{(0)}$ of the family $\mathcal{P}_m : \mathcal{P}_m^{(0)} = \{\mathsf{P}_{n,\theta_0}^* : n \in Z_m\}$, where θ_0 is the "true" value of the parameter θ. Denote the probability on $\mathcal{F}$ inducing the measure $\mathsf{P}_{m,\theta}^*$ on $\mathcal{B}_m$ as $\mathsf{P}_{n,\theta}$. In the considered case we will omit the index θ_0 of the symbols of expectation, variance and covariance, corresponding to the measure P_{n,θ_0} and will simply write $\mathsf{P}_n, \mathsf{E}_n, \mathsf{D}_n$ and Cov_n.

Further we will intensively exploit some auxiliary statements which will be formulated as lemmas. The first statement is well known, and we give it here only for convenience of references.

LEMMA 5.9.1. *Let* $Z_1, \ldots, Z_n$ *be order statistics constructed from a sample of size* n *from a distribution function* Q, *to which corresponds the density* q. *Then the joint density of the random variables* $Z_{r_1}, \ldots, Z_{r_m}$ ($r_1 < r_2 < \ldots < r_m$, $m \le n$) *for* $z_1 \le z_2 \le \ldots \le z_m$ *has the form:*

$$g(z_1, \ldots, z_m) = \tag{9.1}$$
$$\frac{n! Q^{r_1-1}(z_1) q(z_1) (Q(z_2) - Q(z_1))^{r_2-r_1-1} q(z_2) \cdots (1 - Q(z_m))^{n-r_m} q(z_m)}{(r_1 - 1)!(r_2 - r_1 - 1)! \cdots (n - r_m)!}.$$

PROOF of this statement can be found, for example, in (David, 1970).

LEMMA 5.9.2. *Let* $U_1, \ldots, U_n$ *be order statistics constructed from a sample of size* n *from the uniform distribution on* $[0,1]$. *Then*

$$\mathsf{E} \frac{1}{U_i} = \frac{n}{i-1} \qquad (i \ge 2), \tag{9.2}$$

$$\mathsf{D} \frac{1}{U_i} = \frac{n(n-i-1)}{(i-1)^2(i-2)} \qquad (i \ge 3), \tag{9.3}$$

$$\text{Cov}\left(\frac{1}{U_i}, \frac{1}{U_i}\right) = \frac{n(n-1)}{(i-2)(j-1)} - \frac{n^2}{(i-1)(j-1)} \qquad (i > j \ge 2), \tag{9.4}$$

$$\text{Cov}\left(\frac{1}{U_i}, \frac{1}{1-U_{n-i+1}}\right) = -\frac{n}{(i-1)(j-1)} \qquad (n-i+1 > j), \tag{9.5}$$

PROOF. As is known, the i-th order statistic U_i in a sample of size n from the uniform distribution on $[0,1]$ has the beta distribution with the density

$$p_i(x) = \frac{x^{i-1}(1-x)^{n-1}}{B(i, n-i+1)}, \qquad 0 \le x \le 1,$$

where $B(c,b) = \int_0^1 x^{c-1}(1-x)^{b-1} dx$ is Euler's beta function, $c > 0, b > 0$.

Therefore using the properties of beta function, we obtain

$$E\frac{1}{U_i} = \frac{1}{B(i, n-i+1)} \int_0^1 x^{i-2}(1-x)^{n-i} dx = $$

$$\frac{B(i-1, n-i+1)}{B(i, n-i+1)} = \frac{n}{i-1}.$$

$$D\frac{1}{U_i} = E\frac{1}{U_i} - \left(E\frac{1}{U_i}\right)^2 = $$

$$\frac{1}{B(i, n-i+1)} \int_0^1 x^{i-3}(1-x)^{n-i} dx - \frac{n^2}{(i-1)^2} = $$

$$\frac{B(i-2, n-i+1)}{B(i, n-i+1)} - \frac{n^2}{(i-1)^2} = \frac{n^2}{(i-1)^2(i-2)} - \frac{n}{(i-1)(i-2)}.$$

When $i > j$, by Lemma 5.9.1 we have

$$\text{Cov}\left(\frac{1}{U_i}, \frac{1}{U_j}\right) = E\frac{1}{U_i U_j} - \frac{n^2}{(i-1)(i-2)} = $$

$$\frac{n!}{(j-1)!(i-j-1)!(n-i)!} \int_0^1 \int_0^x \frac{1}{xy} y^{j-1}(x-y)^{i-j-1}(1-x)^{n-i} dy dx -$$

$$\frac{n^2}{(i-1)(j-1)} = \frac{B(i-2, n-i+1)B(j-1, i-1)n!}{(j-1)!(i-j-1)!(n-i)!} - \frac{n^2}{(i-1)(j-1)} = $$

$$\frac{n(n-1)}{(i-2)(j-1)} - \frac{n^2}{(i-1)(j-1)}.$$

Further,

$$\text{Cov}\left(\frac{1}{U_j}, \frac{1}{1-U_{n-i+1}}\right) = E\frac{1}{U_j(1-U_{n-i+1})} - E\frac{1}{U_j} \cdot E\frac{1}{1-U_{n-i+1}}. \qquad (9.6)$$

But by the symmetry arguments, $1 - U_{n-i+1} \stackrel{d}{=} U_i$. Hence, due to (9.2) we have

$$E\frac{1}{U_j} \cdot E\frac{1}{1-U_{n-i+1}} = E\frac{1}{U_j} \cdot E\frac{1}{U_i} = \frac{n^2}{(i-1)(j-1)}. \qquad (9.7)$$

Using Lemma 5.9.1 we get

$$E\frac{1}{U_j(1-U_{n-i+1})} = $$

$$\frac{n!}{(j-1)!(n-i-j)!(j-1)!} \int_0^1 \int_0^x y^{j-2}(x-y)^{n-i-j}(1-x)^{i-2} dy dx = $$

$$\frac{n!}{(j-1)!(n-i-j)!(j-1)!} \times$$

$$\int_0^1 (1-x)^{i-2} x^{n-i-1} \, dx \int_0^1 t^{j-2}(1-t)^{n-i-j} \, dt =$$

$$\frac{n! B(n-i, i-1) B(j-1, n-i-j+1)}{(j-1)!(n-i-j)!(j-1)!} = \frac{n(n-1)}{(i-1)(j-1)}. \tag{9.8}$$

Substituting (9.7) and (9.8) in (9.6) we obtain (9.5). The lemma is proved.

Lemma 5.9.1 implies that the joint density of the failures times $Y_1, \ldots, Y_m$ is equal to

$$g^*(y_1, \ldots, y_m) = \frac{n!}{(n-m)!} (1 - H(y_m))^{n-m} \prod_{j=1}^m h(y_j), \tag{9.9}$$

where $y_1 < \ldots < y_m$ and h is the density corresponding to the distribution function H (see (6.1)). Hence, in the situation under consideration Y_m (that is, the time of the last observed failure) is a sufficient statistic for n.

Using representation (9.9), we will find a maximum likelihood estimator for n. Further everywhere $[x]$ denotes the integer part of the number x.

THEOREM 5.9.1. *With the distribution function H being known, the maximum likelihood estimator for n is given by the relation*

$$\hat{n}_{ML} = \max(m, \hat{n}_I), \tag{9.10}$$

where

$$\hat{n}_I = \begin{cases} [\hat{n}], & \text{if } L([\hat{n}]; Y_m, m) \geq L([\hat{n}] + 1; Y_m, m), \\ [\hat{n}] + 1, & \text{if } L([\hat{n}]; Y_m, m) < L([\hat{n}] + 1; Y_m, m), \end{cases} \tag{9.11}$$

$$\hat{n} = \frac{m}{H(Y_m)} \tag{9.12}$$

$$L(n; y, m) = \sum_{j=n-m+1}^n \log j + n \log(1 - H(y)). \tag{9.13}$$

PROOF. It follows from (9.9) that

$$\arg\max_{m \leq n} g^*(Y_1, \ldots, Y_m) = \arg\max_{m \leq n} \log g^*(Y_1, \ldots, Y_m) =$$

$$\arg\max_{m \leq n} L(n; Y_m, m),$$

where the function $L(n; y, m)$ is defined in (9.13). Consider the increments of this function in n. We have

$$L(n; y, m) - L(n-1; y, m) = \log n - \log(n-m) + \log(1 - H(y)) =$$

$$= \log \frac{n(1 - H(y))}{n - m}.$$

Find those n for which the given increments are nonnegative, i.e.,

$$n(1 - H(y)) > n - m. \tag{9.14}$$

Solving inequality (9.14), we obtain $n < m/H(y)$. Similarly we verify that for $n \geq m/H(y)$ the increments of the function $L(n; y, m)$ in n are nonpositive. Hence, since the parameter n takes only integer values, we obtain the required assertion. The proof is completed.

A direct investigation of analytical properties of the maximum likelihood estimator $\hat{n}_{ML}$ on the base of Theorem 5.9.1 seems difficult. However, already when $m \geq 20$ the distribution of $\hat{n}_{ML}$ is practically indistinguishable from negative binomial, whatever the distribution function H is. This fact is experimentally established by the statistical module of the program system QUEST which is designed for distribution identification (Kitaev et al., 1992). An explanation of this fact is that a negative binomial distribution for certain values of parameters may be well approximated by a gamma distribution (and vice versa), and the gamma distribution appears as the limit for order statistics, in terms of which the variable $\hat{n}$ is expressed (9.12). The latter variable differs from $\hat{n}_{ML}$ no more than by one (more explicitly see below). Unlike $\hat{n}_{ML}$, the properties of the variable $\hat{n}$ may be easily established.

THEOREM 5.9.2.   *Whatever the known distribution function H is, the following relation holds*

$$\hat{n} \stackrel{d}{=} m Z_{(n-m+1)} \tag{9.15}$$

where $Z_{(n-m+1)}$ is the $(n-m+1)$-th order statistic constructed from a sample of size n from the Pareto distribution with the distribution function

$$W(x) = \begin{cases} 0, & x < 1; \\ 1 - x^{-1}, & x \geq 1, \end{cases} \tag{9.16}$$

that is, $P_n(\hat{n} < x) = 0$ for $x < m$ and for $x \geq m$

$$P_n(\hat{n} < x) = \left(\frac{m}{x}\right)^n \sum_{i=n-m+1}^{n} C_n^i \left(\frac{x}{m} - 1\right)^i. \tag{9.17}$$

As this is so,

$$E_n \hat{n} = \frac{mn}{m-1}, \tag{9.18}$$

$$D_n \hat{n} = \frac{m^2 n(n-m+1)}{(m-1)^2(m-2)}. \tag{9.19}$$

PROOF. It is well known that if Y is a random variable with a continuous distribution function H, then the random variable $H(Y)$ has the uniform distribution on $[0, 1]$. Hence, $H(Y_m) \stackrel{d}{=} U_m$, where U_m is the m-th order statistic in a sample of size n from the mentioned uniform distribution. If the random variable U has the uniform distribution on $[0, 1]$, then the random variable U^{-1} has the Pareto distribution with distribution function (9.16). Therefore $U_m^{-1} \stackrel{d}{=} Z_{(n-m+1)}$, where $Z_{(n-m+1)}$ is the $(n - m + 1)$-th order statistic constructed from a sample of size n from the mentioned Pareto distribution, since the function $f(y) = y^{-1}$ monotonically decreases for $y > 0$. Thus, relation

(9.17) directly follows from the representation of the distribution function of an order statistic via the distribution function of a sample element (see, e.g., (David, 1980)). Relations (9.18) and (9.19) are the implications of formulas (9.2) and (9.3). The proof is completed.

As we have mentioned, $\hat{n}_{ML}$ and $\hat{n}$ differ no more than by one. Therefore due to Theorem 5.9.2 we can formulate the following statement.

COROLLARY 5.9.1. *The following inequalities hold*

$$\left(\frac{m}{x-1}\right)^n \sum_{i=n-m+1}^{n} C_n^i \left(\frac{x-1}{m} - 1\right)^i \le P_n(\hat{n}_{ML} < x) \le$$

$$\left(\frac{m}{x+1}\right)^n \sum_{i=n-m+1}^{n} C_n^i \left(\frac{x+1}{m} - 1\right)^i, \quad x \ge m, \tag{9.20}$$

$$\left|\frac{E_n \hat{n}_{ML}}{E_n \hat{n}} - 1\right| \le \frac{1}{n}, \tag{9.21}$$

$$\left|\frac{D_n \hat{n}_{ML}}{D_n \hat{n}} - 1\right| \le \frac{4(m-2)}{n-m+1}. \tag{9.22}$$

PROOF. Inequalities (9.20) follow from (9.17) and the inequalities

$$\hat{n} - 1 \le \hat{n}_{ML} \le \hat{n} + 1. \tag{9.23}$$

Relation (9.21) follows from (9.23) and (9.18). Finally, to prove (9.22) notice that by virtue of (9.23),

$$|D_n \hat{n}_{ML} - D_n \hat{n}| \le 4E_n \hat{n} = \frac{4mn}{m-1}$$

Dividing both parts of this inequality by $D_n \hat{n}$ and taking account of (9.19) we obtain (9.22). The corollary is proved.

As can be seen from (9.18) and (9.21), the maximum likelihood estimator $\hat{n}_{ML}$ is not unbiased. However, relation (9.18) shows how the bias may be reduced: we should put

$$\hat{n}_{ML}^* = \frac{m-1}{m} \hat{n}_{ML}. \tag{9.24}$$

As this is so, the variance of the estimator also decreases,

$$D_n \hat{n}_{ML}^* = \left(\frac{m-1}{m}\right)^2 D_n \hat{n}_{ML} < D_n \hat{n}_{ML}.$$

Nevertheless, alas, the properties of estimator (9.24) are not easy to investigate analytically. However, pay attention that transform (9.24) applied to $\hat{n}$ instead of the estimator $\hat{n}_{ML}$ gives an unbiased estimator. Put

$$\hat{n}_0 = \left(\frac{m-1}{m}\right) \hat{n}.$$

It is easy to see that due to (9.18), $E\hat{n}_0 = n$. Moreover, (9.22) implies the inequality

$$\left| \frac{D_n \hat{n}^*_{ML}}{D_n \hat{n}_0} - 1 \right| \leq \frac{4(m-2)}{n-m+1}.$$

However, this inequality is not enough to decide which of the estimators $\hat{n}^*_{ML}$ or $\hat{n}_0$ is more preferable. Later we will discuss the results of simulation, which clarify it all to a certain extent.

Pay attention that the estimator $\hat{n}_0$ may be obtained with the help of the least squares ideology. Let $H_n(x)$ be a left-continuous empirical distribution function constructed from the complete set of order statistics $Y_1, \ldots, Y_m$, $Y_{m+1}, \ldots, Y_n$. Then $H_n(Y_j) = (j-1)n^{-1}$. But it follows from the Glivenko theorem that $H_n(x) \approx H(x)$ for any x. Therefore $H(Y_j)$ should be close to $(j-1)n^{-1}$. Hence, n should be close to $(j-1)/H(Y_j)$ for all j. Realizing the requirement of closeness of n and $(j-1)/H(Y_j)$ for all j simultaneously, we will find the estimator of n according to the condition of minimum in n of the form

$$S_k(n) = \sum_{j=k}^{m} \left[n - \frac{j-1}{H(Y_j)} \right]^2.$$

Taking account of relation (9.2) and Theorem 5.9.2, we come to the family of estimators

$$\left\{ \tilde{n}_k = \frac{1}{m-k+1} \sum_{j=k}^{m} \frac{j-1}{H(Y_j)} \quad k = 1, 2, \ldots, m \right\}.$$

Moreover, Lemma 5.9.2 implies that the variances of $\tilde{n}_1$ and $\tilde{n}_2$ do not exist, and furthermore, $\tilde{n}_1$ have no expectation. The rest estimators of this family are unbiased, and Lemma 5.9.2 implies that for $k \geq 3$,

$$D_n \tilde{n}_k = \frac{n^2}{(m-k+1)^2} \left(\frac{1}{k-2} + \sum_{j=k+1}^{m} \frac{1+2j-2k}{j-2} \right) -$$

$$\frac{n}{(m-k+1)^2} \left(\frac{k-1}{k-2} + \sum_{j=k+1}^{m} \frac{(j-1)(1+2j-2k)}{j-2} \right). \qquad (9.25)$$

Denote $\eta_j = (j+1)/H(Y_{j-2})$, $1 \leq j \leq m-2$. For a fixed k, $3 \leq k \leq m-1$, the random variable $\tilde{n}_k$ would have been an optimal linear with respect to η_j estimator of n, if the random variables η_j, $k-2 \leq j \leq m-2$, were uncorrelated and had equal variances. However, as follows from (9.3) and (9.4), this is not so and hence, for the mentioned k the estimators $\tilde{n}_k$ are not optimal at all. At the same time, relations (9.3) and (9.4) show the direction where the ways of improving the estimators $\tilde{n}_k$ may be found. In fact, due to (9.4) we obtain that for $i < j$

$$\text{Cov}(\eta_i, \eta_j) = \frac{n^2}{i} - \frac{n(i+1)}{i} = n^2 \left(\frac{1}{i} + O\left(\frac{1}{n} \right) \right), \qquad (9.26)$$

as $n \to \infty$, and it follows from (9.3) that

$$D_n \eta_j = \frac{n^2}{i} - \frac{n(i+1)}{i} = n^2 \left(\frac{1}{i} + O\left(\frac{1}{n} \right) \right) \tag{9.27}$$

as $n \to \infty$. Denote $\eta = (\eta_1, \ldots, \eta_{m-2})$. One can see from (9.26) and (9.27) that the $(m-2) \times (m-2)$-matrix $\text{Cov}_n(\eta)$ has the form

$$\text{Cov}_n(\eta) = n^2 C_n, \tag{9.28}$$

moreover,

$$\lim_{n \to \infty} C_n = C \tag{9.29}$$

where $C = (c_{ij})$, $c_{ij} = \min(i^{-1}, j^{-1})$ and the limit is assumed elementwise. The idea of the improvement of the estimator $\tilde{n}_k$ is based on the possible due to (9.29) approximation of the matrix C_n in representation (9.28) by the matrix C and application of the generalized least squares method, according to which if the matrix $\text{Cov}(\eta)$ was precisely equal to $n^2 C$, then the best unbiased linear with respect to η_j estimator of the parameter n would be

$$\tilde{n}_0 = \left[\sum_{i=1}^{m-2} \sum_{j=1}^{m-2} c_{ij}^{(-1)} \eta_j \right] \left[\sum_{i=1}^{m-2} \sum_{j=1}^{m-2} c_{ij}^{(-1)} \right]^{-1}, \tag{9.30}$$

where $c_{ij}^{(-1)}$ are the elements of the matrix C^{-1}, inverse to C (see, e.g., (Seber, 1977)). In order to simplify (9.30), find the matrix C^{-1} in an explicit form. By induction on m, where to prove the inductive passage the bordering method is used (see, e.g., (Faddeev and Faddeeva, 1960)), we make sure that the matrix C^{-1} is tridiagonal:

$$c_{ij}^{(-1)} = \begin{cases} 2i^2, & \text{if } i = j < m - 2, \\ (m-2)^2, & \text{if } i = j = m - 2, \\ -ij, & \text{if } |i - j| = 1, \\ 0 & \text{in other cases.} \end{cases}$$

It is not difficult to see that the sum of the elements of the matrix C^{-1} in any line or in any column (except the last ones) is equal to zero. By virtue of this circumstance expression (9.30) takes a surprisingly simple form:

$$\tilde{n}_0 = \eta_{m-2} = \frac{m-1}{H(Y_m)}. \tag{9.31}$$

In other words, the improved least squares estimator $\tilde{n}_0$ coincides with the improved maximum likelihood estimator $\hat{n}_0$.

Exactly as Theorem 5.9.2 was proved, one can obtain the following statement concerning the properties of the estimator $\hat{n}_0$.

THEOREM 5.9.3. *Whatever the known distribution function H is, the following relation holds:*

$$\hat{n}_0 \stackrel{d}{=} (m-1)Z_{(n-m+1)}, \tag{9.32}$$

where $Z_{(n-m+1)}$ is the $(n-m+1)$-th order statistic constructed from a sample of size n from the Pareto distribution with distribution function (9.16), i.e., $P_n(\hat{n}_0 < x) = 0$ for $x < m-1$, and for $x \geq m-1$

$$P_n(\hat{n}_0 < x) = \left(\frac{m-1}{x}\right)^n \sum_{i=n-m+1}^{n} C_n^i \left(\frac{x}{m-1} - 1\right)^i. \qquad (9.33)$$

As this is so,

$$E_n \hat{n}_0 = n, \qquad D_n \hat{n}_0 = \frac{n(n-m+1)}{m-2}.$$

Now we can return to the comparison of the estimators $\hat{n}_0$ and $\hat{n}_{ML}^*$. Simulation shows that, generally speaking, the estimators $\hat{n}_0$ and $\hat{n}_{ML}^*$ are incomparable. However, we can mention that for $n \geq 400$ the estimator $\hat{n}_0$ has smaller variance practically for all m. Thus, we came to the conclusion that the estimator $\hat{n}_0$ is more preferable than the maximum likelihood estimator $\hat{n}_{ML}^*$ since, first, it is unbiased, second, it has a considerably simpler form, and third, it is more effective for large n. Therefore further we will concentrate on the properties of $\hat{n}_0$.

Consider the asymptotic properties of the estimator $\hat{n}_0$.

THEOREM 5.9.4. *The estimator $\hat{n}_0$ is asymptotically optimal among unbiased estimators of the parameter n that linearly depend on η_j, $j \geq 1$, in the sense that for each fixed $m \geq 3$*

$$\lim_{n \to \infty} n^{-2} D_n \hat{n}_0 \leq \lim_{n \to \infty} n^{-2} D_n^*,$$

where n^* is an arbitrary unbiased estimator of n of the form

$$n^* = \sum_{j \geq 1} p_j \eta_j, \qquad p_j = p_j(n) \in \mathbb{R}^1.$$

This statement is an implication of relations (9.28) and (9.29) and optimality properties of generalized least squares estimators.

Rigorously speaking, Theorem 5.9.4 does not establish asymptotic optimality of the estimator $\hat{n}_0$ in the traditional sense. In fact, asymptotics of estimators is usually meant with respect to an infinitely increasing sample size, which is interpreted as a volume of available information. In our case the size m of the available sample remains fixed and asymptotics is considered with respect to an unknown parameter n. The situation is interesting because traditional asymptotic settings of problems are inadmissible here, since with the growth of the volume of available information, the unknown parameter itself inevitably changes by virtue of the obvious condition $n \geq m$. Though Theorem 5.9.4 implies that

$$\lim_{m \to \infty} \lim_{n \to \infty} n^{-2} D_n \hat{n}_0 \leq \lim_{m \to \infty} \lim_{n \to \infty} n^{-2} D_n n^*,$$

this result cannot be considered exhausting either. The variable

$$\sup_{n \geq m} n^{-2} D_n \hat{n}_0$$

is of greater interest as well as its limit behavior as $m \to \infty$. Investigation of this variable in connection with optimality properties of $\hat{n}_0$ among all unbiased estimators of the parameter n, linearly depending on η_j, represents a special problem. Since $\hat{n}_0 = \tilde{n}_m$, we will consider a more restricted problem and will compare the estimators $\hat{n}_0 = \tilde{n}_m$ and $\tilde{n}_k$ for $k < m$.

For clarity, according to the argument of unification of dimensionality, it is convenient to consider the variable

$$\delta_n(m,k) = n^{-1}\sqrt{D_n \tilde{n}_k},$$

which characterizes the relative error of approximation of the parameter n by the random variable $\tilde{n}_k$. Denote

$$\delta(m,k) = \sup_{n \geq m} \delta_n(m,k).$$

THEOREM 5.9.5. *For all* $k = 3, \ldots, m$,

$$\delta(m,k) \leq$$

$$\frac{1}{m-k+1}\sqrt{\frac{1}{k-2} + 2(m-k) + \log\left(\frac{m-2}{k-2}\right)\left(\frac{2k-5}{m}+3\right)}.$$

PROOF. The inequality

$$D_n \tilde{n}_k \leq$$

$$\frac{n^2}{(m-k+1)^2}\left(\frac{1}{k-2} + 2(m-k) + \left(3 - 2k - \frac{5-2k}{n}\right)\sum_{j=k-1}^{m-2}\frac{1}{j}\right)$$

follows from (9.25) after simple calculations. Now the required follows from the inequalities

$$\sum_{j=k-1}^{m-2}\frac{1}{j} \leq \int_{k-2}^{m-2}\frac{1}{x}\,dx \leq \log\left(\frac{m-2}{k-2}\right).$$

The proof is completed.

In particular,

$$\begin{array}{ll}
\delta(20,3) \approx 0.295 & \delta(20,20) \approx 0.236 \\
\delta(60,3) \approx 0.177 & \delta(60,60) \approx 0.132 \\
\delta(100,3) \approx 0.139 & \delta(100,100) \approx 0.101 \\
\delta(200,3) \approx 0.090 & \delta(200,200) \approx 0.071.
\end{array}$$

Note that for all m

$$\delta_n(m,m) \leq \delta(m,m) = \lim_{n \to \infty} n^{-1}(D_n \hat{n}_0)^{1/2} = (m-2)^{-1/2}, \qquad (9.34)$$

that is, the relative error of the estimator $\hat{n}_0$ is $O(m^{-1/2})$ as $m \to \infty$.

Thus, the estimator $\hat{n}_0$, depending on Y_m only, is essentially more precise than the estimators $\tilde{n}_k$ depending on a greater number of the observations $Y_k, \ldots, Y_m$. In this case the benefit in δ is 20–30%.

Inequality (9.34) may be used for the construction of Chebyshev-type interval estimators of n. Namely, due to Chebyshev's inequality we have from (9.34) for any $\gamma \in (0, 1)$

$$P_n(|\hat{n}_0 - n| \leq n[(m-2)(1-\gamma)]^{-1/2}) \geq \gamma$$

with any $n \geq m$. Therefore a simple (but rather inaccurate) 100γ-percent confidence interval for n has the form

$$\left[\hat{n}_0 \cdot \frac{\sqrt{(1-\gamma)(m-2)}}{\sqrt{(1-\gamma)(m-2)} + 1}, \quad \hat{n}_0 \cdot \frac{\sqrt{(1-\gamma)(m-2)}}{\sqrt{(1-\gamma)(m-2)} - 1} \right].$$

Precise confidence intervals may be constructed by the use of the explicit form of distribution (9.33). However, due to some awkwardness of expression (9.33), the construction of exact confidence intervals is possible only with the help of numerical procedures. At the same time due to (9.33) one can easily obtain asymptotic distributions of $\hat{n}_0$, which may be used for the construction of approximate confidence intervals.

Denote the standard normal distribution function as $\Phi(x)$, and denote the gamma-distribution function with parameters λ and m as $\Gamma_{\lambda,m}(x)$,

$$\Phi(x) = \frac{1}{\sqrt{2\pi}} \int_{-\infty}^{x} e^{-\frac{1}{2}v^2} \, dy, \quad \Gamma_{\lambda,m}(x) = \frac{\lambda^m}{(m-1)!} \int_{0}^{x} y^{m-1} e^{-\lambda y} \, dy.$$

THEOREM 5.9.6. *Let m be fixed. Then for any $x > 0$*

$$\lim_{n \to \infty} P_n(\hat{n}_0/n < x) = 1 - \Gamma_{1,m}\left(\frac{m-1}{x}\right),$$

$$\lim_{n \to \infty} P_n(n/\hat{n}_0 < x) = \Gamma_{1,m}((m-1)x).$$

PROOF of this theorem is based on the well-known fact: if $Z_{(n-m+1)}$ is the $(n-m+1)$-th order statistic in a sample of size n from the continuous distribution function W, then under the conditions of the theorem,

$$\lim_{n \to \infty} P_n(n(1 - W(Z_{(n-m+1)})) < x) = \Gamma_{1,m}(x). \tag{9.35}$$

Substituting expression (9.16) for W into (9.35), we obtain the required. The proof is completed.

Let $g_{1,m}(\beta)$ be the β-quantile of the gamma distribution with parameters 1 and m. Then under conditions of Theorem 5.9.6 the approximate confidence interval for n with the confidence probability γ has the form

$$\left[\hat{n}_0 g_{1,m}\left(\frac{1-\gamma}{2}\right), \hat{n}_0 g_{1,m}\left(\frac{1+\gamma}{2}\right) \right].$$

THEOREM 5.9.7. *Let* $m \to \infty$ *(and therefore* $n \to \infty$*) so that* $m = [n\alpha]$, $\alpha \in (0,1)$. *Then for any* $x \in \mathbb{R}$

$$\lim_{m \to \infty} P_n \left(\sqrt{\frac{n\alpha}{1-\alpha}} \left(\frac{\hat{n}_0}{n} - 1 \right) < x \right) = \Phi(x).$$

PROOF of this statement is also based on representation (9.32). But this time we should use the well-known result about asymptotic normality of order statistics: if $Z_{(r)}$ is the r-th order statistic in a sample of size n from the distribution function W with the density w, $r = [n\alpha]$ ($\alpha \in (0,1)$), $\nu_\alpha = W^{-1}(\alpha)$ and w is continuous and is not equal to zero at the point ν_α, then as $n \to \infty$ $Z_{(r)}$ is asymptotically normal with mean ν_α and variance

$$\frac{\alpha(1-\alpha)}{n[w(\nu_\alpha)]^2}$$

(see, e.g., (Cox and Hinkley, 1975)). Note that for the distribution function W defined by relation (9.16), the value of ν_α is equal to $(1-\alpha)^{-1}$, and the density w in our case has the form

$$w(x) = x^{-2}I(x \geq 1).$$

Substituting these values into the expression for the asymptotic mean and variance, we obtain the required. The proof is completed.

Let $u(\beta)$ be the β-quantile of the standard normal distribution. Then under the conditions of Theorem 5.9.7 the approximate confidence interval for n with the confidence probability γ has the form $[n_l, n_r]$, where

$$n_l = \frac{1}{2} \left(m \left(2\hat{n}_0 - u^2 \left(\frac{1+\gamma}{2} \right) \right) - R \right) \left(m - u^2 \left(\frac{1+\gamma}{2} \right) \right)^{-1},$$

$$n_r = \frac{1}{2} \left(m \left(2\hat{n}_0 - u^2 \left(\frac{1+\gamma}{2} \right) \right) + R \right) \left(m - u^2 \left(\frac{1+\gamma}{2} \right) \right)^{-1},$$

$$R = u \left(\frac{1+\gamma}{2} \right) \left(4m\hat{n}_0(\hat{n}_0 - m) + m^2 u^2 \left(\frac{1+\gamma}{2} \right) \right)^{1/2}.$$

The case $m \to \infty$, $n \to \infty$ so that $n - m = \text{const}$, is easily reduced to the one considered in Theorem 5.9.6. In fact, denoting $n - m + 1 = r$, we obtain that in this case

$$P_n(\hat{n}_0 < x) = P_n \left(\frac{m-1}{U_m} < x \right) = P_n \left(U_{n-r+1} > \frac{m-1}{x} \right) =$$

$$P_n \left(n(1 - U_{n-r+1}) < (r + m - 1) \left(1 - \frac{m-1}{x} \right) \right),$$

where U_j is the j-th order statistic in a sample of size n from the uniform on $[0,1]$ distribution. Hence with the help of the above-mentioned statement about the behavior of extreme order statistics we have

$$P_n(\hat{n}_0 < x) \approx \Gamma_{1,r} \left((m+r-1) \left(1 - \frac{m-1}{x} \right) \right), \qquad x \geq m-1. \qquad (9.36)$$

Unfortunately, in this situation the confidence interval does not have such a simple form as in the previous two, due to the complicated dependence of the approximating distribution on the unknown parameter r.

In practice, the conditions of Theorem 5.9.6, apparently, may be considered fulfilled if, first, the value of $\hat{n}_0$ is large, second, rather many failures have been fixed and, third, time intervals between failures are relatively small and approximately equal.

The conditions of Theorem 5.9.7, apparently, may be considered fulfilled in practice if, first, the value of $\hat{n}_0$ is large, second, rather many failures have been fixed and, third, time intervals between failures are approximately equal, but they essentially exceed time intervals between first failures.

Finally, the conditions leading to approximation (9.36) may be considered fulfilled if the value of $\hat{n}_0$ is large, rather many failures have been fixed and very many tests were held between the last failures.

So assume that on the first stage of the analysis of reliability by order-statistics-type models the estimator $\hat{\theta}$ of the parameter θ of the distribution function $H(x,\theta)$ has been found; in other words, the distribution function H has been identified. If H is identifiable by the estimator $\hat{\theta}$ precisely enough, then the results given above give us a reason to use

$$\hat{\hat{n}}_0 = \frac{m-1}{H(Y_m;\hat{\theta})}$$

as an estimator of n. However, in this case the estimator $\hat{\hat{n}}_0$ is, generally speaking, not obliged to inherit good properties (for instance, unbiasedness) of the random variable $\hat{n}_0$. The properties of $\hat{\hat{n}}_0$ are mainly determined by the properties of the estimator $\hat{\theta}$.

Now we turn to truncation of the first type, when $\mu(t)$, which is the number of failures fixed until the moment t, is a random variable. If we knew only H and $\mu(t)$, then the best estimator of n would be

$$n' = n'(t) = \mu(t)/H(t).$$

This estimator, as is not difficult to see, is unbiased, and its variance equals

$$Dn'(t) = n(1 - H(t))/H(t),$$

whence it follows that for small t, when $H(t)$ is close to zero, this estimator is not accurate. To use all information, in the same way it was done in the situation considered before, we can consider the estimator

$$n^*(t) = \frac{\mu(t) - 1}{H(Y_{\mu(t)};\hat{\theta}_t)},$$

where $\hat{\theta}_t$ is some estimator of the parameter θ. Simulation shows that statistical properties of this estimator for $t = H^{-1}(m/n)$ are similar to the properties of the estimator $\hat{n}(m)$. Therefore for moderate t, the estimator $n^*(t)$ is hitherto preferable to $n'(t)$.

The constructed estimators of the distribution function H and n should be substituted, for example, into the expression for the conditional reliability function $R_t^{(m+1)}(x)$, given in Sect. 5.6, with the help of which main reliability characteristics of the investigated system can be determined.

Appendix 1

Information properties of probability distributions

A1.1 Mathematical models of information and uncertainty

The subject of this appendix lies aside from the main direction of the book. We are not going to present new material here; the results given below are very well known. The character of this appendix is primarily methodical but not theoretical. We undertake an attempt here to formulate some general principles aimed at facilitation of the construction of mathematical (first of all, probability) models in specific practical situations. In Sect. 3.4, 3.5, 4.4 and Chapter 5 we have already demonstrated how important the choice of an appropriate probability distribution modeling real phenomena is in some practical problems. The purpose of the material presented here is to point out some prompts on those distributions which should be first suspected in being mathematical models in more or less uncertain situations in which complete information about the phenomena under study is not available. We also intend to direct the attention of readers of this book to the ideas of Yu. V. Linnik, A. Rényi and others who suggested looking at limit theorems of probability theory as manifestations of the universal principle of non-decrease of uncertainty which, in particular, also reveals itself in the second law of thermodynamics.

We begin with the construction of mathematical models of information and uncertainty themselves. The ideas of connecting the notions of uncertainty and probability due to Hartley and Shannon proved to be very reasonable and fruitful.

Let A and B be events with positive probabilities.

DEFINITION A1.1.1. The Shannon information contained in an event B concerning an event A is

$$I(A|B) = \log \frac{P(A|B)}{P(A)}.$$

If $B = A$, then obviously $I(A|A) = -\log P(A)$. Thus we appear at the following important definition.

DEFINITION A1.1.2. The information contained in an event A is

$$I(A) = -\log P(A). \tag{1.1}$$

The meaning of these definitions is easily clarified by the elementary properties of information defined in this way. They directly resemble those properties which should be inherent to information as assumed by the common sense.

1) If an event with small probability occurs, then, usually, it carries more information than the occurrence of an event with probability close to 1. For example, it seemed almost improbable to catch a live fish later called latimeria. And when at last it was caught, a revolution in ichthyology burst out. At the same time, catching of such a common fish as a herring does not bring much new scientific information.

2) If events A and B are independent, then, obviously, the occurrence of B does not give any information concerning A. Indeed, in this case we have
$$I(A|B) = \log 1 = 0.$$

3) If events A and B are independent, then their simultaneous occurrence carries as much information as occurrence of both of them. Indeed, in this case we have

$$I(AB) = -\log P(AB) = -\log(P(A)P(B)) = \\ -\log P(A) - \log P(B) = I(A) + I(B).$$

The definition of information of logarithmic type (1.1) goes back to Hartley (Hartley, 1928). The base of the logarithm is not very significant and determines only the choice of a unit of information. Most commonly used are logarithms with base 2, for which a unit information (bit) is contained in an event whose probability is $1/2$.

Let $\mathcal{E}$ be an experiment in which only one of n disjoints outcomes $A_1,\ldots,$ A_n can occur. Denote $P(A_i) = p_i$, so that $p_1 + \ldots + p_n = 1$. Then we can regard the information obtained as a result of this experiment as a random variable taking values $I(A_1),\ldots,I(A_n)$ with probabilities $p_1,\ldots,p_n$, respectively. Denote this random variable as $Q(\mathcal{E})$. Introduce the following aggregate information characteristic of $\mathcal{E}$.

DEFINITION A1.1.3. The entropy $H(\mathcal{E})$ of is defined as

$$H(\mathcal{E}) = EQ(\mathcal{E}) = -\sum_{i=1}^{n} p_i \log p_i. \tag{1.2}$$

The entropy of an experiment can serve as a measure of its uncertainty, as can be seen from the well-known properties of $H(\mathcal{E})$.

1) $H(\mathcal{E}) \geq 0$ with equality attained if and only if there is an $i \in \{1, \ldots, n\}$ such that $p_i = 1$.

2) Let $\mathcal{E}_0$ be an experiment with n equiprobable outcomes. Then $H(\mathcal{E}) \leq H(\mathcal{E}_0)$ for any experiment $\mathcal{E}$ with n outcomes.

3) Let $\mathcal{E}_1$ be an experiment with $n - 1$ outcomes constructed from $\mathcal{E}$ by the unification of two outcomes, say, A_i and A_j $(i \neq j)$ and let $\mathcal{E}_2$ be an experiment with outcomes A_i and A_j whose probabilities (within $\mathcal{E}_2$) are, respectively, equal to $p_i/(p_i + p_j)$ and $p_j/(p_i + p_j)$. Then

$$H(\mathcal{E}) = H(\mathcal{E}_1) + (p_i + p_j)H(\mathcal{E}_2).$$

4) $H(\mathcal{E})$ does not depend on $A_1, \ldots, A_n$, but depends only on $p_1, \ldots, p_n$ being a symmetrical function of $p_1, \ldots, p_n$.

5) $H(\mathcal{E})$ is a continuous function of $p_1, \ldots, p_n$.

The proofs of these properties as well as the discussion of other ones can be found, e.g., in (Goldman, 1953). D. K. Faddeev proved that the system of properties 1-5 uniquely determines the function (1.2), see (Faddeev, 1956). Other measures of uncertainty obtained by relaxing some of conditions 1-5 are reviewed, e.g., in (Csiszár, 1977).

It is obviously evident that instead of experiments we can consider discrete random variables taking values $x_1, \ldots, x_n$ with probabilities $p_1, \ldots, p_n$ by assigning the value x_i of a discrete random variable X to the outcome A_i of $\mathcal{E}$. Therefore by complete analogy with (1.2) we can define the entropy of a simple random variable X as

$$H(X) = -\sum_{i=1}^{n} p_i \log p_i,$$

or introducing the density $p(x)$ of X with respect to a counting measure

$$p(x) = \begin{cases} p_i, & x = x_i, \ i = 1, \ldots, n; \\ 0, & x \notin \{x_1, \ldots, x_n\}, \end{cases}$$

as

$$H(X) = -\mathsf{E} \log p(X). \tag{1.3}$$

By analogy with (1.3), if X is an absolutely continuous random variable with Lebesgue density $p(x)$, then we formally define the entropy $H(X)$ of X also as (1.3). The analogy of the definitions of entropies of discrete and absolutely continuous random variables is just formal. As is known, each random variable X can be represented as a pointwise limit of simple random variables X_n. Then, of course, $X_n \Rightarrow X$. But if we have an attempt to use the passage to the limit as the number n of possible values of an approximate simple random variable infinitely increases to obtain the entropy of a limit absolutely continuous random variable in the form (1.3), we will inevitably fail, because the limit for the entropies of prelimit simple random variables in this case is infinite. But (1.3) is the limit for "standardized" entropies of approximating simple random variables in the following sense. Actually the entropy (1.3) of a continuous random variable measures the average information content of X apart from an "infinitely large additive constant". To illustrate this split the range of X into disjoint intervals Δ_i of equal length δ. Define a corresponding discrete approximation X_δ of X so that X_δ equals a fixed element of an interval Δ_i if X falls into the same interval. Then $X_\delta \Rightarrow X$ ($\delta \to 0$) and under some regularity conditions on the density $p(x)$ of X for some points $x_i^* \in \Delta_i$ we have

$$H(X_\delta) = -\sum_i P(X \in \Delta_i) \log P(X \in \Delta_i) = -\sum_i p(x_i^*)\delta \log(p(x_i^*)\delta) =$$

$$-\sum_i p(x_i^*)\delta \log p(x_i^*) - \sum_i p(x_i^*)\delta \log \delta =$$

$$-\sum_i p(x_i^*)\delta \log p(x_i^*) - \log \delta. \tag{1.4}$$

The first term in the right-hand side of (1.4) is an integral sum for $H(X)$ defined by (1.3). Therefore we can write

$$H(X) = \lim_{\delta \to 0}[H(X_\delta) + \log \delta].$$

In accordance with Definition A1.1.2 $-\log \delta$ can be interpreted as the information content of an event whose probability equals δ and thus it can characterize the increase of uncertainty of X_δ due to quantization.

The concept of information can be formalized in many ways. Besides the approach due to Shannon considered above, below we will also consider the Fisher information $J_0(X)$ of a random variable X with continuously differentiable density $p(x)$ which is defined as

$$J_0 = E\left[\frac{p'(X)}{p(X)}\right]^2.$$

A1.2 Limit theorems of probability theory and the universal principle of non-decrease of uncertainty

The entropies (1.2) and (1.3) appear to be formally closely connected with the thermodynamic entropy of a closed physical system defined by Boltzman as

$$H = -\sum_{i=1}^{N} \frac{m_i}{M} \log \frac{m_i}{M}$$

where M is the number of molecules in the system and m_i is the number of molecules with velocity $v \in [v_i, v_i+\delta)$, $i = 1, \ldots, N$. The famous second law of thermodynamics declares that $\Delta H \geq 0$, that is, the entropy of a closed system does not decrease in time. The formal coincidence of definitions of entropies in physics and in information theory gives rise to the question, whether analogs of the second law of thermodynamics exist in probability theory.

Another reason for the search of relationships between the results of probability theory and the second law of thermodynamics is as follows. This physical law also reveals itself as a principle of non-decrease of uncertainty of not necessarily purely physical closed real systems (that is, not necessarily described at a "physical level"). Since probability theory is, in general, an applied science which studies the mathematical models of real phenomena influenced by uncertain or stochastic factors, there should be mathematical, more precisely, probabilistic models of the universal principle of non-decrease of uncertainty itself.

Before we turn to the main conclusion of this appendix and formulate some general principles and problems based on this conclusion, we recall extremal entropy properties of some probability distributions.

LEMMA A1.2.1. (i) *Let Y_1 be a random variable with uniform distribution on $[-a, a]$ for some real $a > 0$. Then for any random variable X such that* $P(-a \leq X \leq a) = 1$, *we have $H(X) \leq H(Y_1)$.*

(ii) *Let Y_2 be a random variable with exponential distribution,* $EY_2 = \alpha > 0$. *Then for any random variable X such that* $P(X \leq 0) = 1$, $EX = \alpha$ *we have $H(X) \leq H(Y_2)$.*

(iii) *Let Y_3 be a normal random variable with parameters $0, \sigma^2$. Then for any random variable X with $EX = 0$ and $EX^2 = \sigma^2$ we have $H(X) \leq H(Y_3)$.*

For the PROOF see, e.g., (Goldman, 1953).

Thus, uniform distribution has maximum entropy among all distributions with bounded support, exponential distribution has maximum entropy among all distributions concentrated on the positive halfline and possessing finite expectations, normal distribution is the law with maximum entropy among all distributions with finite variances.

We see that the distributions mentioned in Lemma A1.2.1 play most important roles in many limit theorems of probability theory. Indeed, the central limit theorem, in full accordance, with its name plays the central role in probability theory and mathematical statistics. Different modifications of this theorem describe convergence to the normal law of distributions of many random variables which additively depend on other "atomic" random variables. Exponential distribution, first, plays an important role appearing as a limit law in limit theorems for random variables also obtained from "atomic" ones but with operation of summation replaced by that of taking maximum. Second, being closely connected with Poisson distributions (exponentially distributed interarrival times characterize Poisson flows), it also implicitly participates in limit theorems for sums of random variables. So the comparison of limit laws appearing in limit theorems of probability theory with the distributions mentioned in Lemma A1.2.1 brings us to the important conclusion that *the universal principle of non-decrease of uncertainty manifests itself in probability theory in the form of limit theorems when the limit is taken with respect* to infinitely increasing number of "atomic" random variables involved in a model.

Although this conclusion is rather a general conjecture than a strictly proved assertion, we can give some evidences in favor of its validity. Indeed, we often see that the mathematical models of real phenomena obtained as approximations constructed on the base of limit theorems of probability theory turn out to be satisfactorily adequate, while the phenomena being modeled, obey the second law of thermodynamics or its generalizations.

Moreover, not long ago new proofs and refinements of some classical limit theorems were obtained using the information technique. As an example we give a central limit theorem for densities and some corollaries. We will follow the ideas of A. R. Barron (Barron, 1986) who in turn followed the results of Yu. V. Linnik (Linnik, 1959), A. Rényi (Rényi, 1970), L. D. Brown (Brown, 1982), I. Csiszár (Csiszár, 1967), S. Kullback (Kullback, 1967) and others.

Linnik (Linnik, 1959) used the information measures of Shannon and Fisher in a proof of weak convergence of standardized sums of independent identically distributed random variables. Rényi (Rényi, 1970, p. 601) states that Linnik established convergence of $\int f_n(x) \log(f_n(x)/\phi(x)) \, dx$ to zero where f_n is the density of a standardized sum and ϕ is the standard normal density. A reading of Linnik reveals that convergence was established only for densities of truncated random variables smoothed by the addition of independent normal random variables. Brown (Brown, 1982) gave an elegant proof of weak convergence based on the decrease of Fisher informations. Barron (Barron, 1986) extended Brown's argument to show that the Fisher informations converge to the reciprocal of the variance (as suggested by the Cramér-Rao bound). He also established a relationship between Fisher information and Shannon entropy described by a rather unexpected identity presented in Lemma A1.2.2 below.

Let X be a random variable with finite variance. Instead of Shannon entropy introduced in (1.3), in the proof of the central limit theorem it is more convenient to deal with relative entropy which is defined in the following way. If $p(x)$ is the density of X (with respect to Lebesgue measure), then the relative entropy of X is

$$D(X) = \int p(x) \log \left[\frac{p(x)}{\phi(x)} \right] dx$$

where ϕ is the normal density with the same mean and variance as p. Otherwise $D(X) = \infty$. It is easily seen that if $\mathsf{E}X = a$, $\mathsf{D}X = \sigma^2$, then

$$D(X) = \int p(x) \log p(x) \, dx - \int p(x) \log \phi(x) \, dx =$$

$$-H(X) + \int p(x) \left[\frac{1}{2} \log(2\pi\sigma^2) + \frac{(x-a)^2}{2\sigma^2} \right] dx =$$

$$-H(X) + \frac{1}{2} \left[\log(2\pi\sigma^2) + 1 \right],$$

so that $H(X) = \frac{1}{2}[\log(2\pi\sigma^2) + 1] - D(X)$ with $\frac{1}{2}[\log(2\pi\sigma^2) + 1]$ obviously being the entropy of a normal law with variance σ^2. Hence, relative entropy

$D(X)$ is the difference between Shannon normal entropy and the entropy of X. By concavity of logarithm, D is nonnegative and equals zero only if $p(x) = \phi(x)$ almost everywhere. (Consequently we obtain that the normal law has maximum entropy for a given variance, i.e., (iii) of Lemma A1.2.1.)

Let Y be a random variable with continuously differentiable density $g(y)$, $EY = a$, $DY = \sigma^2 < \infty$. Define the standardized Fisher information as

$$J(Y) = \sigma^2 E \left[\frac{g'(Y)}{g(Y)} - \frac{\phi'(Y)}{\phi(Y)} \right]^2,$$

where ϕ is the normal density with the same mean and variance as g. The Fisher information $J_0(Y) = E[g'(Y)/g(Y)]^2$ satisfies the relation

$$\sigma^2 J_0(Y) = J(Y) + 1.$$

Since $J \geq 0$ with equality only if $g = \phi$, the normal has minimum Fisher information for a given variance (whence the Cramér-Rao inequality $J_0 \geq \sigma^2$). The standardized informations D and J are translation and scale invariant.

Shannon entropy and Fisher information are linked by the following statement.

LEMMA A1.2.2. *Relative entropy $D(X)$ is an integral of Fisher information in the sense that if X is a random variable with finite variance, then*

$$D(X) = \int_0^1 J(\sqrt{t}X + \sqrt{1-t}Y)\frac{dt}{2t} \qquad (2.1)$$

where Y is an independent normal random variable with the same mean and variance as X.

For the PROOF see (Barron, 1986).

In terms of J and D the central limit theorem is formulated in the following way. Let $\{X_j\}_{j \geq 1}$ be independent and identically distributed random variables with zero mean and variance σ^2 and let $S_n = (X_1 + \ldots + X_n)/\sqrt{n}$ be the standardized sum. Let Y be a normal random variable with zero mean and variance σ^2 independent of $\{X_j\}_{j \geq 1}$. The following theorem states convergence of the Fisher informations.

THEOREM A1.2.1 (Barron, 1986). *Let $S'_n = \sqrt{t}S_n + \sqrt{1-t}Y$ or a fixed $t \in [0, 1]$. Then $J(S'_n) \geq J(S'_{2n})$ for each $n \geq 1$. Furthermore,*

$$\lim_{n \to \infty} J(S'_n) = 0. \qquad (2.2)$$

Equivalently, the sequence of Fisher informations $J_0(S'_n)$ converges to the Cramér-Rao bound σ^2.

Since $J(X) \geq 0$ with equality only if the density of X is normal, relation (2.2) means that the law of $\sqrt{t}S_n + \sqrt{1-t}Y$ becomes "more and more" normal as $n \to \infty$ and hence, due to the Lévy-Cramér theorem on decomposability of a normal law only into normal components, the law of S_n becomes "more and more" normal.

THEOREM A1.2.2 (Barron, 1986). *For each $n \geq 1$ we have $D(S_{2n}) \leq D(S_n)$. Furthermore, the relative entropy converges to zero*

$$\lim_{n \to \infty} D(S_n) = 0 \qquad (2.3)$$

if and only if $D(S_n)$ is finite for some n. Equivalently, Shannon entropy $H(S_n)$ converges to the normal entropy $\frac{1}{2}[\log(2\pi\sigma^2)+1]$ provided the entropy is finite for some n.

Actually Theorem A1.2.2 states something stronger than the classical central limit theorem which describes weak convergence of the laws of normalized sums to the normal distribution. Indeed, I. Csiszár (Csiszár, 1967) proved the inequality

$$\int |p(x) - q(x)| \, dx \leq \sqrt{2D}$$

with $D = \int p(x) \log(p(x)/q(x)) \, dx$, $p(x)$ and $q(x)$ being any probability densities. Hence, since the L_1 convergence of the densities is equivalent to the uniform setwise convergence of the distributions, from (2.3) it follows that

$$\lim_{n \to \infty} \sup_A |P(S_n \in A) - \Phi(A)| = 0 \qquad (2.4)$$

provided the sums S_n satisfy some regularity conditions, where the supremum is taken over all Borel sets A and $\Phi(A)$ is the $(0, \sigma^2)$-normal measure of A. Convergence (2.4) is stronger than the weak convergence for which the supremum in (2.4) should be taken over all sets with zero boundary measure. Moreover, the distance between the law of S_n and Φ in the sense of total variation

$$v_n = \sup_A |P(S_n \in A) - \Phi(A)|$$

with supremum over all Borel sets being directly estimated via standardized Fisher information. Namely, it can be shown that under some rather weak regularity conditions,

$$v_n \leq \sqrt{\frac{\pi}{2} J(S_n)}, \qquad (2.5)$$

see (Papadatos and Papathanasiou, 1995).

If we look at limit theorems of probability theory (e.g., at those for random sums of independent random variables) from an information point of view and agree with the above conclusion that these theorems describe the way the universal principle of non-decrease of uncertainty reveals itself in probability theory, then we immediately notice that the list of possible limit laws is not exhausted by the laws mentioned in Lemma A1.2.1 at all. Immediate examples are mixtures of normal, exponential or other infinitely divisible laws encountered in this book. This may mean, first, that the systems whose mathematical models are random sums are not assumed closed in the sense that the randomness is due not only to summands, but also to indices. As this is so, aggregated is only the uncertainty due to summands since their number is assumed to tend to infinity in some sense while the uncertainty due to indices should not necessarily increase because, in general, no structural assumptions are made concerning them. Second, in accordance with our main conclusion we should expect that the limit laws must possess some extremal entropy properties under certain conditions.

Hence we come against the following problems.

1) Find conditions under which certain limit laws appearing in limit theorems of probability theory possess extremal entropy properties. Immediate candidates to be subjected to such analysis are, of course, stable laws or laws which are limiting for extreme values. If this problem is solved, then, of course one more justification of our main conclusion will be obtained and, which is not less important, the role of these distributions as mathematical models in certain real situations will be clarified.

2) Give information proofs of limit theorems of probability theory such as those on convergence of distributions of sums of independent random variables to nonnormal stable laws or those on convergence of random sums.

3) Since maximum entropy distributions have very favorable characteristic properties such as lack-of-memory property of an exponential law, the equivalence of independence and non-correlatedness of normal random variables, decomposability of normal law only into normal components, a similar property of a Poisson distribution which is closely connected with an exponential law, it is worth to seek for the estimates of stability of characterizations of maximum entropy laws by their properties expressed in terms of information characteristics of type (2.5).

From the practical point of view we should note the following. The principle of maximum entropy is very well known and is widely used for construction of mathematical models. According to this principle, under an uncertainty of the choice of a specific form of a distribution to be used as a model of real phenomena, maximum entropy distributions should be tested first. In accordance with what has been said above this principle can be extended.

1) Under a lack of information the range of the search for appropriate models should be extended to combinations of distributions with maximum entropy, e.g., their mixtures, which may correspond to the models of open systems which do not involve all factors determining their behavior. Indeed, the distributions with maximum entropy are adequate models in ideal situations which assume the closedness of the system being modeled. At the same time no mathematical model, by definition, can take into account all factors that influence the phenomena under consideration. Therefore any modeled system is not closed and the fit of any distribution mentioned in Lemma A1.2.1 to experimental data is rather an exception than a rule. But it is quite natural to begin the

modeling of a system and making corresponding models more and more adequate. It is natural to expect that ideal maximum entropy models should somehow participate in more adequate models.

2) Conversely, if a distribution fits a real phenomena well, then its relationship with maximum entropy distributions should be investigated. For example, a possibility of the representation of this distribution as a mixture of maximum entropy laws should be investigated. An illustration of this principle is the example with Gaussian/inverse Gaussian distribution in Sect. 4.4. This relationship, provided it is established, can allow a researcher to find a setting of an appropriate limit theorem which in its turn can propose better understanding of the phenomena under consideration.

Appendix 2

Asymptotic behavior of generalized doubly stochastic Poisson processes

A2.1 General information on doubly stochastic Poisson processes

In this appendix, we show how the theorems proved in Chapter 3 as well as their extensions work if applied to generalized double stochastic Poisson processes which are natural and practically useful generalizations of well-known Poisson processes that are, of course, most frequently used in applied problems among all counting processes. Recall the definition of an ordinary Poisson process.

DEFINITION A2.1.1. A homogeneous stochastic process $N(t)$, $t \geq 0$, with independent increments which satisfies the conditions

1) $N(0) = 0$;

2) $P(N(t) = 0) = 1 - \lambda t + o(t)$,

$P(N(t) = 1) = \lambda t + o(t)$,

$P(N(t) \geq 2) = o(t)$ as $t \to 0$, for some $\lambda > 0$,

is called a Poisson process with intensity λ.

A Poisson process has step-wise non-decreasing trajectories with the height of each step equal to 1. For any integer $k \geq 1$

$$P(N(t) = k) = e^{-\lambda t} \frac{(\lambda t)^k}{k!},$$

whence it follows that $EN(t) = DN(t) = \lambda t$, $t > 0$. The time intervals between successive jumps of a Poisson process are independent random variables with one and the same exponential distribution: if T_j, $j \geq 1$ are times when the jumps of a Poisson process occur, then

$$P(T_j - T_{j-1} < x) = 1 - e^{-\lambda x}, \qquad x \geq 0 \tag{1.1}$$

where, for definiteness, $T_0 = 0$.

Property (1.1) of a Poisson process is characteristic, i.e., it may be used as the definition of a Poisson process. As we have already seen in Appendix 1, the exponential distribution is a kind of attractor in the framework of reasoning based on the law of non-decrease of entropy. Therefore relation (1.1) is an explanation of the fact that Poisson processes are satisfactory mathematical models in lots of situations which deal with counting of random events, arrivals of customers, etc.

We can say that they are the best initial approximations in many situations. By initial approximation we mean here a model which assumes the situation being modeled to be closed in the sense that the influence of many factors is ignored (see Appendix 1). But for more precise inference one should involve more and more factors into consideration resulting in that the model becomes more and more complicated.

To illustrate this, we consider a generalization of a Poisson process obtained by relaxing the assumption that the intensity is constant. In some sense, we open our model and introduce one more source of randomness, namely, the one related to the intensity of a Poisson process. The resulting model is called a doubly stochastic Poisson process or a Cox process. Now we shall give its formal definition. A homogeneous Poisson process with intensity one will be called a standard Poisson process and denoted $N_1(t)$.

DEFINITION A2.1.2. A random process $\Lambda(t)$, $t \geq 0$, with $\Lambda(0) = 0, \Lambda(t) < \infty$ $(0 < t < \infty)$ almost surely and non-decreasing trajectories is called a random measure.

DEFINITION A2.1.3. Let $\Lambda(t)$ be a random measure. Then the process $N(t) = N_1(\Lambda(t))$ is called a doubly stochastic Poisson process (or a Cox process). In this case we shall say that the Cox process $N(t)$ is controlled by the process $\Lambda(t)$ (or that the process $\Lambda(t)$ controls the Cox process $N(t)$).

General properties of Cox properties are well known. In full detail they are described in (Grandell, 1976). In this section we recall those properties of Cox processes which are connected with the results presented in our book.

Cox processes appear to be closely connected with the operation of rarefaction of point processes described in Chapters 2 and 3. We introduced the operation of elementary rarefaction of a point process (p-rarefaction) in the following way. Let $p \in (0, 1]$ and let $N(t)$ be a point process. The operation of elementary p-rarefaction leaves each point of $N(t)$ unchanged with probability p and omits it with probability $1 - p$. All points are left or omitted independently of each other. The p-rarefied process $N(t)$ will be denoted $N^{(p)}(t)$. Let $\mathcal{P}$ be the set of all point processes, $\mathcal{C}$ be the set of Cox processes. By D_p we denote the operator of p-rarefaction, $D_p : \mathcal{P} \to \mathcal{P}$; $\mathcal{D} = \{D_p N : N \in \mathcal{P}\}$ is the set of point processes obtained by p-rarefaction.

If $N(t)$ is a Poisson process with intensity λ, then $N^{(p)}(t)$ is also a Poisson process, but with intensity $p\lambda$. If by $\mathcal{P}$ we mean the set of distributions of point processes, then we can notice that the operator of p-rarefaction is invertible. In particular, if $N(t)$ is a Poisson process with intensity λ, then $D_p^{-1} N(t)$ is a Poisson process with intensity λ/p. The set $\mathcal{C}$ of Cox processes is closed with respect to p-rarefaction: $D_p N(\cdot) \in \mathcal{C}$ and $D_p^{-1} N(\cdot) \in \mathcal{C}$, if $N(\cdot) \in \mathcal{C}$.

The following theorem was proved by O. Kallenberg.

THEOREM A2.1.1. *Let $\{N_k(\cdot)\}_{k \geq 1}$ be a sequence of point processes and let $\{p_k\}_{k \geq 1} \subset (0,1)$ so that $p_k \to 0$ $(k \to \infty)$. There exists a point process $N(\cdot)$ such that the relation*

$$D_{p_k} N_k(\cdot) \Rightarrow N(\cdot) \qquad (k \to \infty)$$

holds if and only if there exists a random measure $\Lambda(\cdot)$ such that

$$p_k N_k(\cdot) \Rightarrow \Lambda(\cdot) \qquad (k \to \infty).$$

This process $N(\cdot)$ is a Cox process controlled by $\Lambda(\cdot)$.

For the proof see (Kallenberg, 1975).

The following characterization of Cox processes in terms of p-rarefaction is due to J. Mecke.

THEOREM A2.1.2. *A point process $N(\cdot)$ can be obtained by a p-rarefaction for any $p \in (0,1)$ if and only if $N(\cdot)$ is a Cox process. In other words,*

$$\mathcal{C} = \bigcap_{p \in (0,1)} \mathcal{D}_p.$$

For the proof see (Mecke, 1968).

The relationship between Cox processes and renewal processes has a very interesting form. Recall the definition of a renewal process. Let $N(\cdot)$ be a point process with independent distances Y_j, $j \geq 1$, between points. If $\{Y_j\}_{j \geq 1}$ are identically distributed with some distribution function H, then $N(\cdot)$ is called a renewal process. Put $V_k = Y_1 + \ldots + Y_k$, $k \geq 1$.

Assume that $N(t)$, $t \geq 0$, is a Cox process. We are interested in the conditions we should impose on $N(t)$ to guarantee that it is also a renewal process. Let $N(t)$ be controlled by a random measure $\Lambda(t)$ such that $\Lambda(0) = 0$, $\Lambda(\infty) = \infty$. Then $N_1(t) = N(\Lambda^{-1}(t))$, where $\Lambda^{-1}(t) = \sup\{s : \Lambda(s) \leq t\}$. This means that $N_1(t) = \sup\{k \geq 1 : \Lambda(V_k) \leq t\}$. Therefore, the standard Poisson process $N_1(t)$ has $V_k^* = \Lambda(V_k)$ as its jump points and hence, $N(t)$ is a Cox process if and only if $V_1^*, V_2^* - V_1^*, V_3^* - V_2^*, \ldots$ are independent and have one and the same exponential distribution.

The converse result is as follows.

THEOREM A2.1.3. *Let $N(\cdot)$ be a renewal process with H being the distribution function of times between renewals,*

$$h(s) = \int\limits_0^\infty e^{-sx}\, dH(x), \qquad s \geq 0.$$

$N(\cdot)$ *is a Cox process if and only if*

$$h(s) = \frac{1}{1 - \log g(s)} \tag{1.2}$$

where $g(s)$ is the Laplace-Stieltjes transform of some infinitely divisible distribution function $g(x)$ with $G(0) < 1$. Moreover,

$$g(s) = \mathsf{E} \exp\{s\Lambda^{-1}(1)\} \quad \text{and} \quad \Lambda^{-1}(0) = 0,$$

where $\Lambda(t)$ is the random measure controlling $N(t)$.

For the original proof see (Kingman, 1964) and (Grandell, 1976).

Notice that representation (1.2) means that the distribution function H is geometrically infinitely divisible (see Chapters 2 and 4). Therefore the first part of Theorem A2.1.3 (representation (1.2)) follows from Theorem A2.1.2, the definition of a geometrically infinitely divisible distribution (see Definition 4.6.2 with N_θ being a geometrically distributed random variable) and representation (4.6.30) with $\phi(s) = (1+s)^{-1}$. Considering the results of Chapter 2 and Section 4.6 we can give another formulation of Theorem A2.1.3: a renewal process generated by a distribution function H is a Cox process if and only if H is geometrically infinitely divisible. Note that the original proof was obtained without use of the concept of a geometrically infinitely divisible distribution.

Some applications of Cox processes to insurance problems are described in (Embrechts and Klueppelberg, 1993) as well as the reasons why they are good models of the processes of claims arrivals in insurance.

Consider a particular case of a Cox process where the controlling process has the form $\Lambda(t) = \Lambda t$ for some nonnegative random variable Λ. In this case the Cox process is called a mixed Poisson process and Λ is called a structural random variable, the distribution of Λ is respectively called structural. By setting different structural distributions of a mixed Poisson process we obtain a wide class of distributions of the random variable $N(t)$ with fixed $t > 0$. For example, if Λ is gamma distributed, then $N(t)$ has a negative binomial distribution (see (Kendall and Stuart, 1969)). In this case $N(t)$ is called a Polya process. As we will see below, mixed Poisson processes play an important role in convergence rate estimates for general Cox processes.

In what follows we shall consider the asymptotic properties of Cox processes. Our reasoning will be based on a general limit theorem for superpositions of random processes. Due to its importance, we extract this theorem into a separate section.

A2.2 A general limit theorem for superpositions of random processes

According to Definition A2.1.1, a Cox process is a superposition of the standard Poisson process and some nonnegative stochastic process with finite nondecreasing trajectories. In other words, a Cox process is a Poisson process with random time. Therefore we can study its asymptotic behavior with the help of some general results for superpositions of stochastic processes presented in this section.

Of course, an attentive reader has already noticed that in the proof of Theorem 3.2.1 the assumption that the sequence $\{S_n\}$ is constructed by successive sums of independent random variables was used only to prove the tightness of the sequence $\{b_{N_n}/d_n\}$, since, as we have already noted in Chapter 3, transfer Theorem 3.1.2 is valid for arbitrary random sequences $\{S_n\}$ with independent random indices (for brevity, by tightness here and in what follows we will mean weak relative compactness since in the finite-dimensional case these properties of the families of random variables coincide). Therefore, including the assumption of the tightness of the sequence $\{b_{N_n}/d_n\}$ in the conditions of the theorem, we obtain the following statement.

Let $L_1(\cdot,\cdot)$ and $L_2(\cdot,\cdot)$ be probability metrics which metrize weak convergence in the spaces of one- and two-dimensional random variables, respectively. For example, L_1 is the Lévy metric, L_2 is the Lévy-Prokhorov metric.

LEMMA A2.2.1. *Let $\{S_k\}$ be a sequence of random variables, and $\{N_n\}$ be a sequence of integer-valued random variables such that for each n, N_n and $\{S_k\}_{k\geq 1}$ are independent. Let the sequences of numbers $\{a_n\}, \{b_n\}, \{d_n\}$ $(b_n > 0, d_n > 0, b_n \to \infty$ and $d_n \to \infty$ as $n \to \infty)$ provide the tightness of the families of random variables*

$$\left\{\frac{S_n - a_n}{b_n}\right\}_{n\geq 1} \quad \text{and} \quad \left\{\frac{b_{N_n}}{d_n}\right\}_{n\geq 1}.$$

Then

$$\frac{S_{N_n} - c_n}{d_n} \Rightarrow \quad (\text{some}) \quad Z \quad (n \to \infty)$$

for some real c_n if and only if there exists a tight sequence of triples of random variables $\{(Y_n, U_n, V_n)\}_{n\geq 1}$ such that:

1) *$Z \overset{d}{=} U_n Y_n + V_n$ for each $n \geq 1$, where Y_n and the pair (U_n, V_n) are independent;*

2) *$L_1\left(\dfrac{S_n - a_n}{b_n}, Y_n\right) \to 0 \quad (n \to \infty)$;*

3) *$L_2\left(\left(\dfrac{b_{N_n}}{d_n}, \dfrac{a_{N_n} - c_n}{d_n}\right), (U_n, V_n)\right) \to 0 \quad (n \to \infty)$.*

This lemma easily allows us to obtain a general limit theorem for super-positions of independent random processes.

Let $S(t)$ and $M(t)$, $t \geq 0$, be independent random processes such that $P(M(t) < \infty) = 1$ for any $t > 0$. Let $a(t), b(t), d(t)$ be real functions, $b(t) > 0$, $d(t) > 0$.

THEOREM A2.2.1. *Assume that $b(t) \to \infty$ and $d(t) \to \infty$ as $n \to \infty$ and the families of random variables*

$$\left\{ \frac{S(t) - a(t)}{b(t)} \right\}_{t>0} \quad \text{and} \quad \left\{ \frac{b(M(t))}{d(t)} \right\}_{t>0}$$

are tight. Then one-dimensional distributions of appropriately centered and normalized superpositions of the processes $S(t)$ and $M(t)$ weakly converge to the distribution of some random variable Z as $t \to \infty$, i.e.,

$$\frac{S(M(t)) - c(t)}{d(t)} \Rightarrow Z \quad (t \to \infty)$$

for some real function $c(t)$, if and only if there exists a tight family of triples of random variables $\{(Y(t), U(t), V(t))\}_{t \geq 0}$ such that:

1) *$Z \stackrel{d}{=} Y(t)U(t) + V(t)$ for each $t > 0$, where the random variable $Y(t)$ and the pair $(U(t), V(t))$ are independent;*

2) *$L_1 \left(\dfrac{S(t) - a(t)}{b(t)}, Y(t) \right) \to 0 \quad (t \to \infty)$;*

3) *$L_2 \left(\left(\dfrac{b(M(t))}{d(t)}, \dfrac{a(M(t)) - c(t)}{d(t)} \right), (U(t), V(t)) \right) \to 0 \quad (t \to \infty)$.*

On the one hand, this theorem generalizes the famous Dobrushin's lemma (Dobrushin, 1955) from random sequences to random processes and on the other hand, it sharpens the Dobrushin lemma by presenting not only sufficient, but also necessary conditions for the weak convergence of one-dimensional distributions of superpositions of independent random processes. To prove Theorem A2.2.1 it suffices to pass to arbitrary infinitely increasing sequences of time moments and to apply Lemma A2.2.1 along each of these sequences.

A2.3 Limit theorems for Cox processes

The aim of this section is to demonstrate that the asymptotic properties of a Cox process are entirely determined by the asymptotic properties of its controlling process.

LEMMA A2.3.1. *Let $N(t)$ be a Cox process controlled by the process $\Lambda(t)$. Then $N(t) \stackrel{P}{\to} \infty \ (t \to \infty)$ if and only if $\Lambda(t) \stackrel{P}{\to} \infty \ (t \to \infty)$.*

PROOF. First we show that $\Lambda(t) \overset{P}{\to} \infty$ implies $N(t) \overset{P}{\to} \infty$ $(t \to \infty)$. For arbitrary m and n we have

$$P(N(t) \le m) = P(N(t) \le m; \Lambda(t) \le n) + P(N(t) \le m; \Lambda(t) > n) \le$$

$$P(\Lambda(t) \le n) + \int_n^\infty \left(\sum_{k=0}^m e^{-\lambda} \frac{\lambda^k}{k!} \right) dP(\Lambda(t) < \lambda) \equiv$$

$$J_1(t, n) + J_2(t, n, m). \tag{3.1}$$

Let $\varepsilon > 0$ be arbitrary. Consider $J_2(t, n, m)$. Let N_λ be a Poisson random variable with parameter λ and Q_λ be its concentration function, $Q_\lambda(l) = \sup_x P(x \le N_\lambda \le x + l)$, $l \ge 0$. Then from inequality (7.9) given below it follows that, whatever $t > 0$ is,

$$J_2(t, n, m) \le \int_n^\infty Q_\lambda(m + 1) \, dP(\Lambda(t) < \lambda) \le$$

$$3(m + 2) \int_n^\infty \frac{1}{\sqrt{\lambda}} \, dP(\Lambda(t) < \lambda) \le \frac{3(m + 2)}{\sqrt{n}}.$$

Therefore with a fixed m, we can choose $n = n(\varepsilon)$ so that

$$J_2(t, n(\varepsilon), m) < \frac{\varepsilon}{2}. \tag{3.2}$$

Now choose $t = t(\varepsilon)$ so that

$$J_1(t, n(\varepsilon)) < \frac{\varepsilon}{2} \tag{3.3}$$

for all $t \ge t(\varepsilon)$, which is possible due to the assumption $\Lambda \overset{P}{\to} \infty$. But (3.1), (3.2) and (3.3) imply

$$P(N(t) \le m) < \varepsilon$$

for $t \ge t(\varepsilon)$, which means that $N(t) \overset{P}{\to} \infty$ since m and ε are arbitrary.

Now assume that $N(t) \overset{P}{\to} \infty$ and prove that $\Lambda(t) \overset{P}{\to} \infty$ as $t \to \infty$. For arbitrary m and n we have

$$P(\Lambda(t) \le m) = P(\Lambda(t) \le m; N(t) \le n) + P(\Lambda(t) \le m; N(t) > n) \le$$

$$P(N(t) \le n) + \int_0^\infty \left(\sum_{k=n+1}^\infty e^{-\lambda} \frac{\lambda^k}{k!} \right) dP(\Lambda(t) < \lambda) \equiv$$

$$I_1(t, n) + I_2(t, n, m). \tag{3.4}$$

Consider $I_2(t, n, m)$. The function $\psi_k(\lambda) = e^{-\lambda} \lambda^k$ increases in λ for $0 < \lambda < k$. If we choose n in such a way that $n + 1 > m$, then due to the form of the limits of summation and integration in $I_2(t, n, m)$, in all summands we shall have $k > \lambda$. Then

$$I_2(t, n, m) = \sum_{k=n+1}^{\infty} \int_0^m e^{-\lambda} \frac{\lambda^k}{k!} \, dP(\Lambda(t) < \lambda) \leq$$

$$\sum_{k=n+1}^{\infty} e^{-m} \frac{m^k}{k!} \int_0^m dP(\Lambda(t) < \lambda) \leq P(N_m \geq n+1), \tag{3.5}$$

where N_m is a Poisson random variable with parameter m. Let $\varepsilon > 0$ be arbitrary. With a fixed m, by virtue of (3.5), it is possible to choose $n = n(\varepsilon)$ so that

$$I_2(t, n(\varepsilon), m) < \frac{\varepsilon}{2} \tag{3.6}$$

for all $t > 0$. Now choose $t = t(\varepsilon)$ so that

$$I_1(t, n(\varepsilon)) < \frac{\varepsilon}{2} \tag{3.7}$$

for all $t \geq t(\varepsilon)$, which is possible due to the assumption $N(t) \overset{P}{\to} \infty$. Now the desired implication follows from (3.4), (3.6) and (3.7).

THEOREM A2.3.1. Let $N(t)$ be a Cox process controlled by the process $\Lambda(t)$. Let $d(t) > 0$ be a function such that $d(t) \to \infty$ $(t \to \infty)$. Then the following conditions are equivalent:

I. The family of random variables $\{\Lambda(t)/d(t)\}_{t>0}$ is tight and one-dimensional distributions of a normalized Cox process weakly converge to the distribution of some random variable Z as $t \to \infty$:

$$\frac{N(t)}{d(t)} \Rightarrow Z \qquad (t \to \infty). \tag{3.8}$$

II. One-dimensional distributions of the controlling process $\Lambda(t)$, appropriately normalized, converge to the same distribution:

$$\frac{\Lambda(t)}{d(t)} \Rightarrow Z \qquad (t \to \infty). \tag{3.9}$$

PROOF. To prove this theorem we will use Theorem A2.2.1. It is well known that the standard Poisson process $N_1(t)$ is asymptotically degenerate in the sense that

$$\frac{N_1(t)}{t} \Rightarrow 1 \qquad (t \to \infty). \tag{3.10}$$

Therefore in Theorem A2.2.1 we put $S(t) \equiv N_1(t)$, $M(t) \equiv \Lambda(t)$, $b(t) \equiv t$, $a(t) \equiv c(t) = 0$. As we have done so, from (3.10) it follows that any triple of random variables $(Y(t), U(t), V(t))$ considered in Theorem A2.2.1 inevitably has the form $(1, U(t), 0)$ and hence conditions 2) and 3) of Theorem A2.2.1 are reduced to

$$L_1\left(\frac{\Lambda(t)}{d(t)}, U(t)\right) \to 0 \qquad (t \to \infty). \tag{3.11}$$

But due to the form of the triples mentioned above and condition 1) of Theorem A2.2.1, the random variables $U(t)$ for each $t > 0$ should satisfy the relation

$$U(t) \overset{d}{=} Z.$$

Therefore in this case (3.11) is equivalent to (3.9). The theorem is proved.

Theorem A2.3.1 is a sort of a law of large numbers for Cox processes (compare it with Theorem 3.3.5). The following statement dealing with non-randomly centered Cox process may be regarded as a sort of a central limit theorem.

THEOREM A2.3.2. *Let $N(t)$ be a Cox process controlled by the process $\Lambda(t)$. Assume that for some function $d(t) > 0$ such that $d(t) \to \infty$ $(t \to \infty)$, the family of random variables $\{\Lambda(t)/d^2(t)\}_{t>0}$ is tight. Then one-dimensional distributions of a nonrandomly centered and normalized Cox process weakly converge to the distribution of some random variable Z:*

$$\frac{N(t) - c(t)}{d(t)} \Rightarrow Z \qquad (t \to \infty) \tag{3.12}$$

with some real function $c(t)$, if and only if

$$\limsup_{t \to \infty} \frac{c(t)}{d^2(t)} = k^2 < \infty \tag{3.13}$$

and there exists a random variable V such that $Z \overset{d}{=} kW + V$ where W is a random variable with the standard normal distribution independent of V and

$$L_1 \left(\frac{\Lambda(t) - c(t)}{d(t)}, V(t) \right) \to 0 \qquad (t \to \infty), \tag{3.14}$$

where

$$\mathsf{E} \exp\{isV(t)\} = \exp\left\{ -\frac{s^2}{2} \left[k^2 - \frac{c(t)}{d^2(t)} \right] \right\} \mathsf{E} \exp\{isV\}, \quad s \in \mathbb{R}.$$

PROOF. The "only if" part. We will again use Theorem A2.2.1. It is well known that the standard Poisson process $N_1(t)$ is asymptotically normal in the sense that

$$\frac{N_1(t) - t}{\sqrt{t}} \Rightarrow W \qquad (t \to \infty) \tag{3.15}$$

where W has the standard normal distribution. Therefore in Theorem A2.2.1 we put $S(t) \equiv N_1(t)$, $M(t) \equiv \Lambda(t)$, $a(t) \equiv t$, $b(t) \equiv \sqrt{t}$. Then from Theorem A2.2.1 it follows that the family of random variables $\{(\Lambda(t) - c(t))/d(t)\}_{t \geq 0}$ is tight. Let $l_t(q)$ be the left q-quantile of the random variable $\Lambda(t)$. The tightness of the family $\{\Lambda(t)/d^2(t)\}_{t \geq 0}$ implies

$$\sup_t \frac{l_t(q)}{d^2(t)} = c(q) < \infty \tag{3.16}$$

for any $q \in (0,1)$. The tightness of the family $\{(\Lambda(t) - c(t))/d(t)\}$ implies the boundedness of the function $(l_t(q) - c(t))/d(t)$ in t for each $q \in (0,1)$. But

$$\frac{l_t(q) - c(t)}{d(t)} = d(t) \left[\frac{l_t(q)}{d^2(t)} - \frac{c(t)}{d^2(t)} \right], \qquad (3.17)$$

so to guarantee the boundedness of the left-hand side of (3.17) for any $q \in (0,1)$, the difference $l_t(q)/d^2(t) - c(t)/d^2(t)$ should tend to zero as $t \to \infty$ for any $q \in (0,1)$ since $d(t) \to \infty$ as $t \to \infty$. But with account of (3.16) this is possible only if (3.13) holds. Moreover, since

$$\limsup_{t \to \infty} \left| \frac{l_t(q)}{d^2(t)} - \frac{c(t)}{d^2(t)} \right| = 0$$

irrespective of $q \in (0,1)$, we notice that by virtue of (3.12) any triple of random variables $(Y(t), U(t), V(t))$ mentioned in Theorem A2.2.1 should necessarily have the form $(W, \sqrt{c(t)}/d(t), V(t))$, where W is the random variable with the standard normal distribution independent of $V(t)$. Recall that each of these triples should guarantee the possibility of the representation

$$Z \stackrel{d}{=} k(t)W + V(t) \qquad (3.18)$$

for each $t \geq 0$, where for convenience we denoted $k^2(t) = c(t)/d^2(t)$. Consider the family of the random variables satisfying (3.18) in more detail. The boundedness of $k(t)$ and tightness of $\{V(t)\}_{t \geq 0}$ implied by Theorem A2.2.1, allow one to choose a subsequence T_1 from an arbitrary sequence $T = \{t_1, t_2, \ldots\}$ so that

$$k(t) \to k_0 \quad \text{and} \quad V(t) \Rightarrow V_0$$

as $t \to \infty, t \in T_1$, for some number k_0 and some random variable V_0. But then, applying the reasoning we used to prove Theorem 3.2.1, for any $s \in \mathbb{R}$ we obtain

$$Ee^{isZ} = Ee^{isV(t)} \exp\left\{ -\frac{s^2}{2} k^2(t) \right\} \to Ee^{isV_0} \exp\left\{ -\frac{s^2}{2} k_0^2 \right\}$$

as $t \to \infty$, $t \in T_1$, whence it follows that the limit pair (k_0, V_0) also satisfies (3.18). In other words, the set of pairs $(k(t), V(t))$ satisfying (3.18) is closed. Let V be a random variable corresponding to the value $k(t) = k$ in representation (3.18). Then for any $t \geq 0$ we have

$$kW + V \stackrel{d}{=} k(t)W + V(t), \qquad (3.19)$$

where the summands on both sides are independent. Rewrite (3.19) in terms of characteristic functions. We obtain

$$\exp\left\{ -\frac{s^2}{2} k^2 \right\} Ee^{isV} = \exp\left\{ -\frac{s^2}{2} k^2(t) \right\} Ee^{isV(t)} \qquad (3.20)$$

for all $s \in \mathbb{R}$, $t \geq 0$. Expressing the characteristic function of the random variable $V(t)$ from (3.20) we obtain the representation

$$\mathsf{E}\exp\{isV(t)\} = \exp\left\{\frac{s^2}{2}[k^2 - k^2(t)]\right\} \mathsf{E}e^{isV}.$$

Finally, relation (3.14) with $V(t)$ just described, follows from Theorem A2.2.1. The "only if" part is proved.

The "if" part. As we have done in the proof of the "only if" part, we make sure that condition (3.14) together with the tightness of the family $\{\Lambda(t)/d^2(t)\}_{t \geq 0}$ imply

$$L_1\left(\frac{\Lambda(t)}{d^2(t)}, \frac{c(t)}{d^2(t)}\right) \to 0 \qquad (t \to \infty).$$

Now it remains to refer to Theorem A2.2.1 with the account of (3.14). The theorem is proved.

A result similar to Theorem A2.3.2 was proved by Rootzén (Rootzén, 1975). See also (Grandell, 1976).

COROLLARY A2.3.1. *Under conditions of Theorem A2.3.2, a nonrandomly centered and normalized Cox process $N(t)$ is asymptotically normal, i.e.,*

$$\mathsf{P}\left(\frac{N(t) - c(t)}{d(t)} < x\right) \Rightarrow \Phi(x) \qquad (t \to \infty)$$

if and only if

$$\sup_{t \geq 0} \frac{c(t)}{d^2(t)} \leq 1,$$

$$\lim_{t \to \infty} L_1\left(\mathsf{P}\left(\frac{\Lambda(t) - c(t)}{d(t)} < x\right), \Phi\left(\frac{xd(t)}{\sqrt{d^2(t) - c(t)}}\right)\right) = 0.$$

PROOF. The statement follows from Theorem A2.3.2 and the Cramér-Lévy theorem on decomposability of a normal law only into normal components, according to which any random variable $V(t)$ satisfying (3.18) should necessarily be normal with zero mean and variance $1 - k^2(t)$.

Thus, a Cox process is asymptotically normal if and only if so is its controlling process $\Lambda(t)$.

A2.4 Limit theorems for generalized Cox processes

Let $X_1, X_2, \ldots$ be identically distributed random variables and let $N(t)$, $t \geq 0$ be a Cox process controlled by a process $\Lambda(t)$. Assume that for each $t \geq 0$, the random variables $N(t)$, $X_1, X_2, \ldots$ are independent. The process

$$S(t) = \sum_{j=1}^{N(t)} X_j, \qquad t \geq 0 \tag{4.1}$$

will be called a generalized Cox process (for definiteness, we assume $\sum_{j=1}^{0} = 0$). The aim of this section is to present some limit theorems for the one-dimensional distributions of process (4.1). Here our statements will resemble those which have already been formulated in terms of risk processes or stock prices processes in Sections 3.4 and 3.5. In those sections we gave some limit theorems for generalized Cox processes restricted to natural values of the parameter. We shall see here that these theorems remain valid in a general continuous-time case.

For convenience, without essential restriction of generality, we will assume here and in what follows that $E\Lambda(t) = t$, $t \geq 0$. This relation may be interpreted as both the proportionality of the expectation of the controlling process to time and (which is of special importance for asymptotic inference) the parametrization of the process $\Lambda(t)$ by its expectation. Since we consider one-dimensional distributions, this means that we will be interested in the asymptotic behavior of generalized Cox processes under an infinite growth of the expectation of the controlling process. In addition to the above assumptions, first let $EX_1 = 0$ and $DX_1 = \sigma^2$, $0 < \sigma^2 < \infty$. Then it is easily seen that

$$ES(t) = 0, \quad DS(t) = \sigma^2 t, \quad t \geq 0.$$

THEOREM A2.4.1. *Assume that* $\Lambda(t) \overset{P}{\to} \infty \, (t \to \infty)$. *Then one-dimensional distributions of a normalized Cox process weakly converge to the distribution of some random variable* Z:

$$\frac{S(t)}{\sigma\sqrt{t}} \Rightarrow Z \qquad (t \to \infty)$$

if and only if there exists a nonnegative random variable U *such that*

1) $P(Z < x) = \displaystyle\int_0^\infty \Phi(x/\sqrt{y}) \, dP(U < y), \quad x \in \mathbb{R};$

2) $\Lambda(t)/t \Rightarrow U \quad (t \to \infty)$.

PROOF. From Lemma A2.3.1 it follows that the conditions $\Lambda(t) \overset{P}{\to} \infty$ and $N(t) \overset{P}{\to} \infty$ are equivalent. The properties of the random variables $\{X_j\}$ guarantee that

$$P\left(\frac{1}{\sigma\sqrt{k}} \sum_{j=1}^{k} X_j < x\right) \Rightarrow \Phi(x) \qquad (k \to \infty). \tag{4.2}$$

Let $\{t_k\}_{k\geq 1}$ be an arbitrary infinitely increasing sequence of time instants. Put $N_k = N(t_k)$, $b_k = \sigma\sqrt{t_k}$, $d_k = \sigma\sqrt{t_k}$. Then by Theorem 3.3.2 for

$$\frac{1}{\sigma\sqrt{t_k}} S(t_k) = \frac{1}{\sigma\sqrt{t_k}} \sum_{j=1}^{N_k} X_j \Rightarrow Z \qquad (k \to \infty)$$

it is necessary and sufficient that there exists a nonnegative random variable U such that

$$\frac{N_k}{t_k} = \frac{N(t_k)}{t_k} \Rightarrow U \qquad (k \to \infty) \tag{4.3}$$

and

$$P(Z < x) = \mathsf{E}\Phi\left(\frac{x}{\sqrt{U}}\right), \qquad x \in \mathbb{R}. \tag{4.4}$$

But the family of random variables $\{\Lambda(t_k)/t_k\}_{k \geq 1}$ is tight since by the Markov inequality we have

$$\lim_{R \to \infty} \sup_k P\left(\frac{\Lambda(t_k)}{t_k} > R\right) \leq \lim_{R \to \infty} \sup_k \frac{\mathsf{E}\Lambda(t_k)}{t_k R} = \lim_{R \to \infty} \frac{1}{R} = 0.$$

Therefore by Theorem A2.3.1 for (4.3) it is necessary and sufficient that

$$\frac{\Lambda(t_k)}{t_k} \Rightarrow U \qquad (k \to \infty)$$

Now having noted that the sequence $\{t_k\}$ is arbitrary and the distributions of the random variable U in (4.4) for all sequences $\{t_k\}$ coincide due to the identifiability of the family of scale mixtures of normal laws, we complete the proof.

COROLLARY A2.4.1. *Under conditions of Theorem A2.4.1,*

$$P\left(\frac{S(t)}{\sigma\sqrt{t}} < x\right) \Rightarrow \Phi(x) \qquad (t \to \infty)$$

if and only if

$$\frac{\Lambda(t)}{t} \Rightarrow 1 \qquad (t \to \infty).$$

This statement is an immediate consequence of Theorem A2.4.1 and the identifiability of scale mixtures of normal laws.

Now we relax the assumption that X_j have zero means and denote $\mathsf{E}X_1 = a$. Then $\mathsf{E}S(t) = at$, $t > 0$. Note that in this case $DS(t) = \sigma^2 t + a^2 DN(t) \neq \sigma^2 t$, but all the same, for the simplicity of reasoning we will normalize $S(t)$ by $\sigma\sqrt{t}$ but not by $\sqrt{DS(t)}$, thus formally not assuming the existence of the second moment of $\Lambda(t)$.

THEOREM A2.4.2. *Assume that* $\Lambda(t) \overset{P}{\to} \infty$ $(t \to \infty)$. *One-dimensional distributions of a nonrandomly centered and normalized generalized Cox process (4.1) converge weakly to the distribution of some random variable Z:*

$$\frac{S(t) - at}{\sigma\sqrt{t}} \Rightarrow Z \qquad (t \to \infty) \tag{4.5}$$

if and only if there exists a random variable V such that

1) $Z \overset{d}{=} \sqrt{\left(1 + \frac{a^2}{\sigma^2}\right)} W + \frac{a}{\sigma} V$, *where W and V are independent and*

$P(W < x) = \Phi(x)$, $x \in \mathbb{R}$;

2) $\dfrac{\Lambda(t) - t}{\sqrt{t}} \Rightarrow V \qquad (t \to \infty)$.

PROOF. The "only if" part. We will reduce the proof to Theorems 3.3.1 and A2.3.2. With this purpose choose an arbitrary infinitely increasing sequence $\{t_k\}_{k \geq 1}$ of time moments and put $N_k = N(t_k)$, $a_k = ak$, $b_k = \sigma\sqrt{k}$, $c_k = at_k$, $d_k = \sigma\sqrt{t_k}$. Our nearest aim is to show that $N_k/d_k \Rightarrow 1\,(k \to \infty)$. Lemma A2.3.1 yields the equivalence of conditions $\Lambda(t) \overset{P}{\to} \infty$ and $N(t) \overset{P}{\to} \infty\,(t \to \infty)$ whence it follows that $N_k \overset{P}{\to} \infty\,(k \to \infty)$. Now we see that from Theorem 3.3.1 it follows that the families of random variables $\{(N_k - t_k)/\sqrt{t_k}\}_{k \geq 1}$ and $\{N_k/t_k\}_{k \geq 1}$ are tight. Denote the left q-quantile of N_k as $l_k(q)$. The tightness of $\{(N_k - t_k)/\sqrt{t_k}\}_{k \geq 1}$ implies the boundedness of q-quantiles $A_k(q)$ of the random variables $(a_{N_k} - c_k)/d_k = a(N_k - t_k)/(\sigma\sqrt{t_k})$ which are equal to $a(l_k(q) - t_k)/(\sigma\sqrt{t_k})$, that is,

$$\sup_k |A_k(q)| \equiv M(q) < \infty$$

for each $q \in (0,1)$. Therefore we have

$$\left|\frac{l_k(q)}{t_k} - 1\right| = \left|\frac{a(l_k(q) - t_k)}{\sigma\sqrt{t_k}}\right| \left|\frac{\sigma\sqrt{t_k}}{at_k}\right| \leq$$

$$M(q)\frac{\sigma}{|q|}\frac{1}{\sqrt{t_k}} \to 0 \qquad (k \to \infty).$$

Hence, we have $l_k(q)/t_k \to 1\,(k \to \infty)$ for all $q \in (0,1)$. But this means that $N_k/t_k = N(t_k)/t_k \Rightarrow 1$, so our first aim is achieved. In other words, we obtained that $b_{N_k}/d_k \Rightarrow 1$. This fact enables us to directly apply Theorem 3.3.1 because the properties of the random variables $\{X_j\}$ guarantee the normality of the limit law for nonrandom sums whose shift mixtures are identifiable. According to Theorem 3.3.1, for

$$\frac{S(t_k) - at_k}{\sigma\sqrt{t_k}} \Rightarrow Z \qquad (k \to \infty)$$

it is necessary and sufficient that there exists a random variable V_0 such that $Z \overset{d}{=} Y + V_0$ with Y and V_0 independent and Y having the standard normal distribution and

$$\frac{a(N_k - t_k)}{\sigma\sqrt{t_k}} \Rightarrow V_0 \qquad (k \to \infty)$$

Thus, if $a = 0$, then the only possibility for (4.5) to hold is when Z has a normal distribution. This case is covered by Corollary A2.4.1 given above. Let $a \neq 0$. Then from the arbitrariness of the sequence $\{t_k\}$ and the identifiability of shift mixtures of the normal law it follows that if (4.5) holds, then there exists a random variable V_0 such that $Z \overset{d}{=} Y + V_0$ where V_0 and the standard normal random variable Y are independent and

$$\frac{N(t) - t}{\sqrt{t}} \Rightarrow \frac{\sigma}{a}V_0 \qquad (t \to \infty). \tag{4.6}$$

But by Theorem A2.3.2, (4.6) takes place if and only if there exists a random variable V such that $\frac{a}{a}V_0 \stackrel{d}{=} W + V$ where W and V are independent, $P(W < x) = \Phi(x)$, $x \in \mathbb{R}$, and 2) takes place. But this means that $Z \stackrel{d}{=} Y + \frac{a}{\sigma}W + \frac{a}{\sigma}V \stackrel{d}{=} \sqrt{(1 + \frac{a^2}{\sigma^2})}Y + \frac{a}{\sigma}V$. The "only if" part is proved.

The "if" part is now a direct consequence of Theorems A2.3.2 and 3.3.1, because using the same reasoning as in the "only if" part we easily make sure that the conditions of the theorem imply $N_k/t_k \Rightarrow 1$. The theorem is proved.

COROLLARY A2.4.2. *Under conditions of Theorem A2.4.2, one-dimensional distributions of a nonrandomly centered and normalized generalized Cox process (4.1) are asymptotically normal:*

$$P\left(\frac{S(t) - at}{\sigma\sqrt{t}} < x\right) \Rightarrow \Phi\left(\frac{x}{\delta}\right) \qquad (t \to \infty)$$

with some $\delta < \infty$, if and only if $\delta^2 \geq 1$ and

$$P\left(\frac{\Lambda(t) - t}{\sqrt{t}} < x\right) \Rightarrow \Phi\left(\frac{ax}{\sigma\sqrt{\delta^2 - 1}}\right) \qquad (t \to \infty).$$

This statement follows from Theorem A2.4.2 and the Cramér-Lévy theorem on decomposability of the normal law only into normal components.

A2.5 Convergence rate estimates in limit theorems for generalized Cox processes

The aim of this section is to obtain convergence rate estimates in Theorems A2.4.1 and A2.4.2. Since generalized Cox processes are processes of "growing" random sums, this section also serves as an introduction to convergence rate estimates for random sums. The results presented here are an illustration of how general methods of obtaining convergence rate estimates in limit theorems for random sums work. A detailed description of these methods is given in (Kruglov and Korolev, 1990).

First we consider convergence rate estimates in Theorem A2.4.1 which deals with centered summands. So we now assume that $\mathsf{E}X_1 = 0$, $\mathsf{D}X_1 = \sigma^2$, $0 < \sigma^2 < \infty$. From Theorem A2.4.1 it follows that the distribution of the random variable $S(t)/(\sigma\sqrt{t})$ is close to a limit one if and only if the distributions of $\Lambda(t)/t$ and U are close, or which is in some sense the same, the distributions of $\Lambda(t)$ and Ut are close. Let $N^*(t)$ be the Cox process controlled by the process $\Lambda^*(t) = Ut$, $t \geq 0$, (i.e., $N^*(t)$ is the mixed Poisson process with the structural random variable U) and

$$S^*(t) = \sum_{j=0}^{N^*(t)} X_j. \tag{5.1}$$

By $F_t(x)$ and $F_t^*(x)$ we, respectively, denote the distribution functions of the random variables $S(t)/(\sigma\sqrt{t})$ and $S^*(t)/(\sigma\sqrt{t})$. Then from the identity

$$F_t(x) \equiv (F_t(x) - F_t^*(x)) + F_t^*(x) \tag{5.2}$$

it follows that if we know an appropriate estimate of the accuracy of the approximation of the distribution of $S(t)$ by that of $S^*(t)$ and a convergence rate estimate for $S^*(t)$, then we are able to construct an appropriate convergence rate estimate for $S(t)$. Thus, our nearest aim should consist in obtaining an estimate for $\Delta(t) \equiv \sup_x |F_t(x) - F_t^*(x)|$.

LEMMA A2.5.1. *Let $S_1(t)$ and $S_2(t)$ be two generalized Cox processes generated by the same sequence $\{X_j\}_{j\geq 1}$ and controlled by the processes $\Lambda_1(t)$ and $\Lambda_2(t)$, respectively. Then*

$$\sup_x |P(S_1(t) < x) - P(S_2(t) < x)| \leq$$

$$\int_0^\infty \min\left(2, \frac{1}{\sqrt{\lambda}}\right) |P(\Lambda_1(t) < \lambda) - P(\Lambda_2(t) < \lambda)| \, d\lambda.$$

PROOF. Denote $\Psi_t(\lambda) = P(\Lambda_1(t) < \lambda) - P(\Lambda_2(t) < \lambda)$. By the law of total probability we have

$$\sup_x |P(S_1(t) < x) - P(S_2(t) < x)| =$$

$$\sup_x \left| \sum_{k=0}^\infty P\left(\sum_{j=1}^k X_j < x\right) [P(N^{(1)}(t) = k) - P(N^{(2)}(t) = k)] \right| =$$

$$\sup_x \left| \sum_{k=0}^\infty P\left(\sum_{j=1}^k X_j < x\right) \int_0^\infty e^{-\lambda} \frac{\lambda^k}{k!} \, d\Psi_t(\lambda) \right| =$$

$$\sup_x \left| \int_0^\infty \sum_{k=0}^\infty P\left(\sum_{j=1}^k X_j < x\right) e^{-\lambda} \frac{\lambda^k}{k!} \, d\Psi_t(\lambda) \right|,$$

where $N^{(1)}(t)$ and $N^{(2)}(t)$ are the Cox processes controlled by $\Lambda_1(t)$ and $\Lambda_2(t)$, respectively. By integration by parts we obtain

$$\sup_x |P(S_1(t) < x) - P(S_2(t) < x)| =$$

$$\sup_x \left| \int_0^\infty \Psi_t(\lambda) \sum_{k=0}^\infty P\left(\sum_{j=1}^k X_j < x\right) e^{-\lambda} \frac{\lambda^k}{k!} (k - \lambda) \, d\lambda \right| \leq$$

$$\sup_x \left| \int_0^{1/4} \Psi_t(\lambda) \sum_{k=0}^\infty P\left(\sum_{j=1}^k X_j < x\right) e^{-\lambda} \frac{\lambda^k}{k!} (k - \lambda) \, d\lambda \right| +$$

$$\sup_x \left| \int_{1/4}^\infty \Psi_t(\lambda) \sum_{k=0}^\infty P\left(\sum_{j=1}^k X_j < x\right) e^{-\lambda} \frac{\lambda^k}{k!} (k - \lambda) \, d\lambda \right| \equiv$$

$$I_1 + I_2.$$

Consider I_1. Let $\mathcal{E}_0(x)$ be the distribution function of the probability law degenerate at zero. Then

$$I_1 = \sup_x \left| -\int_0^{1/4} \Psi_t(\lambda)\mathcal{E}_0(x)e^{-\lambda}\,d\lambda + \right.$$

$$\left. \int_0^{1/4} \Psi_t(\lambda) \sum_{k=0}^{\infty} P\left(\sum_{j=1}^{k} X_j < x\right)e^{-\lambda}\frac{\lambda^k}{k!}(k-\lambda)\,d\lambda \right| \le$$

$$\int_0^{1/4} |\Psi(\lambda)|\,d\lambda + \int_0^{1/4} |\Psi(\lambda)| \sum_{k=1}^{\infty} e^{-\lambda}\frac{\lambda^{k-1}}{(k-1)!}\,d\lambda = 2\int_0^{1/4} |\Psi(\lambda)|\,d\lambda.$$

At the same time, with the help of the Lyapunov inequality we make sure that

$$I_2 \le \sup_x \int_0^{1/4} |\Psi(\lambda)| \left| \sum_{k=0}^{\infty} P\left(\sum_{j=1}^{k} X_j < x\right)e^{-\lambda}\frac{\lambda^k}{k!}(k-\lambda) \right|\,d\lambda \le$$

$$\int_{1/4}^{\infty} |\Psi_t(\lambda)| \sum_{k=0}^{\infty} e^{-\lambda}\frac{\lambda^k}{k!}\left|\frac{k}{\lambda}-1\right| \le \int_{1/4}^{\infty} \frac{1}{\sqrt{\lambda}}|\Psi_t(\lambda)|\,d\lambda.$$

Now the result is obtained by unifying the estimates for I_1 and I_2. The lemma is proved.

Denote

$$\delta(t) \equiv \int_0^{\infty} \min\left(2, \frac{1}{\sqrt{\lambda}}\right) |P(\Lambda(t) < \lambda) - P(Ut < \lambda)|\,d\lambda. \tag{5.3}$$

Then Lemma A2.5.1 yields the estimate

$$\Delta(t) \le \delta(t). \tag{5.4}$$

As an example, we consider the situation with $\Lambda(t)$ geometrically distributed with parameter $1/t$. It is well known that in this situation $\Lambda(t)/t \Rightarrow U$ $(t \to \infty)$ where $P(U < x) = 1 - e^{-x}$, $x \ge 0$. Put in Lemma A2.5.1 $\Lambda_1(t) = \Lambda(t)$, $\Lambda_2(t) = Ut$ with U just described. In this case the following estimate is known for $|\Psi_t(\lambda)|$ with $t > 1$:

$$|\Psi_t(\lambda)| \le \frac{1}{t-1}\left(\frac{\lambda}{t}+1\right)\exp\left(-\frac{\lambda}{t}\right)$$

(see (Korolev and Selivanova, 1994) and Sect. 5.3). Then as is easily seen, by

Lemma A2.5.1 we have

$$\Delta(t) \leq \delta(t) \leq \frac{1}{t-1} \int_0^\infty \min\left(2, \frac{1}{\sqrt{\lambda}}\right) \left(\frac{\lambda}{t}+1\right) \exp\left(-\frac{\lambda}{t}\right) d\lambda \leq$$

$$\frac{1}{t-1} \int_0^\infty \frac{1}{\sqrt{\lambda}} \left(\frac{\lambda}{t}+1\right) \exp\left(-\frac{\lambda}{t}\right) d\lambda =$$

$$\frac{1}{t(t-1)} \int_0^\infty \sqrt{\lambda} \exp\left(-\frac{\lambda}{t}\right) d\lambda + \frac{1}{t-1} \int_0^\infty \frac{1}{\sqrt{\lambda}} \exp\left(-\frac{\lambda}{t}\right) d\lambda =$$

$$\frac{\sqrt{t}}{t-1} \left(\Gamma\left(\frac{3}{2}\right) + \Gamma\left(\frac{1}{2}\right)\right) = \frac{2\sqrt{\pi t}}{t-1},$$

where $\Gamma(\cdot)$ is Euler's gamma function. This means that in the example under consideration $\Delta(t) = O(1/\sqrt{t})$, which demonstrates the accuracy (or inaccuracy) of the approximation of distributions of generalized Cox processes with geometrically distributed controlling processes by distributions of generalized Polya processes.

REMARK A2.5.1. In statistical applications, when data are observed at some equidistant time instants which is typical for observations of stock prices, it is quite natural to assume from the very beginning that $\Lambda(t) \equiv Ut$ with t being the length of time intervals between observations. It is clear that in this case $\delta(t) \equiv 0$.

LEMMA A2.5.2. Let $\mu_3 = E|X_1|^3 < \infty$ and let N_λ be the Poisson random variable with parameter $\lambda > 0$ independent of the sequence $\{X_j\}_{j\geq 1}$. Then

$$\sup_x \left| P\left(\frac{1}{\sigma\sqrt{\lambda}} \sum_{j=1}^{N_\lambda} X_j < x\right) - \Phi(x) \right| \leq \frac{1}{\sqrt{\lambda}} \left(\frac{\mu_3}{\sigma^3} + 2.5\right).$$

This lemma is a special case of Theorem 6.1.2 in (Kruglov and Korolev, 1990), with the free parameter chosen in such a way that the coefficient at μ_3/σ^3 equals 1.

THEOREM A2.5.1. Let $\mu_3 = E|X_1|^3 < \infty$ and $æ = EU^{-1/2} < \infty$. Then

$$\sup_x \left| F_t(x) - E\Phi\left(\frac{x}{\sqrt{U}}\right) \right| \leq \frac{æ}{\sqrt{t}} \left(\frac{\mu_3}{\sigma^3} + 2.5\right) + \delta(t)$$

with $\delta(t)$ introduced in (5.3).

PROOF. By the law of total probability we have

$$\sup_x \left| P\left(\frac{S^*(t)}{\sigma\sqrt{t}} < x \right) - E\Phi\left(\frac{x}{\sqrt{U}} \right) \right| =$$

$$\sup_x \left| \sum_{k=0}^{\infty} P\left(\frac{1}{\sigma\sqrt{t}} \sum_{j=1}^{k} X_j < x \right) P(N^*(t) = k) - E\Phi\left(\frac{x}{\sqrt{U}} \right) \right| =$$

$$\sup_x \left| \int_0^{\infty} \left[\sum_{k=0}^{\infty} P\left(\frac{1}{\sigma\sqrt{t}} \sum_{j=1}^{k} X_j < x \right) e^{-\lambda t} \frac{(\lambda t)^k}{k!} - \Phi\left(\frac{x}{\sqrt{\lambda}} \right) \right] dP(U < \lambda) \right| \le$$

$$\int_0^{\infty} \sup_x \left| P\left(\frac{1}{\sigma\sqrt{\lambda t}} \sum_{j=1}^{N_{\lambda t}} X_j < x \right) - \Phi(x) \right| dP(U < \lambda),$$

where $N_{\lambda t}$ is a Poisson random variable with parameter λt independent of $\{X_j\}_{j\ge 1}$. Using Lemma A2.5.2 to estimate the integrand, we obtain

$$\sup_x \left| P\left(\frac{S^*(t)}{\sigma\sqrt{t}} < x \right) - E\Phi\left(\frac{x}{\sqrt{U}} \right) \right| \le$$

$$\frac{1}{\sqrt{t}} \left(\frac{\mu_3}{\sigma^3} + 2.5 \right) \int_0^{\infty} \frac{1}{\sqrt{\lambda}} dP(U < \lambda) = \frac{æ}{\sqrt{t}} \left(\frac{\mu_3}{\sigma^3} + 2.5 \right). \tag{5.5}$$

The theorem is proved, since the desired result follows from (5.5), (5.2) and (5.4).

Return to the example we considered before Remark A2.5.1. It follows from Theorem A2.5.1 with the account of the estimate for $\delta(t)$ in the case of geometrically distributed $\Lambda(t)$ and exponential U obtained above, that for these $\Lambda(t)$ and U we have

$$\sup_x \left| P\left(\frac{S(t)}{\sigma\sqrt{t}} < x \right) - E\Phi\left(\frac{x}{\sqrt{U}} \right) \right| = O\left(\frac{1}{\sqrt{t}} \right).$$

Now we turn to the rate of convergence in Theorem A2.4.2 where generalized Cox processes were considered generated by a sequence $\{X_j\}_{j\ge 1}$ of independent identically distributed random variables with nonzero means. Denote $EX_1 = a$ and assume that $a \ne 0$. As above, σ^2 stands for DX_1. According to Theorem A2.4.2, with $t = E\Lambda(t)$ large enough, the distribution of $\frac{1}{\sigma\sqrt{t}}(S(t) - at)$ is close to the limit one, which is $E\Phi\left(\frac{\sigma x - aV}{\sqrt{\sigma^2 + a^2}} \right)$ for some random variable V, or, which is in some sense the same, the distribution of $\frac{1}{\sqrt{t}}(\Lambda(t) - t)$ is close to that of V, or, which is in some sense the same, the distribution of $\Lambda(t)$ is close to that of $\sqrt{t}V + t$.

Let $S^{**}(t)$ be the generalized Cox process generated by the sequence $\{X_j\}_{j\ge 1}$ and controlled by the process $\Lambda^{**}(t) \equiv \max(0, \sqrt{t}V + t)$. We can write the identity similar to (5.2):

$$F_t(x) = (F_t(x) - F_t^{**}(x)) + F_t^{**}(x) \tag{5.6}$$

where F_t and F_t^{**} are the distribution functions of $\frac{1}{\sigma\sqrt{t}}(S(t) - at)$ and $\frac{1}{\sigma\sqrt{t}}(S^{**}(t) - at)$, respectively.

From Lemma A2.5.1 it follows that

$$\sup_x |F_t(x) - F_t^{**}(x)| \le$$

$$\int_0^\infty \min\left(2, \frac{1}{\sqrt{\lambda}}\right) |P(\Lambda(t) < \lambda) - P(\max(0, \sqrt{t}V + t) < \lambda)| \, d\lambda =$$

$$\int_0^\infty \min\left(2, \frac{1}{\sqrt{\lambda}}\right) \left|P\left(\frac{\Lambda(t) - t}{\sqrt{t}} < \frac{\lambda - t}{\sqrt{t}}\right) - P\left(\max(V, -\sqrt{t}) < \frac{\lambda - t}{\sqrt{t}}\right)\right| d\lambda =$$

$$\int_{-\sqrt{t}}^\infty \min\left(2\sqrt{t}, \left(\frac{u}{\sqrt{t}} + 1\right)^{-1/2}\right) \left|P\left(\frac{\Lambda(t) - t}{\sqrt{t}} < u\right) - P(V < u)\right| du \equiv$$

$$\omega(t). \quad (5.7)$$

LEMMA A2.5.3. *Let* $EX_1 = a \ne 0$, $\mu_3 = E|X_1 - a|^3 < \infty$ *and let* N_λ *be a Poisson random variable with parameter* $\lambda > 0$, *independent of the sequence* $\{X_j\}_{j\ge1}$. *Then*

$$\sup_x \left| P\left(\frac{1}{\sigma\sqrt{\lambda}}\left(\sum_{j=1}^{N_\lambda} X_j - a\lambda\right) < x\right) - \Phi\left(\frac{x\sigma}{\sqrt{\sigma^2 + a^2}}\right)\right| \le \frac{1}{\sqrt{\lambda}}\left(\frac{\mu_3}{\sigma^3} + 3.2\right).$$

PROOF. The statement of this lemma follows from Theorem 6.2.1 in (Kruglov and Korolev, 1990) which yields the estimate

$$\sup_x \left| P\left(\frac{1}{\sigma\sqrt{\lambda}}\left(\sum_{j=1}^{N_\lambda} X_j - a\lambda\right) < x\right) - \Phi\left(\frac{x\sigma}{\sqrt{\sigma^2 + a^2}}\right)\right| \le$$

$$\frac{1}{\sqrt{\lambda}}\left(\frac{\mu_3}{\sigma^3} + 2.414\right) + \sup_x \left| P\left(\frac{N_\lambda - \lambda}{\sqrt{\lambda}} < x\right) - \Phi(x)\right|,$$

where the second summand on the right-hand side can be easily estimated as

$$\sup_x \left| P\left(\frac{N_\lambda - \lambda}{\sqrt{\lambda}} < x\right) - \Phi(x)\right| \le \frac{0.766}{\sqrt{\lambda}}$$

by the Berry-Esseen inequality. The lemma is proved.

THEOREM A2.5.2. *Let* $EX_1 = a \ne 0$, $\mu_3 = E|X_1 - a|^3 < \infty$, $E|V| < \infty$. *Then*

$$\sup_x \left| F_t(x) - E\Phi\left(\frac{\sigma x - aV}{\sqrt{\sigma^2 + a^2}}\right)\right| \le \frac{1}{\sqrt{t - 1/2}}\left(\frac{\mu_3}{\sigma^3} + 3.2\right) + \frac{2.3}{\sqrt{t}}E|V| + \omega(t),$$

with $\omega(t)$ *defined in* (5.7).

PROOF. With the account of (5.6) and (5.7) it suffices to estimate

$$\alpha(t) = \sup_x \left| F_t^{**}(x) - E\Phi\left(\frac{\sigma x - aV}{\sqrt{\sigma^2 + a^2}}\right) \right|.$$

Let $N^{**}(t)$ be the Cox process controlled by $\Lambda^{**}(t)$. By $N_{v\sqrt{t}+t}$ we denote a Poisson random variable with parameter $v\sqrt{t} + t$. We have

$$\alpha(t) = \sup_x \left| \sum_{k=0}^{\infty} P(N^{**}(t) = k) P\left(\frac{1}{\sigma\sqrt{t}}\left(\sum_{j=1}^{k} X_j - at\right) < x\right) - \right.$$

$$\left. E\Phi\left(\frac{\sigma x - aV}{\sqrt{\sigma^2 + a^2}}\right) \right| = \sup_x \left| \int_{-\infty}^{-\sqrt{t}} \left[\mathcal{E}_0(x) - \Phi\left(\frac{\sigma x - av}{\sqrt{\sigma^2 + a^2}}\right)\right] dP(V < v) + \right.$$

$$\int_{-\sqrt{t}}^{\infty} \left[\sum_{k=0}^{\infty} e^{-(v\sqrt{t}+t)}\frac{(v\sqrt{t}+t)^k}{k!} P\left(\frac{1}{\sigma\sqrt{t}}\left(\sum_{j=1}^{k} X_j - at\right) < x\right) - \right.$$

$$\left.\left. \Phi\left(\frac{\sigma x - av}{\sqrt{\sigma^2 + a^2}}\right)\right] \right| dP(V < v) \le$$

$$\int_{-\infty}^{-\sqrt{t}} \sup_x \left|\mathcal{E}_0(x) - \Phi\left(\frac{\sigma x - av}{\sqrt{\sigma^2 + a^2}}\right)\right| dP(V < v) +$$

$$\left(\int_{-\sqrt{t}}^{-\sqrt{t}/2} + \int_{-\sqrt{t}/2}^{\infty}\right) \sup_x \left| P\left(\frac{1}{\sigma\sqrt{t}}\left(\sum_{j=1}^{N_{v\sqrt{t}+t}} X_j - at\right) < x\right) - \right.$$

$$\left. \Phi\left(\frac{\sigma x - av}{\sqrt{\sigma^2 + a^2}}\right)\right| dP(V < v) \equiv I_1 + I_2 + I_3.$$

By the Markov inequality we have

$$I_1 + I_2 \le P(V < -\sqrt{t}/2) \le P(|V| > \sqrt{t}/2) \le \frac{2E|V|}{\sqrt{t}}. \tag{5.8}$$

At the same time,

$$I_3 = \int_{-\sqrt{t}/2}^{\infty} \sup_x \left| P\left(\frac{1}{\sigma\sqrt{v\sqrt{t}+t}}\left(\sum_{j=1}^{N_{v\sqrt{t}+t}} X_j - a(v\sqrt{t}+t)\right) < \frac{(\sigma x - av)\sqrt{t}}{\sigma\sqrt{v\sqrt{t}+t}}\right) - \right.$$

$$\left. \Phi\left(\frac{\sigma x - av}{\sqrt{a^2 + \sigma^2}}\right)\right| dP(V < v) =$$

$$\int_{-\sqrt{t}/2}^{\infty} \sup_y \left| P\left(\frac{1}{\sigma\sqrt{v\sqrt{t}+t}}\left(\sum_{j=1}^{N_{v\sqrt{t}+t}} X_j - a(v\sqrt{t}+t)\right) < y\right) - \right.$$

$$\Phi\left(\frac{\sigma y}{\sqrt{a^2+\sigma^2}}\cdot\sqrt{\frac{v}{\sqrt{t}}+1}\right)\bigg|\,d\mathsf{P}(V<v)\le$$

$$\int\limits_{-\sqrt{t}/2}^{\infty}\sup_\nu\left|\mathsf{P}\left(\frac{1}{\sigma\sqrt{v\sqrt{t}+t}}\left(\sum_{j=1}^{N_{v\sqrt{t}+t}}X_j-a(v\sqrt{t}+t)\right)<y\right)-\right.$$

$$\Phi\left(\frac{\sigma y}{\sqrt{a^2+\sigma^2}}\right)\bigg|\,d\mathsf{P}(V<v)+$$

$$\int\limits_{-\sqrt{t}/2}^{\infty}\sup_\nu\left|\Phi(y)-\Phi\left(y\sqrt{\frac{v}{\sqrt{t}}+1}\right)\right|\,d\mathsf{P}(V<v)\equiv I_{31}+I_{32}.$$

Estimate I_{32} by Lemma 6.3.2 in (Kruglov and Korolev, 1990), according to which for $p>0$

$$|\Phi(x)-\Phi(px)|\le x\phi(\min(1,p)x)|p-1|$$

where $\phi(\cdot)$ is the standard normal density. Then

$$I_{32}\le\int\limits_{-\sqrt{t}/2}^{0}\left(\sup_{y>0}y\phi\left(y\sqrt{\frac{v}{\sqrt{t}}+1}\right)\right)\left|1-\sqrt{\frac{v}{\sqrt{t}}+1}\right|\,d\mathsf{P}(V<v)+$$

$$\int\limits_{0}^{\infty}(\sup_{y>0}y\phi(y))\left|1-\sqrt{\frac{v}{\sqrt{t}}+1}\right|\,d\mathsf{P}(V<v)\le$$

$$\frac{\sqrt{2}}{\sqrt{\pi e t}(\sqrt{2}+1)}\int\limits_{-\sqrt{t}/2}^{\infty}|v|\,d\mathsf{P}(V<v)\le\frac{0.21}{\sqrt{t}}\mathsf{E}|V|. \tag{5.9}$$

Estimate I_{31} with the help of Lemma A2.5.3. We have

$$I_{31}\le\left(\frac{\mu_3}{\sigma^3}+3.2\right)\int\limits_{-\sqrt{t}/2}^{\infty}\frac{d\mathsf{P}(V<v)}{\sqrt{v\sqrt{t}+t}}\le\frac{1}{\sqrt{t-1/2}}\left(\frac{\mu_3}{\sigma^3}+3.2\right). \tag{5.10}$$

By unifying (5.8), (5.9) and (5.10) we obtain the desired result. The theorem is proved.

REMARK A2.5.2.   Just as we did in Remark A2.5.1, we note that in statistical applications, if the experimental data are the values of some random process observed at equidistant time moments (time series), then it is quite natural to assume from the very beginning that $\Lambda(t)=\sqrt{t}V+t$ for some random variable V with t being the length of time intervals between successive observations. This assumption results in $\omega(t)=0$.

A2.6 Asymptotic expansions for generalized Cox processes

First consider the case where the sequence of random variables generating a generalized Cox process has zero means. Relation (5.2) allows one to construct an appropriate asymptotic expansion for

$$F_t(x) = \mathsf{P}\left(\frac{S(t)}{\sigma\sqrt{t}} < x\right)$$

with the help of an asymptotic expansion for

$$F_t^*(x) = \mathsf{P}\left(\frac{S^*(t)}{\sigma\sqrt{t}} < x\right)$$

and an estimate for $|F_t(x) - F_t^*(x)|$, for example, given by Lemma A2.5.1. We recall that here $S^*(t)$ is the Cox process generated by the same sequence $\{X_j\}_{j\geq1}$ of independent identically distributed random variables as $S(t)$, but controlled by the process Ut, U being the limit random variable for $\Lambda(t)/t$. So what we are actually going to do is to construct an asymptotic expansion for the distribution of a generalized mixed Poisson process, which is most important from the point of view of statistical analysis by Remark A2.5.1.

We shall say that a random variable Y satisfies the Cramér condition, if

$$\limsup_{|s|\to\infty} |\mathsf{E}\exp\{isY\}| < 1. \tag{6.1}$$

Denote $S_n = X_1 + \ldots + X_n$. It is well known, that if in addition to the above conditions, random variables $\{X_j\}_{j\geq1}$ satisfy the Cramér condition (6.1) and $\mathsf{E}|X_1|^k < \infty$ for some integer $k \geq 3$, then

$$\sup_x \left| \mathsf{P}\left(\frac{S_n}{\sigma\sqrt{n}} < x\right) - \Phi(x) - \sum_{j=1}^{k-2} \frac{Q_j(x)}{n^{j/2}} \right| = o(n^{-(k-2)/2}), \tag{6.2}$$

where

$$Q_j(x) = -\phi(x) \sum H_{j+2l-1}(x) \prod_{m=1}^{j} \frac{1}{k_m!} \left(\frac{\gamma_{m+2}}{(m+2)!\sigma^{m+2}} \right)^{k_m}, \tag{6.3}$$

$j = 1, \ldots, k-2$, summation is carried out over all integer nonnegative solutions $(k_1, \ldots, k_j)$ of the equation $k_1 + 2k_2 + \ldots + jk_j = j$, $l = k_1 + \ldots + k_j$, γ_{m+2} is the semiinvariant of X_1 of order $m+2$ and $H_m(x)$ are Chebyshev-Hermit polynomials of power m, that is,

$$H_m(x)\phi(x) = (-1)^m \phi^{(m)}(x).$$

In particular, if we denote $\alpha_l = \mathsf{E}X_1^l$, $l = 1, 2, \ldots$, then

$$Q_1(x) = -\phi(x)(x^2 - 1)\frac{\alpha_3}{6\sigma^3}, \tag{6.4}$$

$$Q_2(x) = -\phi(x)\left[(x^3 - 3x)\frac{\alpha_4 - 3\sigma^4}{24\sigma^4} + (x^5 - 10x^3 + 15x)\frac{\alpha_3^2}{72\sigma^6}\right]. \qquad (6.5)$$

Our nearest aim is to present an analog of this statement for generalized Cox processes. Denote

$$(\log t)_k = \begin{cases} 1, & k = 3, \\ \log t, & k > 3. \end{cases}$$

THEOREM A2.6.1. *Let* $EX_1 = 0$, $E|X_1|^k < \infty$ *for some integer* $k \geq 3$. *Assume that* X_1 *satisfies the Cramér condition (6.1). Let* U *be the limit random variable for* $\Lambda(t)/t$ *such that* $EU = 1$, $EU^{-(k-2)/2} < \infty$ *and for any* $q \in (0, 1)$

$$(\log t)_k t^{\frac{k-2}{2}} Eq^{Ut} \to 0 \qquad (t \to \infty). \qquad (6.6)$$

Then

$$\sup_x \left| F_t(x) - E\Phi\left(\frac{x}{\sqrt{U}}\right) - \sum_{j=1}^{k-2} \frac{q_j(x)}{t^{j/2}} \right| \leq \epsilon(t),$$

where

$$\epsilon(t) = \delta(t) + o(t^{-(k-2)/2}),$$

$$q_j(x) = EQ_j^*\left(\frac{x}{\sqrt{u}}\right)U^{-j/2}, \quad j = 1,\ldots,k-2,$$

the functions $Q_j^*(\cdot)$ *are defined by formulas (6.3) with the semiinvariants* γ_{m+2} *of order* $m+2$ *replaced by the moments* $\alpha_{m+2} = EX_1^{m+2}$ *of order* $m+2$, $m = 1,\ldots,j$; $j = 1,\ldots,k-2$, *and* $\delta(t)$ *is defined in (5.3).*

Before proving this theorem we make some remarks.

REMARK A2.6.1. *For each* $q \in (0, 1)$ *there is a* $v > 0$ *such that* $q = e^{-v}$. Therefore, denoting the Laplace-Stieltjes transform of U by $\psi_U(s)$, $s \geq 0$, we notice that condition (6.6) is equivalent to

$$\lim_{s \to \infty} (\log s)_k s^{\frac{k-2}{2}} \psi_U(s) = 0, \qquad (6.7)$$

that is, the Laplace-Stieltjes transform of U should decrease faster than $s^{(2-k)/2}$. It is easy to see that condition (6.7) (and hence, (6.6)) is valid for the following distributions: (i) gamma distribution with shape parameter exceeding $(k-2)/2$ (including the exponential distribution for which the only k providing (6.7) is $k = 3$); (ii) Lévy distribution (stable distribution with characteristic exponent $\alpha = 1/2$); (iii) uniform distribution on $[0, 2]$ for which the only k providing (6.7) is $k = 3$; (iv) all distributions concentrated on finite intervals not containing the origin. Condition (6.7) is the analog of Cramér condition for the random variable U.

REMARK A2.6.2. *The first two functions* $Q_1^*(x)$ *and* $Q_2^*(x)$ *have the form*

$$Q_1^*(x) = Q_1(x)$$

$$Q_2^*(x) = -\phi(x)\left[(x^3 - 3x)\frac{\alpha_4}{24\sigma^4} + (x^5 - 10x^3 + 15x)\frac{\alpha_3^2}{72\sigma^6}\right]$$

(compare with (6.4) and (6.5)).

PROOF OF THEOREM A2.6.1. With regard for relations (5.2) and (5.4) it suffices to show that

$$\sup_x \left| F_t^*(x) - \mathsf{E}\Phi\left(\frac{x}{\sqrt{U}}\right) - \sum_{j=1}^{k-2} t^{-j/2} q_j(x) \right| = o(t^{-(k-2)/2}). \qquad (6.8)$$

To prove (6.8) we apply Theorem 5.2 from (Petrov, 1987) according to which if $F(x)$ is a non-decreasing bounded function and $H(x)$ is a differentiable function with bounded variation such that $F(-\infty) = H(-\infty)$ and $\sup_x |H'(x)| \leq c$, then for any $T > 0$ and $b > \frac{1}{2\pi}$ the inequality

$$\sup_x |F(x) - H(x)| \leq b \int_{-T}^{T} \left| \frac{f(s) - h(s)}{s} \right| ds + r(b)\frac{c}{T} \qquad (6.9)$$

holds where $f(s)$ and $h(s)$ are Fourier-Stieltjes transforms of $F(x)$ and $H(x)$, respectively, and $r(b)$ is a positive constant depending only on b. In (6.9) we put

$$F(x) = F_t^*, \quad H(x) = \mathsf{E}\Phi\left(\frac{x}{\sqrt{U}}\right) + \sum_{j=1}^{k-2} t^{-j/2} q_j(x),$$

$$b = \frac{1}{\pi}, \quad T = At^{(k-2)/2}, \quad A > 0.$$

Then we have

$$f(s) = f_t^*(s) = \mathsf{E}\exp\{isS^*(t)/(\sigma\sqrt{t})\},$$

$$h(s) = \int e^{isx} d\left[\mathsf{E}\Phi\left(\frac{x}{\sqrt{U}}\right) + \sum_{j=1}^{k-2} t^{-j/2} q_j(x)\right] =$$

$$\mathsf{E}\left[\int e^{isx} d\Phi\left(\frac{x}{\sqrt{U}}\right) + \sum_{j=1}^{k-2} (tU)^{-j/2} \int e^{isx} dQ_j^*\left(\frac{x}{\sqrt{U}}\right)\right] \equiv$$

$$\mathsf{E}h_t(s, U). \qquad (6.10)$$

It is easy to see that $F(-\infty) = H(-\infty) = 0$ and $\sup_x |H'(x)| < c$. For an arbitrary $\varepsilon > 0$ we choose an $A = A_\varepsilon > 0$ so that $r(1/\pi)\frac{c}{A} < \varepsilon$. Then from (6.9) we have

$$\sup_x \left| F_t^*(x) - \mathsf{E}\Phi\left(\frac{x}{\sqrt{U}}\right) - \sum_{j=1}^{k-2} t^{-j/2} q_j(x) \right| \leq$$

$$\frac{1}{\pi} \int_{-At^{(k-2)/2}}^{At^{(k-2)/2}} \left| \frac{f_t^*(s) - h(s)}{s} \right| ds + \frac{\varepsilon}{t^{\frac{k-2}{2}}} \equiv I + o(t^{-\frac{k-2}{2}}).$$

Therefore to prove (6.8) it suffices to show that

$$I = o(t^{-\frac{k-2}{2}}). \tag{6.11}$$

Prove (6.11). For any $\delta > 0$ the following inequality holds

$$I \le \frac{1}{\pi} \int\limits_{|s| \le \delta\sqrt{t}} \left| \frac{f_t^*(s) - h(s)}{s} \right| ds + \frac{1}{\pi} \int\limits_{\delta\sqrt{t} \le |s| \le At^{\frac{k-2}{2}}} \frac{1}{|s|} |f_t^*(s)| ds +$$

$$\frac{1}{\pi} \int\limits_{\delta\sqrt{t} \le |s| \le At} \frac{1}{|s|} |h(s)| ds \equiv I_1 + I_2 + I_3. \tag{6.12}$$

Estimate the integrals I_1, I_2 and I_3.

Notice that the characteristic function $f_t^*(s)$ of the random variable $S^*(t)/(\sigma\sqrt{t})$ can be represented as

$$f_t^*(s) = \mathsf{E} \exp\left\{ -Ut + Ut f_{X_1}\left(\frac{s}{\sigma\sqrt{t}} \right) \right\} = \mathsf{E}\left(v\left(\frac{s}{\sigma\sqrt{t}} \right) \right)^{Ut} \tag{6.13}$$

where $v(s) = \exp\{f_{X_1}(s) - 1\}$ and $f_{X_1}(s)$ is the characteristic function of X_1. Note that $v(s)$ is the characteristic function of the random variable $Y \equiv \sum_{j=1}^{N_1} X_j$ where N_1 is the standard Poisson random variable independent of $\{X_j\}_{j \ge 0}$. We obviously have $\mathsf{E}Y = 0$, $\mathsf{D}Y = \sigma^2 > 0$, $\mathsf{E}|Y|^k < \infty$. Also note that the semiinvariants of Y coincide with the moments of X_1. So we can apply Lemma 6.2.4 from (Petrov, 1975) to the characteristic function $(v(s/(\sigma\sqrt{ut})))^{ut}$ with n replaced by ut, $u > 0$. From this lemma the following inequality can be deduced: for any $\varepsilon > 0$ there exists a $\delta > 0$ such that for $|s| \le \sqrt{ut}\delta$

$$\left| \left(v\left(\frac{s}{\sigma\sqrt{ut}} \right) \right)^{ut} - e^{-\frac{s^2}{2}}\left(1 + \sum_{j=1}^{k-2} \frac{P_j^*(is)}{(ut)^{j/2}} \right) \right| \le \frac{c_1\varepsilon}{(ut)^{\frac{k-2}{2}}} |s|^{3k} e^{-c_2 s^2} \tag{6.14}$$

where $c_1 > 0$, $c_2 > 0$ and the polynomials $P_j^*(\cdot)$ of power $3j$ are determined by formulas (1.12) in (Petrov, 1975, p. 173) with the semiinvariants γ_m replaced by the moments $\mathsf{E}X_1^m$, $m \le k$. As this is so (see (6.10)),

$$h_t\left(\frac{s}{\sqrt{u}}, u \right) = e^{-s^2/2}\left(1 + \sum_{j=1}^{k-2} \frac{P_j^*(is)}{(ut)^{j/2}} \right). \tag{6.15}$$

We shall apply inequality (6.14) to estimation of I_1. With regard for (6.10), (6.13) and (6.15) we have

$$I_1 \le \frac{1}{\pi}\mathsf{E} \int\limits_{|s| \le \delta\sqrt{t}} \left| \left(v\left(\frac{s}{\sigma\sqrt{t}} \right) \right)^{Ut} - h_t(s, U) \right| \frac{1}{|s|} ds =$$

$$\frac{1}{\pi} \mathrm{E} \int\limits_{|y| \le \delta\sqrt{tU}} \left| \left(v \left(\frac{y}{\sigma\sqrt{tU}} \right) \right)^{Ut} - h_t \left(\frac{y}{\sqrt{U}}, U \right) \right| \frac{1}{|y|} \, dy \le$$

$$\frac{c_1\varepsilon}{\pi t^{\frac{k-2}{2}}} \mathrm{E}U^{-\frac{k-2}{2}} \int\limits_{|y| \le \delta\sqrt{tU}} |y|^{3k-1} e^{-c_2 y^2} \, dy \le$$

$$\frac{2c_1\varepsilon}{\pi t^{\frac{k-2}{2}}} \int\limits_0^\infty y^{3k-1} e^{-c_2 y^2} \, dy \, \mathrm{E}U^{-\frac{k-2}{2}} = o(t^{-\frac{k-2}{2}}), \tag{6.16}$$

since $\mathrm{E}U^{-\frac{k-2}{2}} < \infty$ by the assumption of the theorem.

Now consider the integral I_2 (see (6.12)). Note that the random variable Y satisfies the Cramér condition (6.1) since

$$\limsup_{|s| \to \infty} |v(s)| \le \limsup_{|s| \to \infty} \exp\{|f_{X_1}(s)| - 1\} \le q < 1.$$

Therefore by virtue of (6.6) we have

$$I_2 \le \frac{1}{\pi} \int\limits_{\delta\sqrt{t} \le |s| \le At^{-\frac{k-2}{2}}} \frac{1}{|s|} |f_t^*(s)| \, ds \le \mathrm{E} \int\limits_{\delta\sqrt{t} \le |s| \le At^{-\frac{k-2}{2}}} \frac{1}{|s|} \left| v \left(\frac{s}{\sigma\sqrt{t}} \right) \right|^{Ut} \, ds \le$$

$$2\mathrm{E}q^{Ut} \int\limits_{\delta\sqrt{t}}^{At^{-\frac{k-2}{2}}} \frac{ds}{s} \le C(\log)_k \mathrm{E}q^{Ut} = o(t^{-\frac{k-2}{2}}). \tag{6.17}$$

(Here C is a positive constant.)

Now it remains to estimate I_3 (see (6.12)). With regard for (6.10) and (6.15) we have

$$I_3 \le \frac{1}{\pi} \mathrm{E} \int\limits_{|s| \ge \delta\sqrt{t}} \frac{1}{|s|} |h_t(s, U)| \, ds \le \frac{1}{\pi} \mathrm{E} \int\limits_{|y| \ge \delta\sqrt{t}} \frac{1}{|y|} e^{-y^2/2} \, dy +$$

$$\frac{1}{\pi} \sum_{j=1}^{k-2} t^{-j/2} \mathrm{E}U^{-j/2} \int\limits_{|y| \ge \delta\sqrt{Ut}} \frac{1}{|y|} e^{-y^2/2} |P_j^*(iy)| \, dy \equiv I_{31} + I_{32}. \tag{6.18}$$

According to (1.12) in (Petrov, 1975, p. 173), for $P_j^*(\cdot)$ the estimate

$$|P_j^*(is)| \le \sum_{l=1}^j c_l |s|^{j+2l} \tag{6.19}$$

is valid, $c_l > 0$, $l = 1, \ldots, j$. If $k = 3$, then these inequalities imply that the right-hand side of (6.18) is $o(t^{-1/2})$ by virtue of the dominated convergence theorem and condition $EU^{-1/2} < \infty$. Let $k > 3$. Put $\gamma = Bt^{-1}$ with $B > 0$ sufficiently large to be specified later.

To estimate I_{31} we use the inequality

$$\int_b^\infty e^{-ax^2}\, dx \le \frac{1}{2ab}e^{-ab^2}, \qquad a > 0,\ b > 0,$$

with regard for which we have

$$I_{31} = \frac{2}{\pi\delta\sqrt{t}}EI(0 \le U \le \gamma)\frac{1}{\sqrt{U}}\int_{\delta\sqrt{Ut}}^\infty e^{-v^2/2}\, dy +$$

$$\frac{2}{\pi\delta\sqrt{t}}EI(U > \gamma)\frac{1}{\sqrt{U}}\int_{\delta\sqrt{Ut}}^\infty e^{-v^2/2}\, dy \le$$

$$\frac{2}{\pi\delta\sqrt{t}}\gamma^{\frac{k-3}{2}}EU^{-\frac{k-2}{2}}I(0 \le U \le \gamma) + \frac{2}{\pi\delta\sqrt{t}}EI(U > \gamma)\frac{1}{U}\exp\left\{-\frac{\delta^2 Ut}{2}\right\}. \quad (6.20)$$

Since $EU^{-\frac{k-2}{2}} < \infty$, the first summand in the right-hand side of (6.20) is $o(t^{-(k-2)/2})$, while the second one is no greater than $\frac{2}{\pi\delta^2 B}E\exp\{-\frac{1}{2}\delta^2 Ut\}$ and therefore is also $o(t^{-(k-2)/2})$ by virtue of condition (6.6) (or (6.7)). Hence, $I_{31} = o(t^{-(k-2)/2})$.

In a similar manner we estimate I_{32}. With the account of (6.19) it suffices to estimate the summands of the form ($j = 1, \ldots, k - 2;\ l = 1, \ldots, j$)

$$t^{-j/2}EU^{-j/2}\int_{\delta\sqrt{Ut}}^\infty e^{-v^2/2}y^{j-1+2l}\, dy,$$

which are obviously no greater than the sum

$$Ct^{-j/2}\gamma^{\frac{k-j-2}{2}}EU^{-\frac{k-2}{2}}I(0 \le U \le \gamma) +$$

$$B^{-j/2}EI(U > \gamma)\int_{\delta\sqrt{Ut}}^\infty e^{-v^2/2}y^{j-1+2l}\, dy \qquad (6.21)$$

where $C > 0$. The first summand in (6.21) is $o(t^{-\frac{k-2}{2}})$ due to the assumption $EU^{-\frac{k-2}{2}} < \infty$. Prove that the expectation in the second summand in (6.21) is $o(t^{-\frac{K-2}{2}})$. If we choose B large enough, we easily see that

$$y^{j-1+2l} \le e^{cy}$$

for $y \ge \delta\sqrt{B}$ and some $c > 0$; $j = 1, \ldots, k - 2;\ l = 1, \ldots, j$. Therefore

$$EI(U > \gamma)\int_{\delta\sqrt{Ut}}^\infty e^{-v^2/2}y^{j-1+2l}\, dy \le EI(U > \gamma)\int_{\delta\sqrt{Ut}}^\infty e^{-v^2/2+cy}\, dy \le$$

$$EI(U > \gamma)\frac{1}{\delta\sqrt{Ut}-c}\exp\left\{-\frac{1}{2}\delta^2 Ut + c\delta\sqrt{Ut}\right\} \le$$

$$\frac{1}{2\sqrt{B}-c}E\exp\left\{-Ut\left(\frac{\delta^2}{2} - \frac{c\delta}{\sqrt{B}}\right)\right\} = o(t^{-\frac{k-2}{2}}),$$

since B is large enough and condition (6.6) holds. Now the desired result follows from (6.8), (6.11), (6.16), (6.17) and (6.18). The theorem is proved.

Returning to the example we considered in Sect. A2.5 with generalized Cox process controlled by geometrically distributed process $\Lambda(t)$ with parameter $1/t$, for which $P(\Lambda(t)/t < x) \Rightarrow 1 - e^{-x} = P(U < x)$, we notice that the first (and the only possible due to condition (6.6)) correcting term in the asymptotic expansion for $P(S^*(t)/(\sigma\sqrt{t}) < x)$ with $x > 0$ has the form

$$g(x) = -\frac{\alpha_3}{6\sigma^3\sqrt{t}}\int_0^\infty \phi\left(\frac{x}{\sqrt{u}}\right)\left(\frac{x^2}{u} - 1\right)\frac{1}{\sqrt{u}}\exp(-u)\,du =$$

$$-\frac{\alpha_3}{6\sigma^3\sqrt{2\pi t}}\left[x^2\int_0^\infty u^{-3/2}\exp\left\{-\frac{x^2}{2u} - u\right\}du - \int_0^\infty u^{-1/2}\exp\left\{-\frac{x^2}{2u} - u\right\}du\right] =$$

$$-\frac{\alpha_3}{6\sigma^3\sqrt{2\pi t}}\left[2^{5/4}x^{3/2}K_{-1/2}(\sqrt{2}x) - 2^{3/4}x^{1/2}K_{1/2}(\sqrt{2}x)\right] =$$

$$\frac{\alpha_3\sqrt{x}(1 - \sqrt{2}x)}{3\cdot 2^{3/4}\sigma^3\sqrt{\pi t}}K_{1/2}(\sqrt{2}x)$$

where $K_\nu(y)$ is the cylindric function of an imaginary argument with index ν,

$$K_\nu(y) = \frac{1}{\cos\frac{\nu\pi}{2}}\int_0^\infty \cos(y\sinh(y))\cosh(\nu u)\,du, \quad y > 0,$$

$$K_{-\nu}(y) = K_\nu(y)$$

(see (Gradstein and Ryzhik, 1965)). It is easy to see that $g(-x) = g(x)$. By the way, in this case the limit law $E\Phi(x/\sqrt{u})$ is the Laplace double exponential distribution with density $\frac{1}{\sqrt{2}}\exp\{-\sqrt{2}|x|\}$, $-\infty < x < \infty$.

Now turn to the case where $EX_1 = a \ne 0$. In this situation the limit distribution for nonrandomly centered generalized Cox process is determined by Theorem A2.4.2 and is of the form $E\Phi\left(\frac{\sigma x - aV}{\sqrt{\sigma^2 + a^2}}\right)$ where $\sigma^2 = DX_1$ and V is the limit random variable for the standardized controlling process $(\Lambda(t) - t)/\sqrt{t}$ (where we assume that $E\Lambda(t) \equiv t$). At first we shall formulate a statement on asymptotic expansions for generalized Cox processes with a rather peculiar structure of the random variable V. But as we will see below, this situation turns out to be quite natural for the case under consideration.

THEOREM A2.6.2. *Let the random variable V be of the form $V = V_0 - EV_0$ where V_0 is a nonnegative random variable satisfying the following condition:*

there exist a $\gamma > 0$ and a polynomial $P(h)$ such that for any $h \geq 0$ we have

$$Ee^{hV_0} \leq P(h)e^{\gamma h^2}. \tag{6.22}$$

Assume that $E|X_1|^k < \infty$ for some integer $k \geq 3$ and X_1 satisfies the Cramér condition (6.1). Then for any $t \geq (EV_0)^2$ we have

$$\sup_x \left| P\left(\frac{S(t) - at}{\sigma\sqrt{t}} < x\right) - E\Phi\left(\frac{\sigma x - aV}{\sqrt{\sigma^2 + a^2}}\right) - \sum_{j=1}^{k-2} \frac{w_j(x)}{t^{j/2}} \right| \leq \varepsilon(t)$$

where

$$\varepsilon(t) = \int_0^\infty \min\left(2, \frac{1}{\sqrt{\lambda}}\right) |P(\Lambda(t) < \lambda) - P(\Lambda_0(t) < \lambda)| \, d\lambda + o(t^{-\frac{k-2}{2}}),$$

$$\Lambda_0(t) = \sqrt{t}V + t,$$

$$w_j(x) = \sum_{\substack{l+m=j \\ l,m \geq 0}}^m \sum_{d=0}^x \int_{-\infty}^x \bar{P}_l(-D_y)P_{md}(-D_y)\chi_d(y) \, dy, \quad j = 0, \ldots, k-2;$$

$$\chi_d(y) = \frac{\sigma}{\sqrt{\sigma^2 + a^2}} E\left[V^d \phi\left(\frac{\sigma y - aV}{\sqrt{\sigma^2 + a^2}}\right)\right], \quad d = 0, \ldots, k-2,$$

D_y is the operator of formal differentiation with respect to y, ϕ is the standard normal density, the polynomials $\bar{P}_l(\cdot)$ are defined as

$$\bar{P}_l(it) = \sum \prod_{m=1}^j \frac{1}{k_m!} \left[\frac{(it)^{m+2}\alpha_{m+2}}{(m+2)!\sigma^{m+2}}\right]^{k_m}, \quad j = 1, \ldots, k-2,$$

with summation over all integer nonnegative solutions $k_1, \ldots, k_j$ of the equation $k_1 + 2k_2 + \ldots + jk_j = j$ and $\alpha_{m+2} = EX_1^{m+2}$, $\bar{P}_0(x) \equiv 1$, the polynomials $P_{md}(\cdot)$ are defined by the formal equality

$$\exp\left\{\sum_{l=1}^{k-2} v \frac{x^{l+1}\alpha_{l+1}}{(l+1)!\sigma^{l+1}} t^{-l/2}\right\} = \sum_{m=0}^\infty t^{-m/2} \sum_{d=0}^m v^d P_{md}(x). \tag{6.23}$$

REMARK A2.6.3. It is easy to see that

$$P_{00}(x) \equiv 1, \quad P_{j0} \equiv 0, \quad j = 1, \ldots, k-2;$$

$$P_{11}(x) = \frac{x^2\alpha_2}{2\sigma^2}, \quad P_{21}(x) = \frac{x^3\alpha_3}{6\sigma^3}, \quad P_{22}(x) = \frac{x^4\alpha_2^2}{8\sigma^4},$$

$$P_{31}(x) = \frac{x^4\alpha_4}{24\sigma^4}, \quad P_{32}(x) = \frac{x^5\alpha_2\alpha_3}{12\sigma^5}, \quad P_{33}(x) = \frac{x^6\alpha_2^3}{48\sigma^6}$$

REMARK A2.6.4. The first two functions w_1 and w_2 are of the following form:

$$w_1(x) = -\int \phi\left(\frac{\sigma x - az}{\sqrt{a^2 + \sigma^2}}\right) \left[\frac{\alpha_3}{6\sigma\alpha_2} H_2\left(\frac{\sigma x - az}{\sqrt{a^2 + \sigma^2}}\right) + \right.$$

$$\left. \frac{z}{2} H_1\left(\frac{\sigma x - az}{\sqrt{a^2 + \sigma^2}}\right)\right] d\mathsf{P}(V < z);$$

$$w_2(x) = -\int \phi\left(\frac{\sigma x - az}{\sqrt{a^2 + \sigma^2}}\right) \left[\frac{\alpha_4}{24\sigma^2\alpha_2} H_3\left(\frac{\sigma x - az}{\sqrt{a^2 + \sigma^2}}\right) + \right.$$

$$\frac{\alpha_3}{72\sigma^4\alpha_2} H_5\left(\frac{\sigma x - az}{\sqrt{a^2 + \sigma^2}}\right) +$$

$$z\left(\frac{\alpha_3}{12\sigma^3} H_4\left(\frac{\sigma x - az}{\sqrt{a^2 + \sigma^2}}\right) + \frac{\alpha_3}{6\sigma\alpha_2} H_2\left(\frac{\sigma x - az}{\sqrt{a^2 + \sigma^2}}\right)\right) +$$

$$\left. \frac{z^2\alpha_2^3}{8\sigma^2} H_3\left(\frac{\sigma x - az}{\sqrt{a^2 + \sigma^2}}\right)\right] d\mathsf{P}(V < z),$$

where $H_m(\cdot)$ are Chebyshev-Hermite polynomials.

REMARK A2.6.5. Condition (6.22) holds, if $V_0 = |\xi|$ where ξ is a bounded random variable. It also holds if ξ is normal. It does not hold for exponential or Poisson random variables V_0. V_0 must possess all moments.

Now consider a discrete-time case $t = n = 1, 2, \ldots$ and assume that the controlling process $\Lambda(n)$ has the form

$$\Lambda(n) = \sum_{i=1}^{n} Z_i \qquad (6.24)$$

where $\{Z_i\}$ are independent identically distributed random variables, $Z_1 \geq 0$, $\mathsf{E}Z_1 = 1$ so that $\mathsf{E}\Lambda(n) = n$. Denote

$$\nu_l = \mathsf{E}(Z_1 - 1)^l, \quad l = 1, 2, \ldots$$

Define the formal "semiinvariants" $\ae_j$ by the equality

$$\log \mathsf{E}\exp\{(Z_1 - 1)(f_{X_1}(s) - 1)\} = \sum_{j=2}^{\infty} \frac{\ae_j}{j!}(is)^j.$$

In particular,

$$\ae_2 = \nu_2\alpha_1^2, \quad \ae_3 = 3\nu_2\alpha_2\alpha_1 + \nu_3\alpha_1^3,$$
$$\ae_4 = 3\nu_2\alpha_2^2 + \nu_4\alpha_1^4 + 6\alpha_1^2\alpha_2\nu_3 - 3\nu_2^2\alpha_1^4.$$

THEOREM A2.6.3. Assume that there exist a $\gamma > 0$ and a polynomial $P(h)$ such that for any $h \geq 0$ the random variable Z_1 satisfies the inequality

$$\mathsf{E}e^{hZ_1} \leq P(h)e^{\gamma h}. \qquad (6.25)$$

Let $\mathsf{E}|X_1|^k < \infty$ for integer $k \geq 0$. *Assume that X_1 satisfies the Cramér condition (6.1). Then we have*

$$\sup_x \left| \mathsf{P}\left(\frac{S(n) - an}{\sigma\sqrt{n}} \right) - \Phi\left(\frac{\sigma x}{\sqrt{\ae_2 + \alpha_2}} \right) - \sum_{j=1}^{k-2} \frac{v_j(x)}{n^{j/2}} \right| \leq \varepsilon(n)$$

where $\varepsilon(n)$ was defined in the formulation of Theorem A2.6.2,

$$v_j(x) = \frac{\sigma}{\sqrt{\ae_2 + \alpha_2}} \sum_{\substack{l+m=j \\ l,m \leq 0}} \int_{-\infty}^{x} \overline{P}_l(-D_y) P_m^*(-D_y) \phi\left(\frac{\sigma y}{\sqrt{\ae_2 + \alpha_2}} \right) dy,$$

D_y *is the operator of formal differentiation with respect to y, $P_0^*(x) \equiv \overline{P}_0(x) \equiv 1$, the polynomials $\overline{P}_l(\cdot)$ are defined in Theorem A2.6.2 while $P_m^*(\cdot)$ are defined in the same way as $\overline{P}_l(\cdot)$ but with moments α_{m+2} replaced by semiinvariants $\ae_{m+2}$.*

REMARK A2.6.6. We do not assume the random variables Z_i to satisfy the Cramér condition (6.1) so that they may be lattice.

REMARK A2.6.7. Condition (6.25) is stronger than (6.22). This condition holds for any bounded random variable Z_1, e.g., binomial, but does not hold for exponential or Poisson random variables. Note also that Z_1 possess all moments.

The main stages of the proofs of Theorems A2.6.2 and A2.6.3 resemble those of Theorem A2.6.1.

The structure of the controlling process of type (6.24) is typical for many situations. At the same time it can help us understand better what goes on in Theorem A2.6.2. Return to Theorem A2.6.2 assuming that representations related to (6.24) take place for $\Lambda(n)$ but Z_i are not necessarily positive and consider some examples.

EXAMPLE A2.6.1. Let $\mathsf{E}Z_i = 0$, $\mathsf{D}Z_i = \delta^2$. Put

$$\widetilde{\Lambda}(n) = \frac{1}{\delta}|Z_1 + \ldots + Z_n| + n - \sqrt{\frac{2n}{\pi}}.$$

Then we have $\mathsf{E}\widetilde{\Lambda}(n) \approx n$ and

$$\frac{\widetilde{\Lambda}(n) - n}{\sqrt{n}} \Rightarrow |W| - \mathsf{E}|W| \quad (n \to \infty)$$

where W is the standard normal random variable.

EXAMPLE A2.6.2. Put $Z_0 = 0$. Then under the same conditions on $\{Z_i\}_{i \geq 1}$ as above, putting

$$\widetilde{\Lambda}(n) = \frac{1}{\delta} \max_{0 \leq i \leq n} (Z_0 + Z_1 + \ldots + Z_i) + n - \sqrt{\frac{2n}{\pi}}$$

we also obtain $E\tilde{\Lambda}(n) \approx n$ and

$$\frac{\tilde{\Lambda}(n) - n}{\sqrt{n}} \Rightarrow |W| - E|W| \quad (n \to \infty).$$

EXAMPLE A2.6.3. Under conditions of Example A2.6.1 put

$$\tilde{\Lambda}(n) = \frac{1}{a^2\sqrt{n}}(Z_1 + \ldots + Z_n)^2 + n - \sqrt{n}.$$

Then we have $E\tilde{\Lambda}(n) = n$ and

$$\frac{\tilde{\Lambda}(n) - n}{\sqrt{n}} \Rightarrow W^2 - EW^2 \quad (n \to \infty).$$

In all the three examples we have the same structure of the limit random variable as in Theorem A2.6.2 (the first two examples correspond to $V_0 = |W|$, in the third example $V_0 = W^2$). These examples are nothing more than an illustration of invariance principle and we see that Theorem A2.6.2 corresponds to the situation similar to that in which $\Lambda(t)$ is a function of a Wiener process. As this is so, rather an unexpected structure of the rather V in Theorem A2.6.2 is due to the requirement of positiveness of the controlling process.

Note that in Examples A2.6.1 and A2.6.2 $\sqrt{2n/\pi}$ is the asymptotic expectation of $\frac{1}{\delta}|Z_1 + \ldots + Z_n|$ and $\frac{1}{\delta}\max_{0 \le i \le n}(Z_0 + Z_1 + \ldots + Z_n)$, respectively. By subtracting exact expectations instead of asymptotic ones in the definitions of $\tilde{\Lambda}(n)$ we can provide the exact equality $E\tilde{\Lambda}(n) \equiv n$.

A2.7 Estimates for the concentration functions of generalized Cox processes

In this section we give some estimates for the concentration function of one-dimensional distributions of process (4.1). Recall that the concentration function of a random variable Y is defined as

$$Q_Y(l) = \sup_x P(x \le Y \le x + l), \qquad l \ge 0.$$

Concentration functions of random variables are actually the functions of distribution functions and moreover, they are distribution functions themselves. They were introduced by P. Lévy as informative characteristics of distributions. They were the subjects of many investigations. Their properties are described in some monographs, see, e.g., (Hengartner and Theodorescu, 1973).

It is well known that if $X_1, \ldots, X_n$ are independent random variables with one and the same distribution, then for the concentration function of the sum $S_n = X_1 + \ldots + X_n$ the inequality

$$Q_{S_n}(l) \le C(\varepsilon, \delta)\frac{l+1}{\sqrt{n}}, \qquad l \ge 0, \tag{7.1}$$

holds with $C(\varepsilon, \delta) = (96/95)^2 \max(1, \delta^{-1})\sqrt{\pi/\varepsilon}$, where ε and δ are positive numbers providing

$$|f_{X_1}(s)| \leq 1 - \varepsilon s^2, \quad |s| \leq \delta. \tag{7.2}$$

Here $f_{X_1}(s)$ is the characteristic function of X_1. The existence of a $\delta = \delta(\varepsilon)$ guaranteeing (7.2) for an arbitrary $\varepsilon > 0$ is stated in Theorem 1.2.2 from (Petrov, 1975) which deals with characteristic functions of nondegenerate random variables. We will obtain an analog of (7.1) for the concentration functions of generalized Cox processes (4.1) under conditions $EX_1 = 0$, $DX_1 = \sigma^2$, $0 \leq \sigma^2 < \infty$. Denote $Q_t(l) = Q_{S(t)}(l)$, $l \geq 0$.

THEOREM A2.7.1. *Let $\ae = EU^{-1/2} < \infty$. Then:*

1) *If the random variables $\{X_j\}_{j \geq 1}$ and nondegenerate, then*

$$Q_t(l) \leq \ae C(\varepsilon, \delta)\frac{l+1}{\sqrt{t}} + \delta(t)$$

where $\delta(t)$ is defined in (5.3) and $C(\varepsilon, \delta)$ is the same as in (7.1).

2) *If the distribution of X_1 is degenerate at a point $\alpha \neq 0$, then*

$$Q_t(l) \leq \ae C\left(\frac{11}{24}, 1\right)\frac{l|\alpha|^{-1}+1}{\sqrt{t}} + \delta(t).$$

Before proving this theorem note that there is no analog of item 2) for sums S_n.

PROOF OF THEOREM A2.7.1. From (5.2) and the proof of Lemma A2.5.1 we can conclude that

$$|Q_t(l) - Q_t^*(l)| \leq \sup_x |P(x \leq S(t) \leq x + l) - P(x \leq S^*(t) \leq x + l)| =$$

$$\sup_x \left| \int_0^\infty \left[\sum_{k=0}^\infty P\left(x \leq \sum_{j=1}^k X_j \leq x + l\right) e^{-\lambda}\frac{\lambda^k}{k!}\right] d\Psi_t(\lambda) \right| \leq \delta(t).$$

where $Q_t^*(l) = Q_{S^*(t)}(l)$ and as before, $S^*(t)$ is the generalized Cox process generated by the same sequence $\{X_j\}_{j \geq 1}$ but controlled by the process Ut, $t \geq 0$, with U being the limit random variable for $\Lambda(t)/t$. Therefore

$$Q_t(l) \leq Q_t^*(l) + \delta(t). \tag{7.3}$$

Hence to obtain the assertion of the theorem it suffices to estimate $Q_t^*(l)$. For this purpose we apply Lemma 3.1.3 from (Petrov, 1975), according to which for any random variable Y and any $l \geq 0$, $a > 0$

$$Q_Y(l) \leq \left(\frac{96}{95}\right)^2 \max\left(l, \frac{1}{a}\right) \int_{-a}^a |f_Y(s)| \, ds \tag{7.4}$$

where $f_Y(s)$ is the characteristic function of Y. We have already seen in Section A2.6 that the characteristic function $f_t^*(s)$ of $S^*(t)$ has the form

$$f_t^*(s) = \text{E} \exp\{Ut(f_{X_1}(s) - 1)\}. \tag{7.5}$$

Let X_1 be nondegenerate. Then with account of (7.2) we have

$$|f_t^*(s)| = \text{E} \exp\{Ut(f_{X_1}(s) - 1)\} \le \text{E} \exp\{-Ut\varepsilon s^2\}, \quad |s| \le \delta. \tag{7.6}$$

If $l \ge 1$, then in (7.4) we put $a = \delta/l$. In this case with regard for (7.6) we easily obtain

$$Q_t^*(l) \le \left(\frac{96}{95}\right)^2 l \max\left(1, \frac{1}{\delta}\right) \int_{-\delta/l}^{\delta/l} \text{E} \exp\{-Ut\varepsilon s^2\} \le$$

$$\left(\frac{96}{95}\right)^2 l \max\left(1, \frac{1}{\delta}\right) \text{E} \int_{-\infty}^{\infty} \exp\{-Ut\varepsilon s^2\}\, ds \le C(\varepsilon, \delta)\frac{\ae(l+1)}{\sqrt{t}}. \tag{7.7}$$

For $0 \le l < 1$ it suffices to take $a = \delta$ in (7.4), which yields

$$Q_t^*(l) \le \left(\frac{96}{95}\right)^2 \max\left(l, \frac{1}{\delta}\right) \text{E} \int_{-\infty}^{\infty} \exp\{-Ut\varepsilon s^2\}\, ds \le C(\varepsilon, \delta)\frac{\ae(l+1)}{\sqrt{t}}. \tag{7.8}$$

Now item 1) of the theorem follows from (7.3), (7.7) and (7.8).

Consider the case $P(X_1 = \alpha) = 1$, $\alpha \ne 0$. First let $\alpha = 1$. In this case

$$f_t^*(s) = \text{E} \exp\{Ut(e^{is} - 1)\}$$

and

$$|f_t^*(s)| \le \text{E} \exp\{-Ut(1 - \cos s)\}.$$

Put $a = 1$ in (7.4). Then by virtue of the inequality

$$1 - \cos s \ge \frac{11}{24}s^2, \quad |s| \le 1,$$

and due to the relation $S^*(t) = N^*(t)$ taking place in this case, we obtain the estimate

$$Q_t^*(l) = Q_{N^*(t)} \le \left(\frac{96}{95}\right)^2 \max(l, 1) \int_{-1}^{1} \text{E} \exp\left\{-\frac{11}{24}Uts^2\right\} ds \le$$

$$\left(\frac{96}{95}\right)^2 (l+1)\text{E} \int_{-\infty}^{\infty} \exp\left\{-\frac{11}{24}Uts^2\right\} ds = C\left(\frac{11}{24}, 1\right)\frac{\ae(l+1)}{\sqrt{t}}. \tag{7.9}$$

The case of an arbitrary $\alpha \ne 0$ is considered with the account of (7.9) and the equality

$$Q_t^*(l) = Q_{N^*(t)}\left(\frac{l}{|\alpha|}\right).$$

The theorem is completely proved.

Bibliographical commentary

All the results of the theory of random summation can be conventionally divided into two groups. The first of them includes statements obtained under the assumption of independence of the index and the summands, while the second one involves assertions devoid of such a restriction. In this book, we paid main attention to the results of the first group for many reasons. First, these models are successfully applied to solving practical problems even more often than it might have been expected (and we intended to justify this proposition by many various examples of applied problems considered in this book). Second, the results obtained under the assumption of independence appear to be a good guideline in the investigation of the general case. Third, it is this assumption that made it possible for us to manage without a rather sophisticated apparatus which would have inevitably made this book much more voluminous. But of course, all these are not reasons to ignore deep results of the second group.

As milestones in this direction, along with the results of A. Wald (Wald, 1944) and A. N. Kolmogorov and Yu. V. Prokhorov (Kolmogorov and Prokhorov, 1949) which have already been noted in Chapter 1, we should mention the asymptotic results of F. Anscombe (Anscombe, 1952) who was inspired by the promising research of A. Wald in the field of sequential statistical analysis (Wald, 1947). Major progress in this direction became possible mainly due to the fundamental works of A. Rényi and his followers (Rényi, 1957, 1958, 1960, 1963), (Mogyoródi, 1961, 1962, 1965, 1967), (Blum, Hanson and Rosenblatt, 1963) where mixing sequences were investigated and some variants of the central limit theorem for random sums with non-independent random number of summands were proved. Some special cases, namely, stopped random walks and renewal processes were described in (Gut, 1988).

The total number of scientific publications concerning random summation is now not very far from one thousand, so it is impossible to mention all of them within the limits of this book. The list of references given below includes only those papers that are directly connected with the material of this book. Therefore we apologize to all those whose contribution to the theory of random summation and its applications is not mentioned here.

Along with the sources of the material of Chapter 1 mentioned in the text, we note the paper (Gnedenko, 1941).

Chapter 2 is based on the papers (Gnedenko, 1964a, 1964b), (Solov'ev, 1971) as well as on those mentioned in the text.

The basis of Sections 3.1, 3.2 and 3.3 are the works (Korolev, 1993) and (Korolev, 1994). For multidimensional generalizations wth operator normalization see (Korolev and Kossova, 1995). The material of Sections 3.4, 3.5 and 4.4 was the subject of communication (Korolev, 1995). For historical background and further results on application of methods of probability theory in financial mathematics see (Shiryaev, 1994a, 1994b, 1994c), (Shiryaev, Kabanov, Kramkov and Melnikov, 1994a, 1994b), (Mandelbrot and Taylor, 1967), (Rachev and Ruescheendorf, 1994). A result similar to Theorem 3.4.2 but under slightly different assumptions was recently proved by V. M. Kruglov (Kruglov, 1995). The sourses of Sect. 3.6 are indicated in the text. For some related results see (Szynal, 1976). Other sourses used in Chapter 3 are indicated in the text.

Investigations of random sums in the double array scheme were initiated by the paper (Gnedenko and Fahim, 1969). A major part of Sections 4.1, 4.2 and 4.3 consists of the results of (Korolev and Kruglov, 1993) and (Korolev, 1993). Theorem 4.2.1, along with other results of Sections 4.2 and 4.3 reported by V. Yu. Korolev at the seminar on Stability Problems for Stochastic Models in April, 1993 in Moscow State University was independently proved by M. Finkelstein, H. G. Tucker and J. A. Veeh (Finkelstein, Tucker and Veeh, 1995). Theorem 4.3.1 generalizes a result of (Szász, 1972). A detailed description of the limit behavior of noncentered random sums in the double array scheme was given in (Kruglov and Korolev, 1990). The sourses of Sections 4.5 and 4.6 are noted in the text.

Chapter 5 contains material published in (Korolev, 1992), (Korolev and Petukhov, 1994, 1995), (Korolev and Selivanova, 1994). More references on software reliability can be found in (Musa, Iannino and Okumoto, 1987).

All the sourses of Appendix 1 are indicated in the text.

The basis of Section A2.1 are the works (Grandell, 1976), (Embrechts and Klueppelberg, 1993). Theorems A2.2.1, A2.3.1, A2.3.2 are published in (Korolev, 1996). The rest of the material of Appendix 2 is the content of (Bening and Korolev, 1996).

References

1) Abel, N., *Oeuvres, Vol. 2*, Christiania, 1881.

2) Aivazyan, S. A., Bukhshtaber, V. M., Enyukov, I. S. and Meshalkin, L. D., *Applied Statistics. Classification and Reduction of Dimensionality*, Finansy i Statistika, Moscow, 1989 (in Russian).

3) Ambartsumyan, R. V., Mecke, J. and Stoyan, D., *An Introduction to Stochastic Geometry*, Nauka, Moscow, 1989 (in Russian).

4) Anscombe, F. I., *Large-sample theory of sequential estimation*, Proc. Cambridge Phil. Soc., 48, 600, 1952.

5) Asmussen, S., *Applied Probabilities and Queues*, John Wiley, New York, 1987.

6) Bachelier, L., *Theorie de la speculation*, Ann. Ecole Norm. Sup., 17, 21, 1900 (Reprinted in The Random Character of Stock Market Prices, Coother, Ed., MIT Press, Cambridge, MA, 1967, 17).

7) Bagirov, E. B., *The Method of Mixtures and Its Application to Deducing of Lower Estimates for Distributions of Functions of Normal Random Variables*, Ph.D. Thesis, Steklov Mathematical Institute of Academy of Sciences of USSR, Moscow, 1988 (in Russian).

8) Barndorff-Nielsen, O. E., *Exponentially decreasing distributions for the logarithm of a particle size*, Proc. R. Soc. London, A, 353, 401, 1977.

9) Barndorff-Nielsen, O. E., *Hyperbolic distributions and distributions on hyperbolae*, Scand. J. Stat., 5, 151, 1978.

10) Barndorff-Nielsen, O. E., *Gaussian\\Inverse Gaussian Processes and the Modelling of Stock Returns*, Working paper, Aarhus University, 1994.

11) Barndorff-Nielsen, O. E., Kent, J. and Sørensen, M., *Normal variance-mean mixtures and the z-distribution*, Int. Stat. Rev., 50, 145, 1982.

12) Barron, A. R., *Entropy and the central limit theorem*, Ann. Probab., 14, 336, 1986.

13) Beekman, J. A., *Collective risk results*, Trans. Soc. Actuaries, 20, 182, 1968.

14) Belov, A. G., Galkin, V. Ya. and Ufimtsev, M. V., *Probabilistic and Statistical Problems in Experimental Separation of Multiple Processes*, Moscow University Press, Moscow, 1985.

15) Belyaev, Yu. K., *A limit theorem for rarefied flows*, Theory Probab. Appl., 8, 175, 1963.

16) Beneš, V. E., *On queues with Poisson arrivals*, Ann. Math. Stat., 28, 670, 1957.

17) Bening, V. E. and Korolev, V. Yu., *Asymptotic behavior of generalized Cox processes*, Bull. Moscow University, Ser. 15 Comput. Math. Cybern., to appear, 1996.

18) Blum, J. R., Hanson, D. L. and Rosenblatt, J. I., *On the central limit theorem for the sum of a random number of independent random variables*, Z. Wahrscheinlichkeitstheorie verw. Geb., 1, 389, 1963.

19) Borovkov, A. A., *Probability Theory*, Nauka, Moscow, 1986 (in Russian).

20) Bowers, N. L., Gerber, H. U., Hickman, J. C., Jones, D. A. and Nesbitt, C. J., *Actuarial Mathematics*, Society of Actuaries, Itasca, IL, 1986.

21) Brocklehurst, S., Chan, P. Y., Littlewood, B. and Snell, J., *Recalibrating software reliability models*, IEEE Trans. Software Eng., 16, 458, 1990.

22) Brown, L. D., *A proof of the central limit theorem motivated by the Cramér-Rao inequality*, in Statistics and Probability: Essays in Honor of C. R. Rao, Kallianpur, G., Krishnaiah, P. R. and Ghosh, J. K., Eds., North-Holland, Amsterdam, 1982.

23) Chow, Y. S. and Teicher, H., *Probability Theory: Independence, Interchangeability, Martingales*, Springer, Berlin-Heidelberg-New York, 1988.

24) Clark, P. K., *A Subordinated Stochastic Process Model of Cotton Futures Prices*, Ph. D. Thesis, Harvard University, Cambridge, MA, 1970.

25) Clark, P. K., *A subordinated stochastic process model with finite variance for speculative prices*, Econometrica, 41, 135, 1973.

26) Cox, D. R. and Hinkley, D. V., *Theoretical Statistics*, Halsted Press, New York, 1974.

27) Csiszár, I., *Information-type measures of difference of probability distributions and indirect observations*, Studia Sci. Math. Hung., 2, 299, 1967.

28) Csiszár, I., *Information measures: a critical survey*, in Trans. VII Prague Conf. on Information Theory, Statistical Decision Functions, Random Processes and of the 1974 European Meeting of Statisticians, Prague, August 18-23, 1974, Vol. A, Academia Publishing House of the Czechoslovak Acad. Sci., Prague, 1977.

29) David, H. A., *Order Statistics*, John Wiley, New York, 1981.

30) Dawid, A. P., *Statistical theory: the prequential approach (with discussion)*, J. R. Stat. Soc. Ser. A, 147(2), 278, 1984.

31) Dobrushin, R. L., *A lemma on the limit of a compound random function*, Uspekhi Mat. Nauk, 10, 157, 1955.

32) Einstein, A., *On the motion of particles suspended in a liquid at rest, assumed by the molecular-kinetic theory of heat*, Ann. Phys. (Leipzig), 14, 549, 1905.

33) Eggenberger, F., *Die Wahrscheinlichkeitsanteckung*, Mitt. Verein. Schweiz. Versich. Nachr., 19, 31, 1924.

34) Elder, A., *Trading for a Living*, Financial Trading Seminars, Inc., New York, 1993.

35) Embrechts, P. and Klueppelberg, C., *Some aspects of insurance mathematics*, Theory Probab. Appl., 38, 374, 1993.

36) Faddeev, D. K., *On the concept of entropy of a finite probabilistic scheme*, Uspekhi Mat. Nauk, 11, 227, 1956.

37) Faddeev, D. K. and Faddeeva, V. N., *Computational Methods of Linear Algebra*, Fizmatgiz, Moscow, 1960 (in Russian).

38) Feller, W., *An Introduction to Probability Theory and Its Applications*, John Wiley, New York, 1950.

39) Feller, W., An *Introduction to Probability Theory and Its Applications*, Vol. *1*, 3rd ed., John Wiley, New York, 1968.

40) Feller, W., *An Introduction to Probability Theory and its Applications*, Vol. *2*, 2nd ed., John Wiley, New York, 1971.

41) Finkelstein, M., Kruglov, V. M. and Tucker, H. G., *Convergence in law of random sums with non-random centering*, J. Theor. Probab., 7, 565, 1994.

42) Finkelstein, M. and Tucker, H. G., *A necessary and sufficient condition for convergence in law of random sums of random variables under nonrandom centering*, Proc. Am. Math. Soc., 107, 1061, 1990.

43) Finkelstein, M., Tucker, H. G. and Veeh, J. A., *Convergence of random sums under nonrandom centering*, Theory Probab. Appl., 36, 397, 1991.

44) Finkelstein, M., Tucker, H. G. and Veeh, J. A., *Convergence of nonrandomly centered random sums*, Adv. Appl. Math., to appear, 1995.

45) Galtsov, M. V., *Models of Random Sets Sifting*, Ph.D. Thesis, Moscow State University, Moscow, 1992.

46) Galtsov, M. V. and Solov'ev, A. D., *An elementary model of complex software testing*, Bull. Moscow University, Ser. 1 Math., Mech., 5, 74, 1991.

47) Gnedenko, B. V., *To the theory of Geiger-Mueller counters*, J. of Exp. and Theoret. Phys., 11, 101, 1941.

48) Gnedenko, B. V., *A Course in Probability Theory*, Radyanska Shkola, Kiev-L'vov, 1949 (in Ukrainian).

49) Gnedenko, B. V., *On non-loaded doubling*, Izvestiya AN SSSR, Tech. Cybern., 4, 3, 1964.

50) Gnedenko, B. V., *On doubling with repair*, Izvestiya AN SSSR, Tech. Cybern., 5, 111, 1964.

51) Gnedenko, B. V., *On the relationship between the theory of summation of independent random variables and problems of queueing theory and reliability theory*, Rev. Roumaine Math. Pures Appl., 12, 1243, 1967.

52) Gnedenko, B. V., *Stability theorems for limit distributions of order statistics*, Theory Probab. Appl., 28, 809, 1983.

53) Gnedenko, B. V., *On limit theorems for a random number of random variables*, Lect. Notes Math., 1021, 167, 1983.

54) Gnedenko, B. V., *Theory of Probability*, 6th ed., Nauka, Moscow, 1988 (in Russian).

55) Gnedenko, B. V., Belyaev, Yu. K. and Solov'ev, A. D., *Mathematical Methods in Reliability Theory*, Nauka, Moscow, 1965 (in Russian).

56) Gnedenko, B. V. and Fahim, H., *On a transfer theorem*, Doklady AN SSSR, 187, 15, 1969.

57) Gnedenko, B. V. and Freyer, B., *Some remarks to a paper by I. N. Kovalenko*, Lit. Mat. Sbornik, 9, 463, 1969.

58) Gnedenko, B. V. and Kolmogorov, A. N., *Limit Distributions for Sums of Independent Random Variables*, Addison-Wesley, Reading, MA, 1954.

59) Gnedenko, B. V. and Kovalenko, I. N., *Introduction to Queueing Theory*, 2nd ed., Birkhauser, Boston, 1989.

60) Goel, A. K. and Okumoto, K., *A Markovian model for reliability and other performance measures*, in Proc. Natl. Computer Conf., 1979, 769.

61) Goldman, S., *Information Theory*, Constable and Company, London, 1953.

62) Gradstein, I. S. and Ryzhik, I. S., *Tables of Integrals, Series and Products*, Academic Press, New York, 1965.

63) Grandell, J., *Doubly stochastic Poisson processes*, Lect. Notes Math., 529, 1, 1976.

64) Grandell, J., *Aspects of Risk Theory*, Springer, Berlin-Heidelberg- New York, 1990.

65) Greenwood, M. and Yule, G. U., *An inquiry into the nature of frequency distribution representative of multiple happenings with particular reference to the occurence of multiple attacks of disease or repeated accidents*, J. R. Stat. Soc., 83, 255, 1920.

66) Gut, A., *Stopped Random Walks*, Springer, Berlin-Heidelberg-New York, 1988.

67) Harris, T., *Theory of Branching Processes*, Prentice-Hall, Englewood Cliffs, NJ, 1963.

68) Hartley, R. V. L., *Transmission of information*, Bell System Tech. J., 7, 535, 1928.

69) Heckman, J., Robb, R. and Walker, J. R., *Testing for mixture of exponentials hypothesis and estimating the mixing distribution by the method of moments*, J. Am. Stat. Assoc., 85, 582, 1990.

70) Hengartner, W. and Theodorescu, R., *Concentration Functions*, Academic Press, New York-London, 1973.

71) Henley, E. J. and Kumamoto, H., *Reliability Engineering and Risk Assessment*, Prentice-Hall, Englewood Cliffs, NJ, 1981.

72) Heyde, C. C., *A rate of convergence result for supercritical Galton-Watson process*, J. Appl. Probab., 7, 451, 1970.

73) Heyde, C. C., *Some central limit analogues for supercritical Galton-Watson processes*, J. Appl. Probab., 8, 52, 1971.

74) Heyde, C. C. and Brown, B. M., *An invariance principle and some convergence rate results for branching processes*, Z. Wahrscheinlich-keitstheorie verw. Geb., 20, 271, 1971.

75) Hull, J., *Options, Futures and Other Derivative Securities*, Prentice-Hall, New York-London, 1989.

76) Ivchenko, G. I. and Medvedev, Yu. I., *Mathematical Statistics*, Vysshaya Shkola, Moscow, 1992 (in Russian).

77) Jelinski, Z. and Moranda, P. B., *Software reliability research*, in *Statistical Computer Performance Evaluation*, Freiberger, W., Ed., New York, 1972, 465.

78) Johnson, N. L. and Kotz, S., *Distributions in Statistics: Discrete Distributions*, Houghton Mifflin, Boston, 1969.

79) Kabak, I. S. and Rappoport, G. N., *Software reliability estimation from its mathematical model*, in Flexible Automatic Production Design Problems, Moscow, 1987, 236.

80) Kalashnikov, V. V. and Vsekhsvyatskii, S. Y., *Metric estimates of the first occurrence time in regenerative process*, Lect. Notes Math., 1155, 100, 1985.

81) Kallenberg, O., *Limits of compound and thinned point processes*, J. Appl. Probab., 12, 269, 1975.

82) Kendall, D. G., *Some problems in the theory of dams*, J. R. Stat. Soc., Ser. B, 19, 207, 1957.

83) Kendall, M. G., *The analysis of economic time series. Part I. Prices*, J. Royal Statist. Soc., 96, 11, 1953.

84) Kendall, M. G. and Stuart, A., *The Advanced Theory of Statistics, Vol. 1, Distribution Theory*, 3rd ed., Hafner, New York, 1969.

85) Khinchin, A. Ya., *Mathematical theory of a stationary queue*, Mat. Sbornik, 39, 73, 1932.

86) Khinchin, A. Ya., *Su una legge dei grandi numeri generalizzata*, Giorn. Ist. Ital. Attuari, 7, 365, 1936.

87) Khinchin, A. Ya., *Asymptotic Laws of Probability Theory*, ONTI, Moscow, 1936 (in Russian).

88) Khinchin, A. Ya., *Limit Theorems for Sums of Independent Random Variables*, GONTI, Moscow, 1938 (in Russian).

89) Khinchin, A. Ya., *On unimodal distributions*, Trans. Res. Inst. Math., Mech., University of Tomsk, 2, 1, 1938.

90) Kingman, J. F. C., *On doubly stochastic Poisson processes*, Proc. Cambridge Phil. Soc., 60, 923, 1964.

91) Kitaev, M. V., Korolev, V. Yu., Lavrentyev, V. A., Nazarov, L. V., Petukhov, A. V., Ushakov, V. G. and Zinovyev, P. V., *QUEST (QUEueing Systems Testing), User's Manual*, JV Dialogue-Moscow University, Moscow, 1992.

92) Klebanov, L. B., Manija, T. M. and Melamed, J. A., *One problem of V.M.Zolotarev and analogs of infinitely divisible and stable distributions in the scheme of summation of a random number of random variables*, Theory Probab. Appl., 29, 757, 1984.

93) Klebanov, L. B., Manija, T. M. and Melamed, J. A., *Analogs of infinitely divisible and stable laws for sums of a random number of random variables*, in IV Vilnius Conference on Probab. Theory and Math. Statistics. Abstracts of Communications, Vol. 2, Vilnius, 1985, 40.

94) Klebanov, L. B. and Melamed, J. A., *Analytical problems related to sums of a random number of random variables*, in XX Colloquium on Probab. Theory and Math. Statistics. Abstracts of Communications, Metsnierba, Tbilisi, 1986, 21 (in Russian).

95) Koenigs, G., *Recherches sur les integrales de certaines equations fonctionnelles*, Ann. Sci. Ecole Norm. Sup., 1 (supplement), 2, 1884.

96) Kolchin, A. V., *Sums of Independent Random Variables and Some Combinatorial Problems*, Ph.D. Thesis, Moscow State University, Moscow, 1993 (in Russian).

97) Kolchin, V. F., *Random Mappings*, Optimization Software, New York, 1986.

98) Kolchin, V. F., *On the number of transpositons with restrictions on lengths of cycles*, Discrete Math., 1, 97, 1989.

99) Kolmogorov, A. N. and Prokhorov, Yu. V., *On sums of a random number of random summands*, Uspekhi Mat. Nauk, 4, 168, 1949.

100) Kopylov, V. V., *A method of estimating software reliability from the results of statistical tests*, Kibernetika (Kiev), 5, 123, 1990.

101) Korolev, V. Yu., *On the probability distribution appearing in registration of the output of multiple nuclear processes*, Bull. Moscow Univ., Ser. 15 Comput. Math. Cybern., 2, 36, 1980.

102) Korolev, V. Yu., *The asymptotics of randomly indexed random sequences*, in Stability Problems for Stochastic Models. Proc. of the Seminar, Zolotarev, V. M. and Kalashnikov, V. V., Eds., Institute for Systems Studies, Moscow, 1989, 60.

103) Korolev, V. Yu., *On limit distributions of randomly indexed random sequences*, Theory Probab. Appl., 37, 564, 1992.

104) Korolev, V. Yu., *Probabilistic and statistical methods for predicting complex software reliability*, Bull. Moscow Univ., ser. 15 Comput. Mathem. and Cybern., 47, 3, 1992.

105) Korolev, V. Yu., *Limit Distributions for Random Sequences with Random Indices and Their Applications*, Doct. Sci. Thesis, Moscow State University, Moscow, 1993 (in Russian).

106) Korolev, V. Yu., *Convergence of random sequences with independent random indices. I*, Theory Probab. Appl., 39, 313, 1994.

107) Korolev, V. Yu., *Limit theorems of random summation in some problems of actuarial and financial mathematics*, in XVII Seminar on Stability Problems for Stochastic Models, Kazan, June 19-25, 1995, Abstracts of Communications, Kazan State University, Kazan, 1995, 24.

108) Korolev, V. Yu., *Limit behavior of centered random sums of independent identically distributed random variables*, J. Math. Sci., 76, 2153, 1995.

109) Korolev, V. Yu., *A general theorem on limit behavior of superpositions of independent random processes with applications to Cox processes*, J. Math. Sci., to appear, 1996.

110) Korolev, V. Yu. and Kossova, E. V., *Convergence of multidimensional random sequences with random indices*, J. Math. Sci., 76, 2259, 1995.

111) Korolev, V. Yu. and Kruglov, V. M., *Limit theorems for random sums of independent random variables*, Lect. Notes Math., 1546, 100, 1993.

112) Korolev, V. Yu. and Kruglov, V. M., *Random Sequences with Random Indices*, 1997.

113) Korolev, V. Yu. and Petukhov, A. V., *Estimation of the number of missing observations with applications to software reliability*, J. Math. Sci., 75, 1415, 1995.

114) Korolev, V. Yu. and Petukhov, A. V., *Mathematical methods of reliability growth analysis*, in Stability Problems for Stochastic Models. Proceedings of the Seminar. Perm, 1992, Zolotarev, V. M., Kruglov, V. M. and Korolev V. Yu., Eds., TVP/VSP, Moscow/Utrecht, 1994, 103.

115) Korolev, V. Yu. and Selivanova, D. O., *Conditionally Geometrical and Conditionally Exponential Reliability Growth Models*, Manuscript deposed in VINITI 12 May, 1994. No. 1187-B94 (in Russian).

116) Kovalenko, I. N., *On a class of limit distributions for rarefied flows of homogeneous events*, Lit. Mat. Sbornik, 5, 569, 1965.

117) Kruglov, V. M., *The Method of Accompanying Infinitely Divisible Laws*, Doct. Sci. Thesis, Moscow State University, Moscow, 1975 (in Russian).

118) Kruglov, V. M., *On convergence of distributions of sums of a random number of independent random variables to the normal law*, Bull. Moscow Univ., Ser. 1 Math., Mech., 5, 5, 1976.

119) Kruglov, V. M., *Additional Chapters of Probability Theory*, Vysshaya Shkola, Moscow, 1984 (in Russian).

120) Kruglov, V. M., *Convergence of moments of random sums*, Theory Probab. Appl., 33, 360, 1988.

121) Kruglov, V. M., *Mixtures of probability distributions*, Bull. Moscow Univ., Ser. 15 Comput. Math. Cybern., 2, 4, 1991.

122) Kruglov, V. M., *Weak convergence of random sums under nonrandom centering*, J. Math. Sciences, 76, 2275, 1995.

123) Kruglov, V. M. and Korolev, V. Yu., *Limit Theorems for Random Sums*, Moscow University Press, Moscow, 1990.

124) Kruglov, V. M. and Titov, A. N., *Mixtures of probability distributions*, Lect. Notes Math., 1233, 41, 1986.

125) Kullback, S., *A lower bound for discrimination in terms of variation*, IEEE Trans. Inform. Theory, IT-13, 126, 1967.

126) Kupper, J., *Some aspects of cumulative risk*, Astin Bull., 3, 85, 1965.

127) Linnik, Yu. V., *An information-theoretic proof of the central limit theorem with the Lindeberg condition*, Theory Probab. Appl., 4, 288, 1959.

128) Little, R. A. and Rubin, D. B., *Statistical Analysis with Missing Data*, John Wiley, New York, 1987.

129) Littlewood, B., *Stochastic reliability growth: a model for fault-removal in computer programs and hardware designs*, IEEE Trans. Reliab., R30, 313, 1981.

130) Littlewood, B. and Verrall, J. L., *A Bayesian reliability growth model for computer software*, J. R. Stat. Soc. Ser. C, 22, 332, 1973.

131) Loève, M., *Probability Theory*, 3rd ed., Van Nostrand, Princeton, NJ, 1963.

132) Lukacs, E., *Characteristic Functions*, 2nd ed., Griffin, London, 1970.

133) Lundberg, F. *Ueber die Theorie der Ruckversicherung*, in Trans. VI Int. Congr. of Actuaries, Vol. 1, 1909, 877.

134) Mandelbrot, B., *The variation of certain speculative prices*, J. Business, 36, 394, 1963.

135) Mandelbrot, B., *New method in statistical economics*, J. Political Economy, 71, 421, 1963.

136) Mandelbrot, B., *Long-run linearity, locally Gaussian processes and infinite variances*, Int. Economic Rev., 10, 82, 1969.

137) Mandelbrot, B. and Taylor, H. M., *On the distribution of the stock price difference*, Operat. Res., 15, 1057, 1967.

138) Mecke, J., *Eine charakteristische Eigenschaft der doppelt stochastischen Poissonischen Prozesse*, Z. Wahrscheinlichkeitstheorie verw. Geb., 11, 74, 1968.

139) Miller, D. R., *Exponential Order Statistics Models of Software Reliability Growth*, George Washington University Tech. Report T-496/84, Washington, DC, 1984.

140) Mogyoródi, J., *On limiting distribution for sums of a random number of independent random variables*, Studia Sci. Math. Hung., 6, 365, 1961.

141) Mogyoródi, J., *A central limit theorem for the sum of a random number of independent random variables*, Studia Sci. Math. Hung., 7, 409, 1962.

142) Mogyoródi, J., *On the law of large numbers for the sum of a random number of independent random variables*, Ann. Univ. Sci. Budapest Ser. Math., 8, 33, 1965.

143) Mogyoródi, J., *A central limit theorem for the sums of a random number of independent random variables*, Ann. Univ. Sci. Budapest Ser. Math., 10, 171, 1967.

144) Mogyoródi, J., *Some remarks on the rarefaction of the renewal processes*, Lit. Mat. Sbornik, 11, 303, 1971.

145) Moran, P. A. P., *A probability theory of a dam with a continuous release*, Q. J. Math., Oxford Series, 2, 130, 1956.

146) Moranda, P. B., *Prediction of software reliability during debugging*, in Proc. of Annual Reliability and Maintainability Symp., 1975, 327.

147) Musa, J., *Software reliability measurement*, J. Syst. Software, 1, 223, 1980.

148) Musa, J., Iannino, A. and Okumoto, K., *Software Reliability: Measurement, Prediction, Application*, McGraw-Hill, New York, 1987.

149) Nelson, E., *A Statistical Basis for Software Reliability Assessment*, TRW Software Series TRW-SS-73-03, Redondo Beach, CA, 1973.

150) Neyman, J., *On a new class of "contagious" distributions applicable in entomology and bacteriology*, Ann. Math. Stat., 10, 35, 1939.

151) Novitskii, P. V. and Zograf, I. A., *Estimation of Errors of Measurements*, Energoatomizdat, Leningrad, 1991 (in Russian).

152) Palchun, B. P., *A method of testing programs for reliability, in Functional Stability of Special Software of Automatic Systems*, Moscow, 1989, 111 (in Russian).

153) Papadatos, N. and Papathanasiou, V., *Distance in variation and a Fisher-type information*, Math. Methods of Stat., 4, 230, 1995.

154) Parnov, E. I., *On the Cross of Infinities*, Atomizdat, Moscow, 1967 (in Russian).

155) Petrov, V. V., *Sums of Independent Random Variables*, Springer, Berlin-Heidelberg-New York, 1975.

156) Petrov, V. V., *Limit Theorems for Sums of Independent Random Variables*, Nauka, Moscow, 1987 (in Russian).

157) Poincaré, H., *Sur une class nouvelle de trancendantes uniformes*, J. Math. Pures Appl., 4 Ser., 6, 313, 1890.

158) Pollaczek, F. and Geiringer, H. *Ueber die Poissonische Verteilung und die Entwicklung Willkurlichen Verteilunger*, Z. Angew. Math. Mech., 8, 292, 1928.

159) Pollaczek, F., *Ueber eine Aufgabe der Wahrscheinlichkeitstheorie, Part I*, Math. Zeitschr., 32, 64, 1930.

160) Pollaczek, F., *Ueber eine Aufgabe der Wahrscheinlichkeitstheorie, Part II*, Math. Zeitschr., 32, 729, 1930.

161) Prabhu, N. U., *Stochastic Storage Processes: Queues, Insurance Risk and Dams*, Springer, Berlin-Heidelberg-New York, 1980.

162) Rachev, S. and Ruescheendorf, L., *Models for option prices*, Theory Probab. Appl., 39, 150, 1994.

163) Rényi, A., *A Poisson-folyamat egy jellemzese*, Magyar. Tud. Akad. Mat. Kut. Int. Kozl., 1, 519, 1956.

164) Rényi, A., *On the asymptotic distribution of the sum of a random number of independent random variables*, Acta Math. Acad. Sci. Hung., 8, 193, 1957.

165) Rényi, A., *On mixing sequences of sets*, Acta Math. Acad. Sci. Hung., 9, 215, 1958.

166) Rényi, A., *On the central limit theorem for the sum of a random number of independent random variables*, Acta Math. Acad. Sci. Hung., 11, 97, 1960.

167) Rényi, A., *On stable sequences of events*, Sankhya, Ser. A, 25, 293, 1963.

168) Rényi, A., *Probability Theory*, North-Holland, Amsterdam, 1970.

169) Robbins, H., *The asymptotic distribution of the sum of a random number of random variables*, Bull. Am. Math. Soc., 54, 1151, 1948.

170) Rootzén, H., *A Note on the Central Limit-Theorem for Doubly Stochastic Poisson Processes*, Techn. Report, The University of North Carolina, 1975.

171) Rychlik, Z. and Szynal, D., *On the limit behaviour of the sums of a random number of independent random variables*, Bull. Acad. Pol. Sci., Ser. Math., Phys., Astron., 20, 401, 1972.

172) Rychlik, Z. and Szynal, D., *On the limit behaviour of the sums of a random number of independent random variables*, Colloq. Math., 28, 147, 1973.

173) Samuelson, P. A., *Rational theory of warrant pricing*, Ind. Manag. Rev., 6(spring), 13, 1965.

174) Schneider, H., *Truncated and Censored Samples from Normal Populations*, Marcel Dekker, New York-Basel, 1986.

175) Schroeder, E., *Ueber iterirte Funktionen*, Math. Ann., 3, 296, 1871.

176) Seal, H. L., *Stochastic Theory of a Risk Business*, John Wiley, New York, 1969.

177) Seber, G. A. F., *Linear Regression Analysis*, John Wiley, New York, 1977.

178) Seneta, E., *Regularly varying functions*, Lect. Notes Math., 508, 1, 1976.

179) Sevastyanov, B. A., *Branching Processes*, Nauka, Moscow, 1971 (in Russian).

180) Shannon, C., *A mathematical theory of communication*, Bell Syst. Tech. J., 27, 379, 1948.

181) Shanthikumar, J. G., *Software reliability models: a review*, Microelectron. Reliab., 23, 903, 1983.

182) Shick, G. L. and Wolverton, R. W., *Assessment of software reliability, in Proc. on Operations Research*, Wurzberg-Vien, 1973, 395.

183) Shiryaev, A. N., *On some basic concepts and some basic stochastic models used in finance*, Theory Probab. Appl., 39, 5, 1994.

184) Shiryaev, A. N., *Actuarial and financial business: current state of art and perspectives of development. Notes of communication presented at Constituent Conference of Russian Society of Actuaries, September 14, 1994*, Rev. Ind. Appl. Math., Ser. Financial Actuarial Math., 1, 684, 1995.

185) Shiryaev, A. N., *Stochastic problems of financial mathematics*, Rev. Ind. Appl. Math., Ser. Financial Actuarial Mathematics, 1, 780, 1995.

186) Shiryaev, A. N., Kabanov, Yu. M., Kramkov, D. O. and Melnikov, A. V., *Towards the theory of pricing of options of both European and American types. Part I. Discrete time*, Theory Probab. Appl., 39, 23, 1994.

187) Shiryaev, A. N., Kabanov, Yu. M., Kramkov, D. O. and Melnikov, A. V., *Towards the theory of pricing of options of both European and American types. Part II. Continuous time*, Theory Probab. Appl., 39, 80, 1994.

188) Shooman, M. L., *Probabilistic models for software reliability prediction*, in Statistical Computer Performance Evaluation, Freiberger, W., Ed., New York, 1972, 465.

189) Skellam, J. G., *Studies in statistical ecology. Part I. Spatial pattern*, Biometrica, 39, 346, 1952.

190) Solov'ev, A. D., *Asymptotic behavior of the time of the first occurrence of a rare event in a regenerative process*, Izvestiya AN SSSR, Tech. Cybern., 6, 79, 1971.

191) Sweeting, T. J., *Asymptotically independent scale-free spacings with application to discordancy testing*, Ann. Stat., 14, 1485, 1986.

192) Szász, D., *On classes of limit distributions for sums of a random number of identically distributed random variables*, Theory Probab. Appl., 17, 424, 1972.

193) Szász, D. and Freyer, B., *A problem of the theory of summation with a random index*, Lit. Mat. Sbornik, 11, 181, 1971.

194) Szynal, D., *On limit distribution theorems for sums of a random number of random variables appearing in the study of rarefaction of a recurrent process*, Zastosowania Mat., 15, 277, 1976.

195) Takács, L., *The limiting distribution of the virtual waiting time and the queue size for single-server queue with recurrent input and general service times*, Sankhya, Ser. A, 25, 91, 1963.

196) Takács, L., *On the distribution of the supremum for stochastic processes with interchangeable increments*, Trans. Am. Math. Soc., 199, 367, 1965.

197) Tallis, G. M., *The identifiability of mixtures of distributions*, J. Appl. Probab., 6, 389, 1969.

198) Teicher, H., *Identifiability of mixtures*, Ann. Math. Stat., 32, 244, 1961.

199) Teicher, H., *Identifiability of finite mixtures*, Ann. Math. Stat., 34, 1265, 1963.

200) Tikhonov, A. N., *Solutions of Ill-Posed Problems*, John Wiley, New York, 1977.

201) Tucker, H. G., *Convolutions of distributions attracted to stable laws*, Ann. Math. Statist., 39, 1381, 1968.

202) Umarov, A. Yu., *Sums of a Random Number of Random Variables: a Study of Infinitely Divisible Distributions*, Ph.D. Thesis, Tashkent State University, Tashkent, 1992 (in Russian).

203) Van Pul, M., *Asymptotic properties of statistical models in software reliability*, in 2nd Bernoulli Society World Congress. Abstracts of Communications, Uppsala, 1990, 43.

204) Volkov, L. I., *Aircraft Complexes Operation Control*, Vysshaya Shkola, Moscow, 1981 (in Russian).

205) Volkov, L. I. and Shishkevich, A. M., *Aircraft Reliability*, Vysshaya Shkola, Moscow, 1975 (in Russian).

206) Wagoner, W. L., *The Final Report on a Software Reliability Measurement Study*, Aerospace Corp. Rep. TOR-0074(41112)-1, El Segundo, CA, 1973.

207) Wald, A., *On cumulative sums of random variables*, Ann. Math. Stat., 15, 283, 1944.

208) Wald, A., *Sequential Analysis*, John Wiley, New York, 1947.

209) Yakowitz, S. J. and Spragins, J. D., *On the identifiability of finite mixtures*, Ann. Math. Statist., 39, 209, 1968.

210) Yamada, S. and Osaki, S., *Nonhomogeneous error detection rate models for software reliability growth*, in Stochastic Models in Reliability Theory, Springer, Berlin-Heidelberg-New York, 1984, 120.

211) Yamada, S. and Osaki, S., *Optimal software release policies for a nonhomogeneous error detection rate model*, Microelectron. Reliab., 26, 691, 1986.

212) Zolotarev, V. M., *One-Dimensional Stable Distributions*, American Math. Soc., Providence, RI, 1986.

213) Zolotarev, V. M., *Modern Theory of Summation of Independent Random Variables*, Nauka, Moscow, 1986 (in Russian).

Index

U-plot, 189
$\mathcal{N}$-Gaussian distribution function, 139
$\mathcal{N}$-infinitely divisible laws, 145
p-rarefaction, 218

accompanying infinitely divisible distributions, 2
additively closed family, 60
analysis of computational algorithms, 6
asymptotic expansions, 239

Bayesian reliability growth models, 158
Beekman's convolution formula, 12
Bernstein theorem, 30, 58
Berry-Esseen inequality, 236
beta distribution, 193
binomial distribution, 14
bordering method, 199
branching processes, 14
breakdowns, 4
Brownian motion, 78
busy period, 8

Cauchy distribution, 64
Cauchy-Schwarz inequality, 184
censoring, 192
centers, 114
Chebyshev-Hermite polynomials, 239
Chebyshev-type interval estimators, 202
claims, 6
classical risk process, 67
cold reserve, 17
completely monotone function, 29
concentration function, 223
Conditionally geometric models, 158
conditionally geometric models, 160
contour, 6
controlling process, 218
cosmic particles, 1
Cox process, 75, 218
Cramér condition, 239
Cramér-Rao bound, 212
cross-sections of nuclear reactions, 4
cryptography, 6
customer, 8
cycle, 6
cylindric function of an imaginary argument, 245

de Finetti theorem, 146
deaths, 4
defective products, 13
diffusion equation, 78
discrete reliability growth model, 157
domain of $\mathcal{N}$-attraction, 150
domain of geometric attraction, 34
dominated convergence theorem, 95
double exponential distribution, 64, 245
doubly stochastic Poisson process, 75, 218

economical Brownian motion, 126
elementary rarefaction, 33
empirical distribution function, 198
entropy, 208
European call option, 126
excess coefficient, 82
expectation of a random sum, 2
exponential distribution, 22

Fisher information, 210
Fredholm integral equation, 128
Fubini theorem, 128

Galton–Watson process, 14
gamma distribution, 50
Gaussian/inverse Gaussian distribution, 125
Geiger-Müller counters, 12
general rarefaction, 85
generalized conditionally exponential models, 180
generalized Cox process, 228
generalized homogeneous reliability growth models with independent decrements, 185
generalized hyperbolic distributions, 126
generalized Jelinski-Moranda model, 181
generalized least squares estimators, 200
generalized Poisson laws, 3
generalized Polya processes, 234
generalized risk process, 67
geometric Brownian motion, 126
geometric distribution, 9
Glivenko theorem, 198
globally parametrized renewing model, 161
globally parametrized renewing reliability growth models, 170

Hermite distribution, 5
homogeneous conditionally exponential re-
       liability growth model with in-
       dependent decrements, 174
hot reserve, 17
Hyperexponential distribution, 170

identifiable families, 59
identifiable mixtures, 59
ill-posed problem, 129
infinitely divisible distribution, 98
infinitely divisible distribution function,
       32
information measures, 212
initial capital, 10
insurance, 3
inverse Gaussian distribution, 126
Ito's lemma, 127

Jelinski-Moranda model, 161

Khinchin's theorem about convergence of
       types, 96
Kruglov-Titov theorem, 65, 112, 113
kurtosis, 82

Lévy distribution, 79, 240
Lévy metric, 50
Lévy's inequality, 52
Lévy-Cramér theorem, 73, 75, 76, 107,
       113, 213
Lévy-Prokhorov metric, 50
ladder epochs, 12
Lagrange formula, 183
Laplace distribution, 64
Laplace-Stieltjes transform, 20
law of large numbers, 66
leptokurtosity, 78
light-weight reserve, 17
Lindeberg condition, 78
loaded reserve, 17
loading, 17
local limit theorem, 7
Lyapunov's inequality, 184

main element, 17
maturity, 126
maximum aggregate loss, 11
maximum likelihood method, 187
mean time between failures, 190
method of least squares, 187
method of moments, 188
mixed Poisson process, 220
modified Bessel function, 125
Mosaic reliability growth models, 180

Neyman distribution, 5
normal distribution, 50

order statistics, 193
order-statistics-type models, 175

parametric renewing reliability growth mod-
       els, 170
partially renewing models, 172
photonuclear reaction, 4
Poincaré equation, 88, 140
point process, 218
Poisson distribution, 1
Poisson process, 217
Poisson theorem, 4
Poisson-binomial distribution, 5
Pollaczek-Khinchin formula, 9
Polya process, 220
premiums, 6
primary particles, 4
principle of non-decrease of uncertainty,
       207
pseudosolution, 129

quantile, 94
quantization, 210
queue, 8

Rényi-Mogyoródi theorem, 66
Raikov's theorem on the decomposability
       of a Poisson law, 113
random measure, 218
Randomly infinitely divisible distributions,
       137
rarefaction, 84
rarefied renewal processes, 84
Rayleigh distribution, 171
record values, 12
reference function, 154
regular reliability growth models, 159
relative entropy, 212
relative stability, 39
reliability function, 186
reliability growth model, 157
reliability of technical systems, 17
renewal duration, 18
renewal process, 219
renewing conditionally geometrical relia-
       bility growth model, 160
Renewing models, 169
reproducibility with respect to parame-
       ter, 60
reserve element, 17
risk process, 5, 67
ruin probability, 11

scarlet fever, 4
scatters, 114
second law of thermodynamics, 207
secondary particles, 4
semiinvariant, 239
sequential analysis, 14

Shannon entropy, 212
Shannon information, 207
slowly varying function, 35, 69
smallpox, 4
spatial arrangement of individuals, 4
special rarefaction, 85
specified mappings, 156
stable distribution, 130
standard Poisson process, 225
standardized Fisher information, 213
steam boilers explosions, 4
Stirling-Hermite distribution, 5
stochastic transformation, 156
stochastic volatility model, 128
stock price, 77
stock return, 83
storage control, 10
strict stability, 138
structural random variable, 220
structural distributions, 220
Student distribution, 64
subordinated Wiener process, 79
supercritical branching process, 86
supercritical Galton–Watson process, 15
superpositions of stochastic processes, 221
surplus, 6
symmetrical stable laws, 64
symmetrization inequality, 51

Taylor formula, 28
telephone operator, 8
theory of dams, 10
thermodynamic entropy, 210
total variation, 214
transfer theorem, 41
transposition, 6
trend indicators, 83
trouble-free performance duration, 18
truncation, 192

uncertainty, 207
uniform distribution, 7, 211
unimodal distribution function, 30
unloaded reserve, 17

victims in transport accidents, 4
virtual waiting time, 10
volatility, 126

waiting time, 8
Wald identity, 13
warm reserve, 17
weak relative compactness criterion, 43
Weibull distribution, 171
Wiener process, 78